Overcurrent Relay Advances for Modern Electricity Networks

Arturo Conde Enríquez

ELSEVIER

ACADEMIC PRESS
An imprint of Elsevier

Academic Press is an imprint of Elsevier
125 London Wall, London EC2Y 5AS, United Kingdom
525 B Street, Suite 1650, San Diego, CA 92101, United States
50 Hampshire Street, 5th Floor, Cambridge, MA 02139, United States
The Boulevard, Langford Lane, Kidlington, Oxford OX5 1GB, United Kingdom

ISBN 978-0-323-91777-3

For information on all Academic Press publications
visit our website at https://www.elsevier.com/books-and-journals

Publisher: Charlotte Cockle
Acquisitions Editor: Graham Nisbet
Editorial Project Manager: Ali Afzal-Khan
Production Project Manager: Prasanna Kalyanaraman
Cover Designer: Mark Rogers

Typeset by STRAIVE, India

Dedication

To my treasures, Roxana and Linda

Contents

Preface xiii
Nomenclature xv

1. Mathematical model and application of overcurrent relays 1
   1.1 Overcurrent relays: Overview 1
   1.2 Time curve 2
   1.3 OCR coordination 5
   1.4 DOCR coordination 10
   1.5 General considerations 14
   1.6 Functional structure of digital relays 18
      1.6.1 Phasor estimation 19
      1.6.2 Instantaneous OCR 23
      1.6.3 Inverse time relay 24
      1.6.4 Dynamic performance 27
      1.6.5 Limitations of the OCR 27
   1.7 Practical applications 30
      1.7.1 Coordination of overcurrent devices 31
   1.8 Summary 35
   References 35

2. Adaptive operating time for overcurrent relays 37
   2.1 Introduction 37
   2.2 Adaptive time 38
   2.3 Evaluation of the adaptive operating time 41
   2.4 Functional structure of the proposed relay 43
      2.4.1 Implementation 44
   2.5 Time curve fitting algorithms 46
      2.5.1 Equations of the time-current characteristics 47
      2.5.2 Time curve fitting program 50
   2.6 Curve fitting program 53
      2.6.1 Fuse fitting 55
      2.6.2 Recloser fitting 59
   2.7 Curve fitting using the *H*(*I*) function 60

2.7.1 Comparison between curve fits using the functions $H(I)$ and $T(I)$ 63
2.8 Coordination of overcurrent protection 66
2.8.1 Coordination of an adaptive relay and conventional relay 69
2.8.2 Coordination of a conventional relay, adaptive relay, and fuse 70
2.8.3 Coordination between a conventional relay, adaptive relay, and recloser 70
2.9 Comparison between overcurrent relays 71
2.10 Summary 76
References 77

3. Overcurrent coordination via metaheuristic algorithms 81

3.1 Introduction 81
3.2 DOCR coordination via metaheuristic algorithms 82
3.2.1 Data preparation for relay coordination formulation 83
3.3 Genetic operators 91
3.3.1 Population (P_x) 91
3.3.2 Selection 91
3.3.3 Crossover 94
3.3.4 Mutation 95
3.4 Metaheuristic optimization algorithms 96
3.4.1 Genetic algorithms 97
3.4.2 Invasive weed optimization 99
3.4.3 Differential evolution algorithm 103
3.4.4 Gray wolf algorithm 107
3.4.5 DOCRs coordination through GWO 109
3.4.6 DOCRs coordination via MOGWO 110
3.5 Test systems and results 113
3.6 Results and discussion 113
3.6.1 Test case I 114
3.6.2 Test case II 114
3.7 Summary 117
References 117

4. Design of nonconventional time curves for overcurrent relays 121

4.1 Nonconventional time curves of overcurrent relays 121
4.2 Approach one: Proposed general model 122
4.3 OCR coordination by IWO 127
4.4 OCR coordination by GWO and DE 132
4.4.1 Test systems 133

4.5 **Phase and ground coordination** 137
4.5.1 Academic test system 138
4.5.2 Results 139
4.6 **Approach two: Dynamic *J(I)* function** 141
4.6.1 Proposed model 142
4.6.2 Case studies 148
4.7 **Summary** 157
References 158

5. **Coordination of nonconventional DOCR based on the time interval criterion** 161

5.1 **Introduction** 161
5.2 **Novel method of OCR coordination via time intervals** 163
5.2.1 Proposed optimization problem 164
5.3 **Metaheuristic algorithms for the interval method** 166
5.3.1 DE for proposed coordination method using time intervals 167
5.3.2 Relay coordination verification based on time intervals 169
5.4 **Test systems** 174
5.5 **Results and discussions** 175
5.5.1 Performance of the optimization algorithm with the proposed method 176
5.5.2 Case 1: Comparison of conventional and interval coordination methods 178
5.5.3 Case 2: Evaluation of the proposed method with different intervals 182
5.5.4 Case 3: Performance of the proposed method for topological changes to the nine-bus IEEE system 185
5.6 **Summary** 189
References 191

6. **Overcurrent relay protection for active microgrids** 193

6.1 **Introduction** 193
6.2 **Architecture of a microgrid** 194
6.2.1 Models of renewable sources 195
6.3 **MG protection** 205
6.3.1 Weak source contribution 208
6.3.2 Islanding detection 209
6.3.3 Island saving scheme 213
6.4 **DOCRs coordination proposal for MG protection** 217
6.4.1 Academic test system 219
6.5 **Summary** 229
References 230

7. Sensitivity enhancement via adaptive methods for phase DOCR in active electrical networks with high CIGRs connection 233

7.1 Sensitivity of directional overcurrent relays overview 233
7.2 The lack of sensitivity in active networks with CIGRs connection 235
7.2.1 Negative sequence relays 236
7.3 Sensitivity detection system 238
7.4 Adaptive pickup current 240
7.4.1 Relay setting groups 241
7.4.2 Test of microgrid connected to the utility 242
7.5 Continuous adaptive overcurrent 245
7.5.1 Adaptive characteristic of the overcurrent relay 246
7.5.2 Control logic for adaptive relay 249
7.6 Adaptation of the pickup current with RMS 250
7.6.1 Control logic 255
7.6.2 Evaluation of adaptive pickup current 255
7.7 Adaptation of the pickup current with load demand 258
7.7.1 Demand control logic 259
7.7.2 Evaluation 261
7.8 Comparison between the two current values used in adaptive logic 263
7.9 Summary 263
References 264

8. Distance and overcurrent relay coordination 267

8.1 Overcurrent and distance relay coordination 267
8.2 Use of distance relay to replace a DOCR for lack of sensitivity 269
8.2.1 Distance relay in DOCR coordination 271
8.2.2 Nonconventional time curves with distance relay coordination 273
8.3 Distance relay 274
8.3.1 Distance relay algorithm 275
8.3.2 Validation of relay operation 279
8.4 Optimization formulation for overcurrent and distance relay coordination 281
8.4.1 Test systems 282
8.4.2 Results 283
8.5 Dynamic coordination 287
8.6 Summary 292
References 292

9. Online centralized control for adaptive protection system 295

9.1 Introduction 295
9.2 Dynamic operating conditions of electrical systems 296
9.2.1 Topological changes 297
9.2.2 Input/output of generators 298
9.2.3 Change of demand 299
9.3 Online protection system 300
9.3.1 Coordination algorithm 302
9.4 Special considerations for online coordination system 303
9.4.1 Network topology 303
9.4.2 Load flow and fault current analysis 304
9.4.3 Special considerations 305
9.4.4 Sensitivity analysis 309
9.5 Code of optimization algorithms 310
9.6 Evaluation and comparison among the algorithms in IEEE test systems 313
9.6.1 Test system data 313
9.6.2 Parameter settings of coordination algorithms 314
9.6.3 Results and comparison 315
9.7 Discussion 321
9.7.1 Interconnection of RES 324
9.8 Time interval coordination method 324
9.9 Summary 326
References 326

10. Virtual relay design and lab tests of an adaptive overcurrent relay 329

10.1 Introduction 329
10.2 Relay programming 330
10.3 Virtual OCR 332
10.3.1 Virtual relay 334
10.3.2 Modules of virtual relay 335
10.4 Stand-alone examples 336
10.4.1 Operating time validation 336
10.4.2 Cold load current 338
10.4.3 Subharmonic currents 339
10.4.4 Coordination examples 339
10.5 Laboratory test structure 344
10.5.1 Acquisition of signals in real time 346
10.5.2 Sensitivity 355
10.6 Signal file 359
10.7 Internal signal generation 362
10.8 Summary 365
References 366

Index 369

Preface

This book discusses the application and the improved performance of overcurrent relays to highly dynamic power systems. Advanced solutions such as new adaptive relay designs and new coordination methods are used to solve application problems in electrical networks. This book presents the functional alternatives that allow the relay to be viewed as an active element of the electrical network, modifying the functional structure by incorporating dynamic adjustments according to the quasidynamic state of electrical networks and presenting the formulation of a new relay coordination method.

The operating structure of electrical networks has been modified in recent decades with the interconnection of highly intermittent generation. The application of protection systems presents a challenge for the detection of faults and security, avoiding false operation. Current-based protection is the most widely used in electrical systems, offering phase and ground protection in all voltage networks. Due to its functional simplicity, it is the most affected by highly dynamic operating conditions. The functional improvement of these relays is urgent and highly impactful, incorporating the information available from other locations of the relay to tune in to the current operating conditions of the electrical network.

Current protection systems are functionally passive, with predefined settings that are not adjusted to the dynamic conditions of the electrical network. The use of communication channels has been very favorable to help improve protection performance, either speeding up its operation or avoiding false triggers. Nevertheless, the use of standardized time curves and fixed settings does not always represent the best solution for the diversity of coordination problems of the electrical networks.

The active relay presented in this proposal is conceived as an element that is subject to adaptive adjustment according to the dynamic operating conditions of the electrical network and with optimization strategies for coordination. The implementation of proposed systems for setting and online coordination is aimed at improving the relay operation. Also, the use of nonstandardized time curves offers a wide range of applications that allow to improve coordination. Through it, it is possible to obtain a new coordination method, where the setting of each overcurrent relay is stand-alone, improving operation times and reducing coordination violations.

Because the protection zone of the overcurrent relay is highly dynamic under highly intermittent conditions, there may be protection zones without adequate sensitivity. This has repercussions in undetected faults and in operating times, which, in many cases, impact on the quality of voltage in the electrical networks, triggering a low-voltage load and increasing the operational technical losses of the electrical network. Adaptable schemes that seek protection retuning will manage to monitor and guarantee more sensitive protection zones, reducing fault release times for a better quality of supply. The requirements for the conformation of these monitoring systems of protection zones and the dynamic adjustment of relays are presented.

Finally, the laboratory platform necessary for the functional evaluation of an overcurrent relay is presented. Expensive laboratory and simulation equipment requirements are avoided. It is possible by means of modest equipment to carry out the operational validation of overcurrent relays subject to various operating conditions.

Arturo Conde Enríquez

Nomenclature

$A(I)$	dynamic nonlinear function
A, p, B	time curve constants
ACO	ant colony optimization
AS	adjustable setting
BESS	battery storage systems
CC	conventional time curves
CIGRs	converter interfaced generation resources
CT	current transformer
CTI	coordination time interval
Cx	chromosome
DE	differential algorithms
DER	distributed energy resources
DG	distributed generation
DOCR	directional overcurrent relay
GA	genetic algorithms
G_k	accumulated value of the integrator
GWO	gray wolf optimization
$H(I_k)$	relay functions to represent the dynamic disk displacement
HIF	high impedance fault
I_{pickup}^{Adp}	adaptive load current
I_{load}	load current
I_{pickup}	pickup current of relay
I_{pickup_max}	maximum load current
$I^{primary}$	multiple of current
Ireset	reset current
I_{sc}	short circuit current
IWO	invasive weed optimization
$J(I_k)$	relay functions to represent the tap position
kop	operation sample
MG	microgrid
MOGWO	multiobjective gray wolf optimizer
NCC	nonconventional time curves
OCR	overcurrent relay
OF	objective function

PS	population size
PSM	plug setting multiplier
PT	potential transformers
RES	renewable sources
t_b	operating time of the backup relay
TCC	type of characteristic curves
TDS	time dial setting
t_p	operating time of the primary relay

Chapter 1

Mathematical model and application of overcurrent relays

1.1 Overcurrent relays: Overview

Overcurrent relays (OCRs) have operating times that have made them convenient for the protection of electrical networks for over a century. The operation times of these relays are determined by the operating conditions, such as temporary overload, damage curves, and the presence of other protection devices, rather than functional limitations. OCR can be applied at practically all voltage levels; they are used in distribution circuits as primary protection; in industrial systems, for the protection of equipment such as generators, transformers, and motors; and as backup protection for transmission lines in the form of ground protection or as backup for phase protection. Under operating conditions with fault current supply from both ends, directional operation of the relay is required, and a DOCR therefore has a wider range of applications. Recently, its applications have been evaluated owing to the increase in distributed energy resources (DERs) in the network, which cause bidirectional flow in networks with a radial nature and a reduction in the magnitude of fault currents, causing a lack of sensitivity.

Overcurrent protection is used as primary protection for the own line and as backup protection for faults on adjacent lines, and both functions are performed based on the same time curve. DOCRs use short-circuit current as an indicator of a fault, but the variation in the Thevenin impedance and the fault type causes different current magnitudes for the same fault location. The relay has great uncertainty concerning the fault location and a highly dynamic reach; in view of this, the coordination of OCRs is carried out based on the operation time of a neighboring relay. Thus, for a short-circuit condition, both relays will have the same operating dynamics, and the coordination between the primary (R_1) and backup (R_2) relays is guaranteed by the coordination time

Overcurrent Relay Advances for Modern Electricity Networks. https://doi.org/10.1016/B978-0-323-91777-3.00001-5

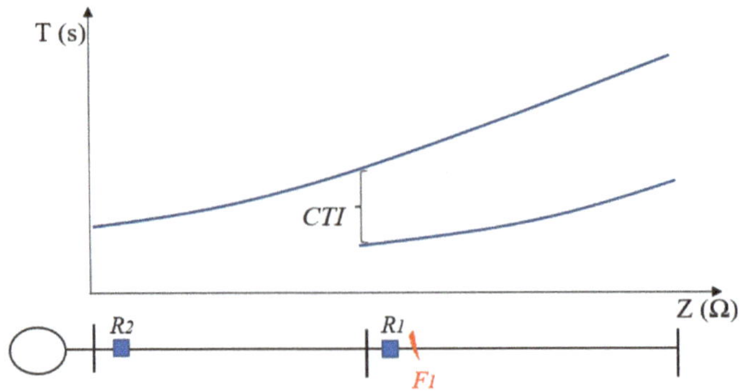

FIG. 1.1 Fault location for overcurrent coordination.

interval (*CTI*), which is determined based on the maximum fault current of both relays (F_1 in Fig. 1.1).

1.2 Time curve

The inverse type of time curves is widely used for line protection, as the operating times are generally adequate for this application. The time curve of the OCR represents an operating time in inverse proportion to the current signal flowing in the relay and is represented on a bi-logarithmic scale to allow for better visualization for adjustment and coordination. Existing standards define the analytical expressions for the time curves of inverse time relays, and these time curves differ because the operating time decreases as the current increases. Fig. 1.2 shows these standardized curves [1]. In electromechanical relays, different types of curves are obtained by means of a magnetic flux saturation process that affects the induction disk. The saturation is controlled and different degrees of inversion of the curve are obtained by means of irregular areas on the disk. In digital relays, the coefficients of the analytical expression for the time curve are defined to obtain the desired family of curves.

The time curve of an OCR is defined by international standards as follows [1,2]:

$$t(s) = \left[\frac{A}{\left(\frac{I_{sc}}{I_{pickup}} \right)^{p} - 1} + B \right] \times TDS \tag{1.1}$$

$$t(s) = \left[\frac{A}{\left(\frac{I_{sc}}{I_{pickup}} \right)^{p} - 1} \right] \times TDS \tag{1.2}$$

where *TDS* is the time dial setting used to set the time operation, I_{sc} is the maximum short-circuit current measured by the coordinated relays, and I_{pickup} is the

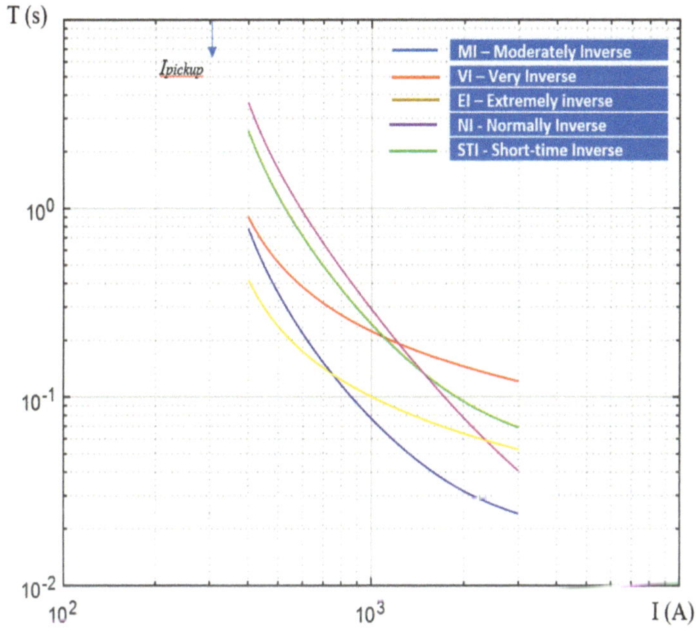

FIG. 1.2 Standardized operating time curves.

setting of the relay and the vertical asymptote of the time curve. These international standards are used to define the type of time curve based on their parameters, where A, B, and p are defined by IEEE Std (1.1) and IEC Std (1.2) defines A and p (see Table 1.1). Specifically, B is the horizontal asymptote of time curve, whereas A and p define the shape of the time curve.

The relay operates when the pickup current (I_{pickup}) is greater than the maximum load current (I_{load}), where a plug setting multiplier (PSM) is used to avoid false operation, and a minimum I_{sc} is applied to ensure adequate fault detection.

$$I_{pickup} = I_{load} \times PSM \tag{1.3}$$

PSM is a factor that prevents operation under normal conditions caused by temporary overloads, power transfer, and errors in current transducers. Typical values are 1.5–2.0 [3], although the specific value will be determined by the application. The variables used for relay coordination are TDS and PSM, and these are defined to set each pair of relays. For different coordination conditions, the time curves must be shifted to achieve the CTI, and this is why the asymptotes of the time curve and their dependence on the values of the time characteristic must be identified.

The asymptotes of the inverse time curve of the OCR are obtained from the settings for I_{pickup} and TDS. The design criteria for the time curves are established based on the electromechanical properties of the relays, where the torque

TABLE 1.1 Parameters of standard time curves.

Standard	Curve type	*A*	*B*	*p*
IEC-60255	SI—Standard inverse (C1)	0.14	0	0.02
	VI—Very inverse (C2)	13.5	0	1
	EI—Extremely inverse (C3)	80	0	2
	STI—Short-time inverse (C4)	0.05	0	0.04
	LTI—Long-time inverse (C5)	120	0	1
ANSI/IEEE	MI—Moderately inverse	0.0515	0.1140	0.02
	VI—Very inverse	19.61	0.491	2.0
	EI—Extremely inverse	28.2	0.1217	2.0
	NI—Normally inverse	5.95	0.18	2.0
	STI—Short-time inverse	0.02394	0.01694	0.02

exerted on the relay disk depends on the magnetic flux that affects it, and this is directly proportional to the ampere-turns. If the number of turns is increased, the current necessary to operate the relay decreases, and vice versa; this allows us to modify the value of I_{pickup} by varying the tap of the number of turns, and the digital relay reproduces this behavior through the modification of I_{pickup}. Fig. 1.3 shows the effect of the tap change, where the operating characteristics move from left to right with a change in I_{pickup}.

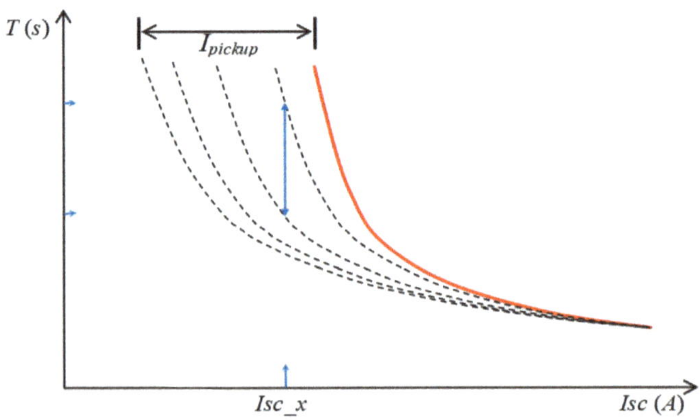

FIG. 1.3 Inverse time curve for change in I_{pickup}.

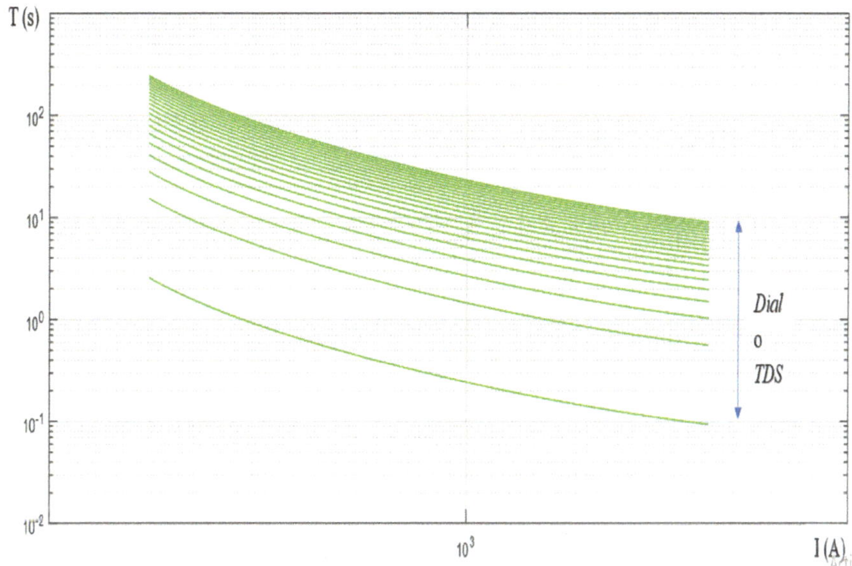

FIG. 1.4 Family of curves created by changing the *TDS* setting.

The inverse time curve has a disadvantage when it is necessary to use large magnitudes for the settings of I_{pickup}; this scenario occurs in electrical networks with high current demand and is very common in actual electrical networks. A high value for I_{pickup} results in a high value for the operating time, and the relay time increases for a fault current lower than the maximum current, as shown in Fig. 1.3. This increase in operating time is appreciable for any current I_{sc_x} between similar time curves, and the time difference is due only to the difference in I_{pickup}.

For a particular type of time curve, the *TDS* setting defines different times and is therefore used to determine the operating time during relay coordination. In an analogous way to an electromechanical relay, the operating time depends on the start position of the disk, if the disk starts from further away, it takes longer to reach the contact position for the same current. This effect is used to change the operating time for the relay based on the *TDS*, thereby generating a family of curves. By means of this parameter, the operating time of the relay can be controlled, as shown in Fig. 1.4.

1.3 OCR coordination

OCR coordination aims to establish a time sequence for the operation of the primary relay and its backup relays. If the primary relay fails to operate, the current contribution to the fault must be eliminated by the operation of the

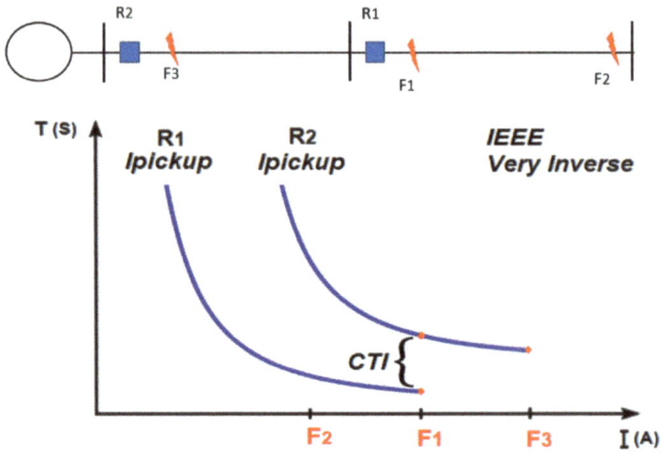

FIG. 1.5 Fault location for overcurrent coordination.

backup relays. To verify the backup function of these lines, fault detection during minimum generation must be guaranteed by considering the analysis of $(n-1)$ contingencies. The coordination criterion is based on compliance with Eq. (1.4) for each of the m pairs of relays. The breaker fault release time, uncertainties in the calculation of the fault current, and the errors in the instrument current transformers (CTs) are considered when determining the operating time of the backup relay (t_b) based on the time of the primary relay (t_p). The CTI must consider the total fault release time of the primary relay to avoid simultaneous tripping, and a time interval of between 0.2 and 0.5 s is usually applied [3].

$$t_b{}^m - t_p{}^m \geq CTI \tag{1.4}$$

The backup relays must detect faults in the remote line-end, and to achieve this, minimum phase-to-phase faults (F_2) are used to check the sensitivity of relay R_2 for the adjacent lines, as illustrated in Fig. 1.5. Minimal generation and arc-phase resistances (usually 5 Ω) are typically used to obtain minimal fault currents. The backup time depends on the primary time, owing to the uncertainty in the relay reach; this causes the coordination of n relays to have a functional dependence, since the time setting of one relay depends on the time setting of another relay.

In the early days of relay production, each manufacturer obtained a time curve as a result of the design of their relay, and protection engineers used acetates showing these time curves to achieve coordination. After the digital revolution, various organizations proposed standards [1,2] in which specific time curves were defined (and coordination software was developed). Traditionally, coordination is done via selection of a time curve from a catalog, meaning that the time curve determines the success of coordination. However, these standardized time curves were not found to solve all coordination problems, and due to the increase in coordination commitments in modern power systems with renewable sources (RESs), many manufacturers have defined their own curves. The process of relay design has therefore returned to the beginning.

Another point of view is that the coordination problem defines the time curve for enhancing the coordination via nonconventional curves (NCCs). These nonstandardized time curves can offer more solutions, since they are more flexible and can be adapted to different coordination problems. However, the curve selection process becomes more complicated, as it involves a greater number of time curves, and this increase in complexity is solved by optimization algorithms, which can obtain the settings for relays that comply with the restrictions on coordination.

Coordination of radial networks. The settings of the relays needed for coordination are based on I_{pickup} and TDS, since the type of curve and the constants (A, p, and in some cases B) must be previously defined. To ensure the protection of distribution and transmission networks, it is general practice that the same type of curve is selected; typically, the very inverse type is used due to its having tolerance times to adequate overload conditions for the operation of the electrical network. However, there is greater difficulty in selecting curves for industrial networks due to the presence of different protection devices such as fuses and low-voltage switches, and the need for protection of cables and transformers, and hence different types of curves are commonly selected. The value of I_{pickup} does not depend on the coordination process, and this is previously determined as shown in Eq. (1.3), using the maximum value of I_{load} from a contingency analysis.

The process of relay coordination is done with pairs of relays, where the relay on the downstream or load side is known as the primary relay, and the relay on the upstream or source side is the backup relay. The TDS settings for primary relays are known, and the objective is therefore to compute the TDS for the backup relays.

General procedure for coordinating relays. The processes of adjustment and relay coordination must be following the same steps for each pair of relays: after the first coordination is complete, the backup relay is considered the primary relay and the next upstream relay becomes the backup relay. The steps are as follows:

(1) Calculate I_{pickup} for all relays (Eq. 1.3).
(2) Determine the primary relay time, as TDS is known.
(3) Determine the backup relay time (Eq. 1.4).
(4) Based on T_{backup} and I_{sc}, determine the TDS for the backup relay.

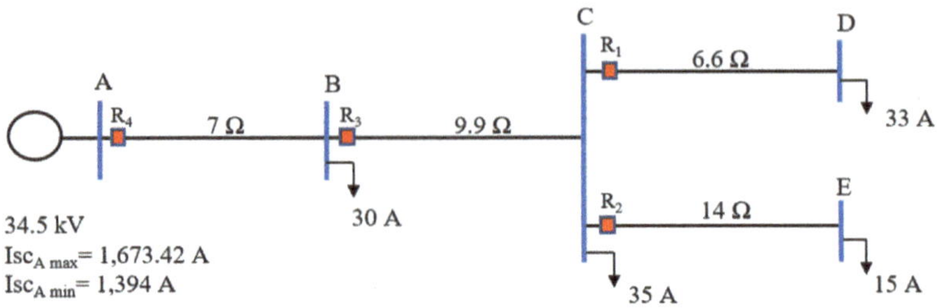

FIG. 1.6 Radial network.

Example. DOCR coordination is performed for the 34.5-kV radial system shown in Fig. 1.6. In this problem, the number of pairs for coordination is $m=3$.
Solution:

Coordination example, input data: Thevenin impedances

$$Z_{th\_max} = \frac{V_{LN}}{I_{sc}} = \frac{34{,}500/\sqrt{3}\,V}{1673\,A} = 11.9\,\Omega, \quad Z_{th\_min} = \frac{34{,}500/\sqrt{3}\,V}{1394\,A} = 14.28\,\Omega$$

Short-circuit currents,

$$I^{B}_{sc\_max} = \frac{V_{LN}}{Z_{th\_max} + Z_{AB}} = 1016.25\,A \quad I^{B}_{sc\_min} = \frac{V_{LN}}{Z_{th\_min} + Z_{AB}} = 906.21\,A$$

Fault currents

Bus	I_{sc_max}	I_{sc_min}
A	1673.42	1394
B	1016.25	906.21
C	675.2	624.79
D	551.76	517.63
E	457.89	434.14

Pickup currents:

$$I^{R_1}_{pickup} = 1.5 \times 33\,A = 49.5\,A$$
$$I^{R_2}_{pickup} = 1.5 \times 15\,A = 22.5\,A$$
$$I^{R_3}_{pickup} = 1.5 \times (35 + 15 + 33) = 124.5\,A$$
$$I^{R_4}_{pickup} = 1.5 \times (30 + 35 + 15 + 33) = 169.5\,A$$

Relay coordination process (Fig. 1.7):

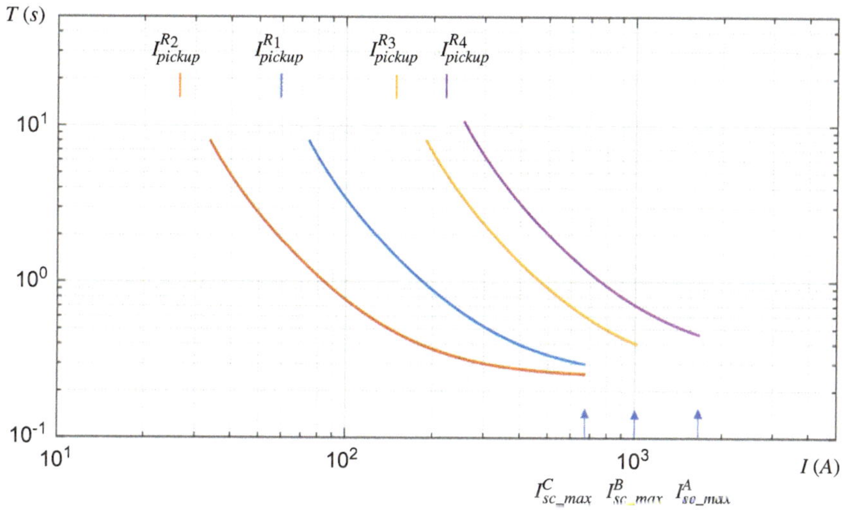

FIG. 1.7 Coordination chart.

Very inverse IEEE. $A = 19.61,\ B = 0.491,\ p = 2$

Relay R_1. Current $TDS_{R_1} = 0.5$ Relay R_2. Current $TDS_{R_2} = 0.5$

$$T_{R_1} = \left[\frac{19.61}{\left(\frac{675.2}{49.5}\right)^2 - 1} + 0.491\right] \times 0.5 = 0.298 \quad T_{R_2} = \left[\frac{19.61}{\left(\frac{675.2}{22.5}\right)^2 - 1} + 0.491\right] \times 0.5 = 0.256$$

Relay R_3. The slowest relay R_1 is selected. $CTI = 0.3$

$$T_{R_3} = T_{R_1} + CTI = 0.598 \quad TDS_{R_3} = \frac{T_{R_3}}{\frac{A}{\left(\frac{I_{sc}}{I_{pickup}}\right)^p - 1} + B} = 0.50$$

Relay R_4. Time coordination criterion

$$T^B_{R_3} = \left[\frac{19.61}{\left(\frac{1016.25}{49.5}\right)^2 - 1} + 0.491\right] \times 0.8 = 0.398 \rightarrow T_{R_4} = T^B_{R_3} + CTI = 0.698$$

$$TDS_{R_4} = 0.664$$

For the same fault location, different topological configurations and network operation conditions can modify the fault current contributions. Fig. 1.8 shows the time curve for an OCR for a range of currents and operation times. Both relays (relays R_3 and R_4 in Fig. 1.16) see the same fault current, so current variations will result in higher *CTIs*; however, the proper sequence of operation is obtained only in the case where the primary relay does not operate.

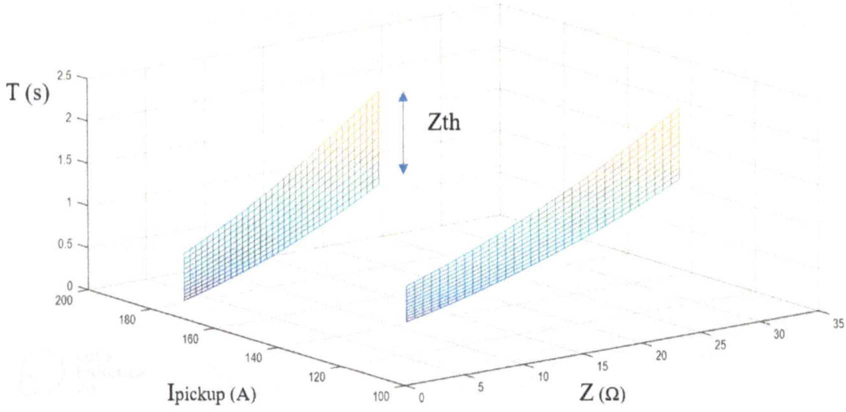

FIG. 1.8 Tridimensional chart of relay pair.

The process of calculating the settings and coordination relays is carried out based on the maximum border values for both I_{sc} and I_{load}. In this way, we ensure that under any other operating condition, coordination criterion (3) is not violated.

Sensitivity. To provide a backup function, minimal faults at the adjacent remote line-end must be detected. To overcome the sensitivity filter, each relay in the network is analyzed as follows:

$$S_b = \frac{Isc_{2\varnothing}}{I_{pickup}} \geq 1.5 \tag{1.5}$$

where $Isc_{2\phi}$ is the minimum phase-to-phase fault current at the remote line-end with a fault resistance of 5 Ω. The backup relay (R_2) must operate for fault F_2 in Fig. 1.5.

A lower limit of 1.5 is considered in Eq. (1.5) since the operation times are large and are therefore useless for protection purposes for multiples of current (M) between 1 and 1.5. This range can be considered a dead zone for protection purposes (Fig. 1.9). The curve of M versus T is generally provided for coordination, and the protection engineer needs to verify that the active zone of the relay is used for the adjustment.

Security. The range of M was established as 1.5–20. A value of 20 was used to avoid a current of more than 100 A in the secondary and hence damage to the relay (Eq. 1.6). When a current of 5 A is used in the secondary CT, safety can be ensured when:

$$\frac{I_{sc\,max}}{20} \leq CT \tag{1.6}$$

1.4 DOCR coordination

DOCRs are designed to detect faulty contribution in the right current contribution and find applications in electrical networks with fault supply on both sides of the line. This contribution can be through other lines with a source connection

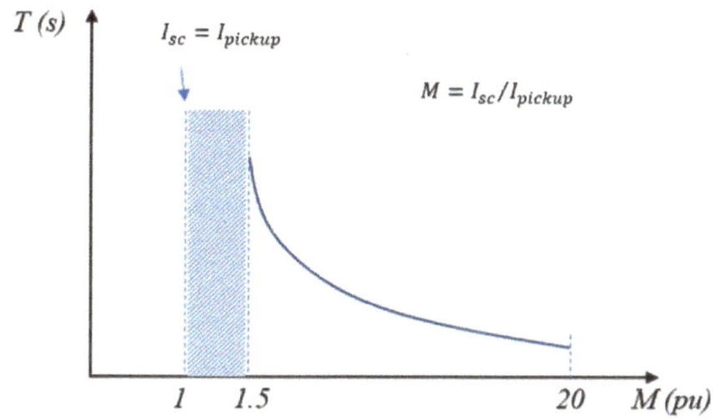

FIG. 1.9 Multiples of current of time curve.

or through the connection of distributed generation (DG) sources. The directionality is calculated based on the use of potential transformers (*PTs*) for the polarization signal and the current signal from the *CT* as the operation signal; the phase difference between these signals defines whether the fault is in front of or behind the relay [4]. When a neutral connection is available, the polarization can be by $3I_0$. For operation of the relay, the fault current must be detected based on I_{pickup} and the fault location must be in front of the relay and activated by the directional function, as illustrated in Fig. 1.10.

The polarization signal must not change phase during the fault. In contrast, the operation signal must shift by around 180 degrees for faults in front of and behind the relay, and this phase change is exploited by the directional function. The most common directional function of a DOCR is illustrated in Fig. 1.11. For relay protection with phase a, the polarizing quantity is V_{bc}. The terms *maximum torque line* and *zero torque line* are inherited from the electromechanical era and are also known as operating lines or thresholds in digital relays.

DOCRs mainly find applications in mesh networks as both backup protection for phase and earth in transmission networks and primary protection for active subtransmission and distribution networks.

Meshed networks. In meshed networks, additional contributions from adjacent lines are frequent, and this infeed current effect causes the primary relay to see more fault current than the backup relay. Coordination is not affected

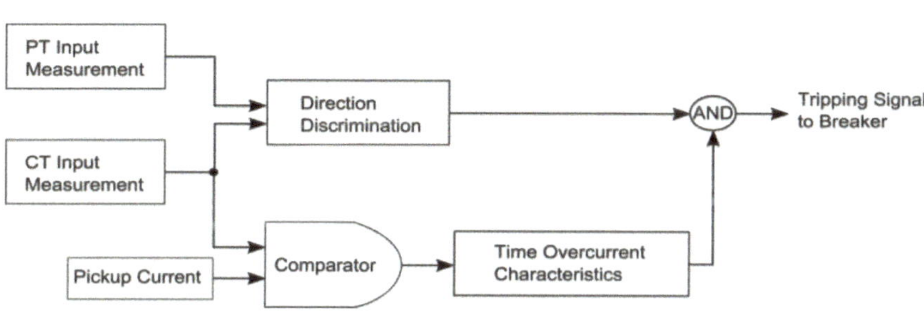

FIG. 1.10 Logic scheme for tripping of a DOCR.

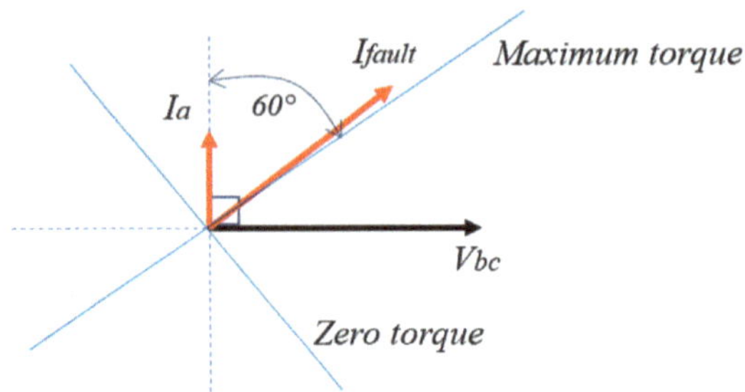

FIG. 1.11 Ninety-degree type connection for DOCR.

because only the primary operation time is reduced, and although the *CTI* is greater, the coordination conditions are fulfilled. In addition, it is likely that the I_{pickup} for a primary relay (R_1) is greater than for a backup relay (R_2), and the crossing of these curves can make coordination difficult (Fig. 1.12). The use of different types of time curves, different short-circuit magnitudes, and different relays (I_{pickup}) can cause crossing of the curves. Relay R_2 is sensitive to a detected fault F_2 (Fig. 1.5), as shown in Fig. 1.12; however, if F_2 is close to the intersection of the curves, coordination may be lost. It is therefore important to verify that the coordination of faults is fulfilled at both ends of the line for primary relays.

The complexity of the coordination problem increases exponentially with the size of the power system. For example, the 14-bus IEEE test system has 40 relays, resulting in 50 coordination pairs, while the 30-bus IEEE test system has 82 relays with 120 coordination pairs. In case of updates to a relay due to a change in the topology or operation of the electrical network, recoordination of all the relays is necessary even if there is no change in the load or fault current of any relay. In the coordination solution provided by optimization methods, this complexity forces the optimization algorithm to perform a more exhaustive search to comply with the restrictions. In the case of nonpermissible operating times for some relays, due to the limitations of the protection principle (maximum demand, low short-circuit current) or a lack of sensitivity, this result may affect other relays, giving long times owing to the coordination requirement.

The coordination criterion establishes the operational sequence of the relays. When a fault occurs on the protected line, the primary relay must operate immediately; however, if it does not operate, the backup relay must be given enough time to provide backup to the adjacent line. The functional dependency between relays means that a modification of the setting for one relay has a domino effect, affecting the coordination of the entire network (Eq. 1.4).

Small generators such as turbo diesel must be included in the fault analysis [5,6], because fault contribution must be increased causing a reduction in the relay operation time, simultaneous trips are possible. As a result, loss of coordination can occur, as shown in Fig. 1.13 between relays R_1 and R_2 in F_1^{DG}. In a

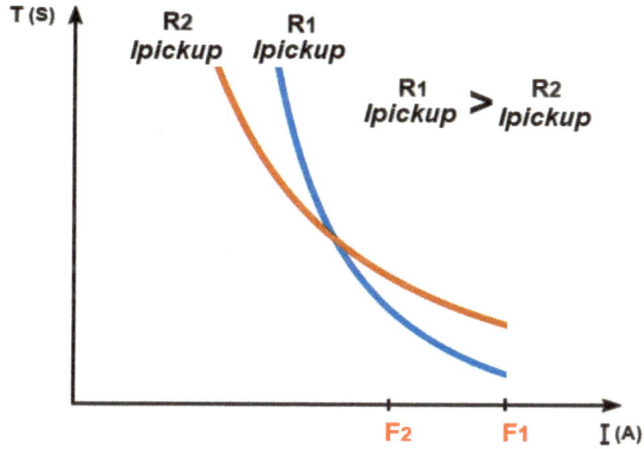

FIG. 1.12 Fault location for overcurrent coordination.

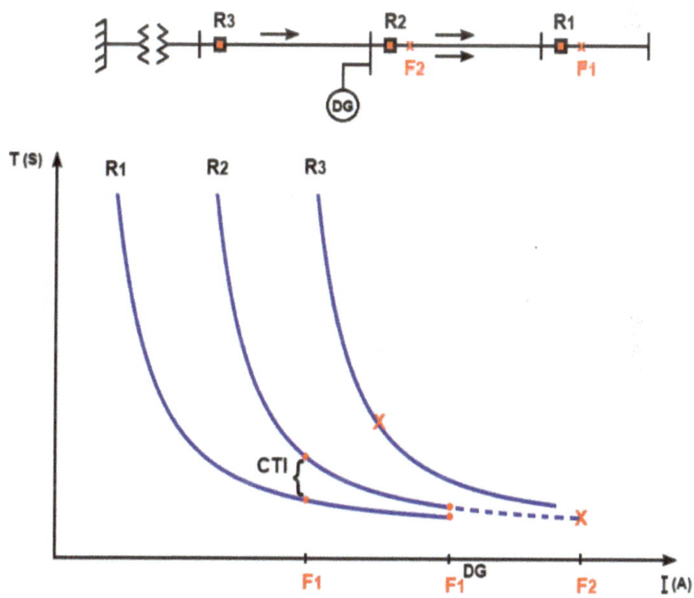

FIG. 1.13 Time OCR coordination.

different sense, the infeed fault contribution of DG (F_2) does not see by relay R_3, and its time is not affected; the coordination between R_2 and R_3 will be maintained, but with a higher *CTI*.

The increase in the fault current due to a DG source can have some effect on the relay timing, as it may be within the time asymptotic region of the time curve. That is, the mere presence of DG does not imply a loss of coordination, and this condition must be evaluated. However, noncontrolled sources connected by converters, such as wind power or photovoltaics, should not be considered, as the fault contribution is poor and short-lived. The sensitivity of DOCRs can therefore be compromised.

1.5 General considerations

The initial data used for coordination of DOCRs are the load flow and fault current. The solution to the nonlinear power flow equations consists of the voltages and load currents of the lines. The magnitude and direction of I_{load} are used to calculate I_{pickup} in order to avoid false tripping of the relay, and a contingency analysis must be carried out to obtain the maximum values. Likewise, a fault current analysis must consider the topologies and sources that result in the maximum fault current, and it is important that solid faults and the open end of the line should be considered. The fault current is used for coordination, while the location of the fault is also very important; depending on the optimization formulation, it may be necessary to calculate phase-to-phase faults for relay sensitivity verification.

Contingence analysis. Operating conditions that increase the current during line or generator operations and that can result in power transfer must be considered. An analysis of $n-1$ contingencies is carried out to compute I_{load}. As shown in Fig. 1.14, when line 1–4 is out of service, power is transferred to line 1–5, giving an increase in current that would cause the R_{15} relay to false operation.

The name of each relay is assigned based on the network topology; for example, relay R_{14} is on a branch line between buses 1 and 4, although the relay itself is located on bus 1. The six-bus IEEE test system [4] in Fig. 1.15 shows the names assigned to the relays.

The formation of coordination pairs uses a simple methodology. For each primary relay, the backup relays are on all adjacent buses in the same direction

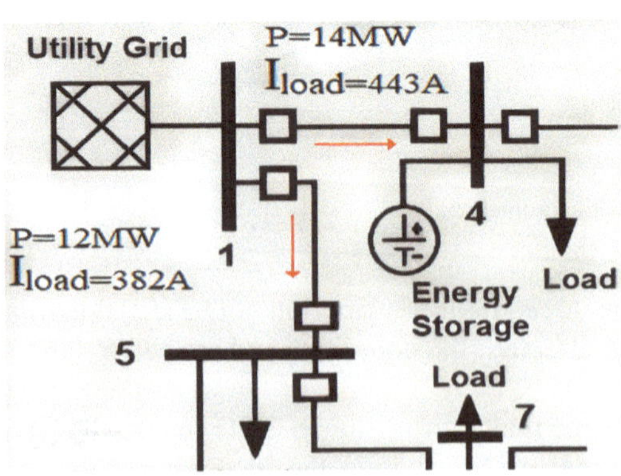

FIG. 1.14 Power transferred during line outage.

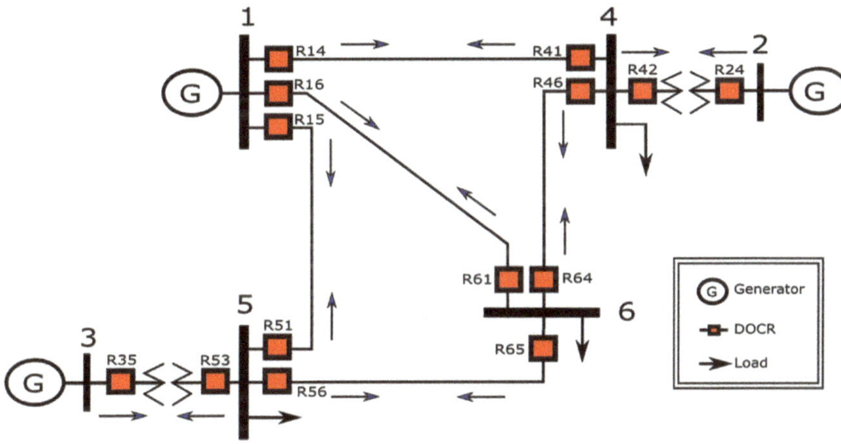

FIG. 1.15 Six-bus IEEE network system.

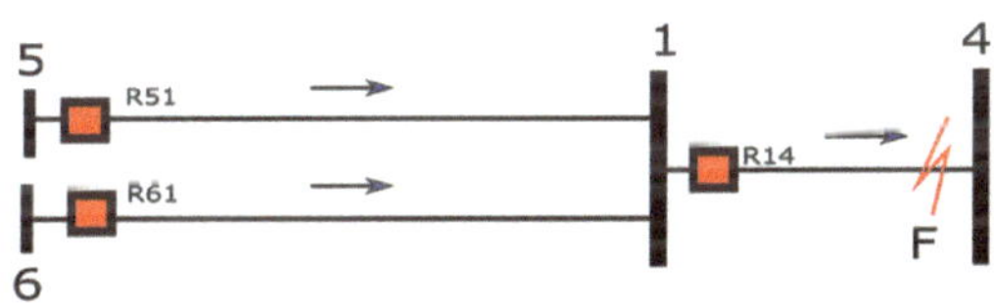

FIG. 1.16 Coordination pairs.

of operation. For example, for clockwise coordination, the lines 5-**1** and 6-**1** are located adjacent to relay **R14**, as shown in Fig. 1.16. Hence, relays R51 and R61 are the backup relays, and the backup-primary coordination pairs are **R51-R14** and **R61-R14**. Counterclockwise coordination is done in the same way. This process has computational advantages and can be carried out automatically. Power transformer protection may also be included, although coordination with the transformer damage curve must be verified.

The maximum fault currents required to calculate the operation time of each relay are determined by a fault current analysis. The solid fault current is located at the bus nearest to the primary relay; for phase relays, three-phase faults are simulated, whereas for ground relays, one-phase ground faults are used. In radial lines, both primary and backup relays have the same fault current, but in meshed networks it often happens that the fault currents for the primary relays are greater than for the backup relays. It is therefore important to compute the fault current at both ends of the line to ensure there is no crossing of the curves. For example, Fig. 1.16 shows that the minimum fault current (*F*) opening the line-end of bus 4 must be seen by relay R_{51} and R_{61} as backups to R_{14}.

Sensitivity. When coordination is verified at both ends of the protected line, a sensitivity analysis can be used to identify relays that cannot be coordinated due to system constraints. This does not mean that nonsensitive relays are not adjusted; instead, they are flagged and excluded from coordination programs as they result in infeasible solutions. Eq. (1.5) is used to analyze each coordination pair to determine if they can be coordinated.

The six-bus IEEE test system is used for the sensitivity evaluation (Fig. 1.15) with a voltage of 20 kV and 14 relays. The number of relays that are involved in the coordination process is simplified when we exclude the relays that protect the transformers. The total number of relays that meet the sensitivity criteria is between 10 and 14 (Table 1.2). Relays that do not meet the sensitivity criteria are shown in red.

A lack of sensitivity must be resolved by the use of negative sequence relays or distance relays, as described as follows.

Approach 1: Negative sequence relays. A solution to the problem of sensitivity (and partially to the problem of high operation times) is to use a negative sequence current in the OCR. This method yields high sensitivity as its setting value is not affected by I_{load}. Phase protection (sec +) and earth protection (*sec* o) are functionally independent and do not need to be coordinated. Although both the phase and the ground can operate in the case of single-phase faults, there is generally no conflict, since the ground protection is faster and more sensitive. However, the negative sequence relay must be coordinated with the positive sequence relays, since phase-to-phase faults will be seen by both relays and coordination may be compromised.

To provide backup for phase-to-phase faults, the negative sequence overcurrent function can be enabled with the following criteria for its application, as illustrated in Fig. 1.17.

TABLE 1.2 Results of a sensitivity evaluation.

Backup relay	I_{pickup} (A)	$Isc_{2\emptyset}$ (A)	Sensitivity
51	1718	2750	1.60
41	1992	2144	1.08
62	2428	3380	1.39
52	2916	4016	1.38
63	909	3345	3.68
43	800	4186	5.23
34	800	3729	4.66
25	2916	4122	1.41
15	1718	3850	2.24
36	909	3496	3.85
14	1992	4948	2.48
26	2428	3122	1.29

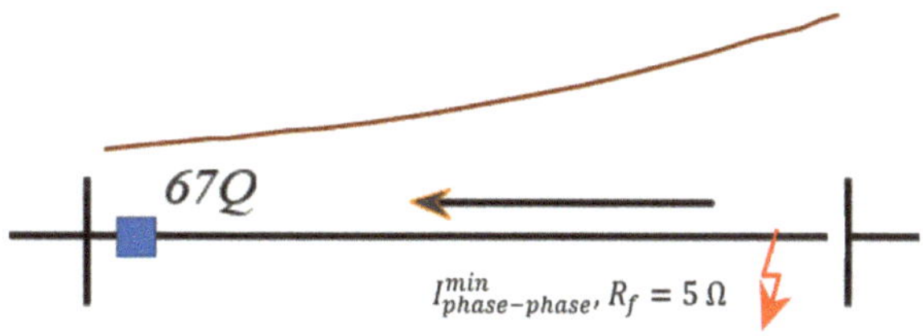

FIG. 1.17 A negative sequence relay.

(a) It must act as a backup for the primary relay of the protected line for phase-to-phase and phase-to-phase ground faults (Eq. 1.7).

$$I^{Q}_{pickup} = 0.5 \times I^{min}_{phase-phase} \tag{1.7}$$

(b) A coordination margin must be maintained with regard to the remote bus line protection, for phase-to-phase "close-in" faults.

(c) Its application should be evaluated in networks with high imbalances and asymmetric contributions of the RES.

Approach 2: Distance relay and DOCR coordination. An alternative solution to the problem of sensitivity of OCRs in networks with low levels of short-circuit current is the use of distance relays. An impedance measurement allows for better discrimination between the load and fault states for two basic reasons: (a) the magnitude of the impedance varies more than the current when a fault occurs, because the voltage is reduced; and (b) the impedance angle varies as the fault occurs, which serves as an additional indicator of the fault status. Since the coordination of overcurrent and distance relays is difficult, distance elements have been used in distribution networks to monitor overcurrent elements.

Four coordination points are used, based on the three zones of the distance relay. The most critical coordination points are located where the time curve of the DOCR and the characteristic curve of distance relay are closest (points *b* and *d* in Fig. 1.18). If only the first and second zones of the distance relay are considered, coordination is simplified, as point *c* will not be involved. The setting and coordination criteria are described in detail in Chapter 8. The performance of DOCRs and distance relays is different, and crossing of the curves is possible, meaning that the restrictions must be included in the coordination process.

The dynamic performance of the OCR is determined by the dynamics of the fault current, and the operating time until the moment at which the contacts are closed is unknown. The integration process inherent to the OCR must be considered when carrying out coordination with relays that rely on another operating principle, such as distance relays. The zone of protection of DOCRs is very dynamic, as it depends only on the current signal. In addition, other factors such as phasor estimation, the design parameters, and the functional structure of the relay must be considered.

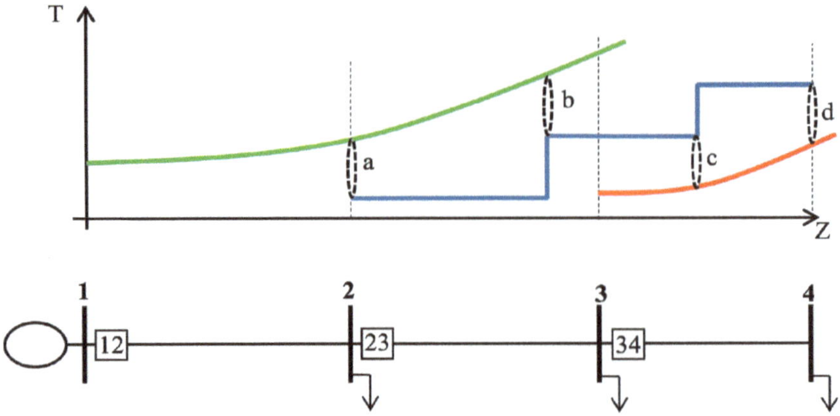

FIG. 1.18 Coordination points for distance relays and OCRs.

1.6 Functional structure of digital relays

When designing digital OCRs, the objective is to emulate the operation of overcurrent protection devices, in terms of both their dynamic response to operation in the face of time-varying fault currents and the state of restoration when the fault is eliminated, either by the same device or by another form of protection.

Fig. 1.19 shows a functional diagram of a DOCR, where the analog processing and the analog-digital signal conversion are not shown for simplicity. The digital filter receives digitized samples $(i_r)_k$ of the input signal and delivers at its output the modulus of the phasor $(I_r)_k$ for each sampling instant. The instantaneous and inverse time elements of the relay process the input signal and emit trip signals where appropriate, which are processed and amplified by the output element.

Due to the simplicity of OCRs, the RMS current can be used for protection proposes; however, for many applications, the phase of the signal is required. For example, a polarization signal is needed for the directional function, to allow for a comparison between the operational signal (current) and the polarization signal (V or I). For phasor calculations, the size window of one cycle is commonly used to reach a compromise between speed and accuracy.

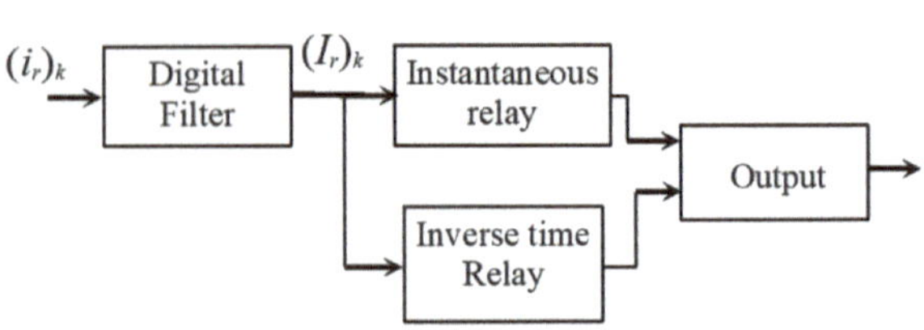

FIG. 1.19 Simplified functional diagram of a digital overcurrent relay.

The relay receives a current signal with various periodic and nonperiodic components. Although these are caused by the fault, they are not useful for locating the fault due to their random nature. Phasor estimation must be applied to remove these components using analog and digital filters [7,8], and DOCR operation must be performed using only the fundamental component of the signal, since it is affected by the location of the fault.

1.6.1 Phasor estimation

The process of phasor estimation relies on analog and digital filters (Fig. 1.20). The analog filter used is a Butterworth, as it has a flat response in the passband and a monotonic decrease in the stop band [9,10] (Fig. 1.21). An analog-to-digital converter (ADC) is also used in which the sampling frequency must

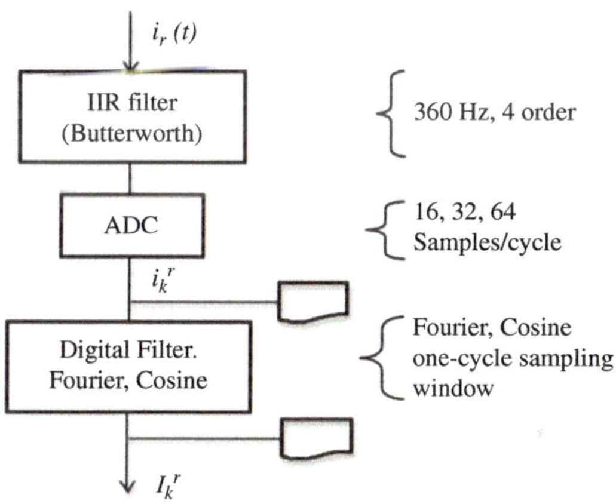

FIG. 1.20 Phasor module estimation.

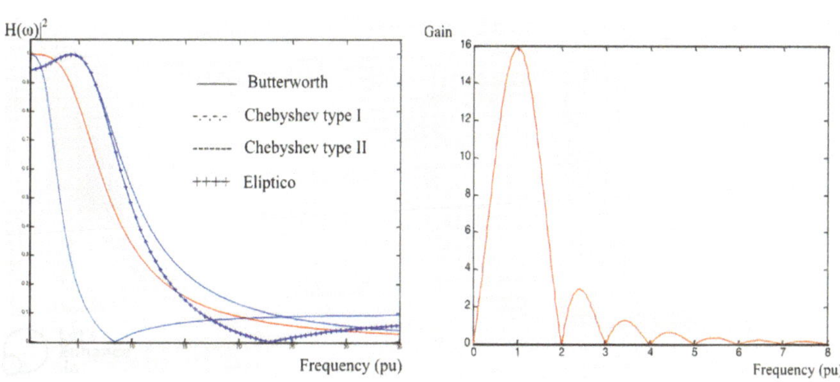

FIG. 1.21 Frequency spectra for IIR and FIR filters.

be defined to represent the original signal, to avoid aliasing. For protection applications, a minimum sampling frequency of 960 Hz allows for good characterization of the signal and frequency overlap with the digital filter. It is also possible to remove the analog filter by oversampling the signal [11].

Digital filtering is performed with finite impulse response (FIR) filters, since they do not involve recursion. In addition, infinite impulse response (IIR) filters generally produce phase distortion, unlike the FIR type that has a linear phase response, and this allows the frequency response to be zero at the natural harmonic frequencies, thus enabling the rejection of these components from the signal (Fig 1.21B). A cosine filter is preferred as it yields better rejection of the direct current component.

The phasor computation is done using a Fourier algorithm, where the real (cosine) and imaginary (sine) components are used. Fig. 1.22 shows the phasor estimation process. The sliding window effect is important for protection purposes, as the relay response times are reduced. One phasor is estimated at each position (sample k) of the data window of size K, and for protection it is important that the window is shifted for each sample, to avoid delays in the operation of the relay. This leads to a processing of each protection unit in the intersample time, for that, the computational overload of the relay must be evaluated, since in the event of an asymmetric fault condition, all the protection units can be activated.

The obtained phasors are nonrecursive; it is possible to compensate for the phasor speed to make them recursive, but this has no practical use. Phasor processing can be simplified by determining only the real (cosine) or imaginary

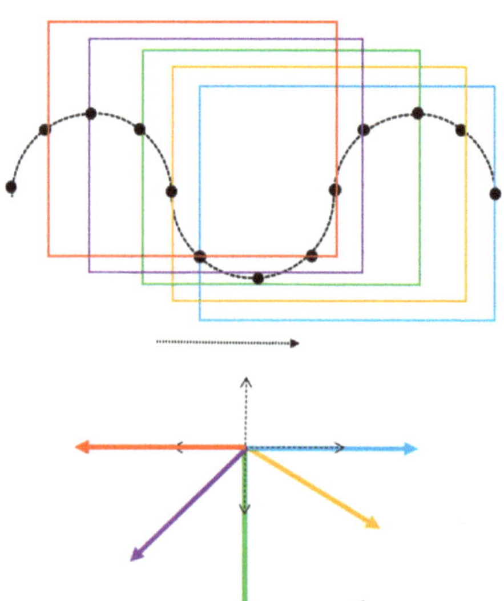

Fourier coefficients

$$Y_c^{(i)} = \frac{2}{K}\sum_{k=1}^{K} x_k \cos(k\theta)$$

$$Y_s^{(i)} = \frac{2}{K}\sum_{k=1}^{K} x_k \sin(k\theta)$$

$$|Y^{(i)}| = \sqrt{\left[Y_c^{(i)}\right]^2 + \left[Y_s^{(i)}\right]^2}$$

$$\varphi^{(i)} = arctan\frac{Y_s^{(i)}}{Y_c^{(i)}}$$

FIG. 1.22 Phasor estimation for DOCRs.

(sine) part of the phasor. The cosine filter is determined by considering a component of the phasor with a delay of a quarter of a cycle and assuming that this is the value of the imaginary component (Eq. 1.8). This has the benefit of rejecting the direct current component, which is very advantageous for protection purposes; however, it has the disadvantage that the lateral lobes have greater magnitude, meaning that there is less rejection of interharmonic components than with the Fourier filter.

$$\left|Y^{(i)}\right| = \sqrt{\left[Y_c^{(i)}\right]^2 + \left[Y_c^{(i-K/4)}\right]^2} \quad \varphi^{(i)} = arctan\frac{Y_c^{(i-K/4)}}{Y_c^{(i)}} \tag{1.8}$$

For protection applications, estimation algorithms need to respond quickly to discontinuous signals caused by a fault. Phasor estimation for a current signal during a fault is shown in Fig. 1.23, and it can be observed that during the passage of the data window, there is an error in the estimated phasor due to the discontinuity caused by the fault. The duration of this error is critical in terms of the operation of the relay, since the delay in the operation of the relay is proportional to the duration of the window. In the case of fractional windows, the estimate is generally poor and the one that is processing asymmetric signals.

In active electrical networks with RES connection via converter-interfaced generation resources (CIGRs), signals with nonfundamental frequency components are present, such as harmonics and subharmonic or interharmonic components. The use of a Fourier filter gives natural zeros in the harmonic frequencies, allowing for the rejection of these components. However, components that are not integer multiples of the fundamental frequency give rise to an error in the estimate. To identify these components by means of a Fourier method, it is necessary to estimate several cycles of the signal, and the estimate

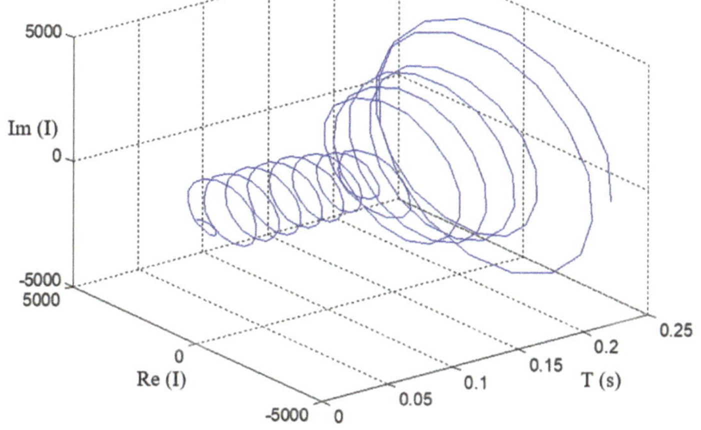

FIG. 1.23 Current phasor during a fault.

of the interharmonics is carried out based on a relationship where the required number of cycles is obtained by dividing the fundamental frequency by the required frequency of the interharmonics [12]. For example, for a resolution of 5 Hz in a 60-Hz signal, 12 cycles are necessary. As previously stated, this condition cannot be used for protection purposes. As an alternative to using a signal filter, an estimate of the signal is performed to identify the frequency components present in the signal.

Signal estimation by the Prony method. Although this was developed by Gaspard Riche de Prony in 1795, its main practical applications were discovered in the digital computer age. Like the Fourier transform, the Prony method extracts valuable information from a uniformly sampled signal and builds a series of damped complex exponentials or damped sinusoids (Fig. 1.24). This allows us to estimate the frequency, amplitude, phase, and damping components of a signal.

The extended Prony method fits a linear combination of exponential terms to a signal $y(t)$, where each term is composed of four components: amplitude A_n, damping factor σ_n, frequency f_n (hertz), and angle θ_n (radians) [13]. The interval between samples is defined as N. Each exponential term with a different frequency is treated as a unique mode of the original signal $y(t)$, and the four elements of each mode can be identified from the representation of the state space of a data record with fixed sampling (Eq. 1.9). The procedure used to estimate signals for transmission and distribution networks is described in [13].

$$y(t) = \sum_{n=1}^{N} A_n e^{\sigma_{nt}} \cos\left(2\pi f_n t + \theta_n\right) \tag{1.9}$$

Application of the Prony method to protection. Conventional relays use a cosine filter to eliminate components other than the fundamental signal; however, they are not capable of eliminating the subharmonic and interharmonic components present in the signal. In contrast, the Prony method can detect

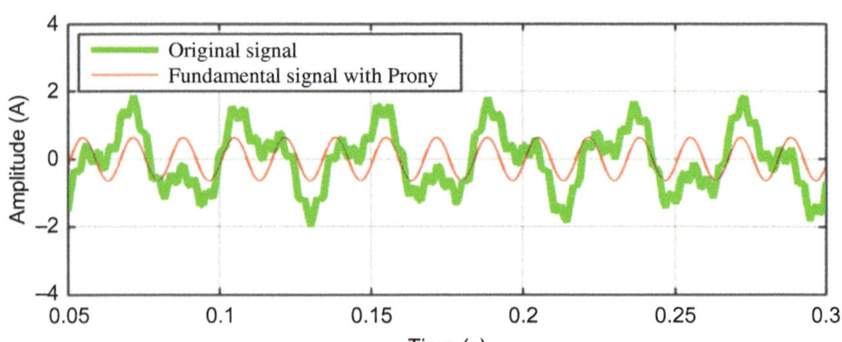

FIG. 1.24 Frequency estimation of signals with the Prony method.

any frequency component present in the signal, where the estimates will be equal to $M-1$ (e.g., for $M=16$ there will be 15 estimates of the signal (t)). One way to select the optimal estimate is by determining the error for each sample and then obtaining the average error percentage for the estimate.

The generation of nonfundamental frequencies in active networks with a strong presence of sources with CIGRs is a concern in the fields of protection and measurement. In the case of protection, the relays reject harmonic components through digital filtering; however, subharmonic and interharmonic frequencies affect the performance of the protection. In the specific case of OCRs, the error introduced by the presence of these components is not of great concern, as the protection operates over several cycles, but in the case of distance or differential relays, these errors can be significant.

1.6.2 Instantaneous OCR

The instantaneous OCR is designed to operate without an intentional time delay when the current exceeds the relay setting [14]. It is used only as primary protection to reduce the trip time of the inverse time OCR, and its protection zone is set to 120% of the maximum fault current value at the remote node. A reduction in the short-circuit magnitude decreases the operating range of an instantaneous relay, and it can even be lost under certain operating conditions.

Fig. 1.25 shows a functional diagram of the instantaneous element of the relay, which basically consists of an amplitude comparator and a counter. The comparator emits a trip signal for each sampling instant within which the input signal $(I_r)_k$ exceeds an operating threshold I^i_{pickup}, which represents the pickup current of the instantaneous element:

$$(\mathrm{I}_r)_k \geq I^i_{pickup} \tag{1.10}$$

Once the comparator has been activated, it continues to emit trip signals as long as $(I_r)_k$ does not fall below its reset threshold I^i_{rep}, which constitutes the reset current of the relay:

$$(I_r)_k \geq I^i_{rep} \tag{1.11}$$

The comparator has a low level of hysteresis, which ensures reliable operation even for variable currents close to the pickup current. The counter compares the number of consecutive trigger signals received based on a certain

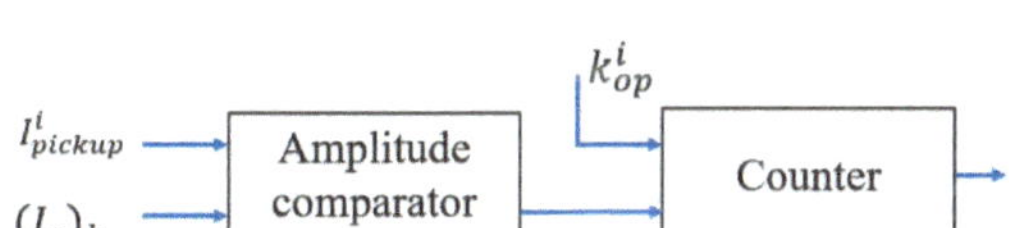

FIG. 1.25 Functional diagram of an instantaneous relay.

threshold k_{op}^{i}, and when this threshold is reached, it emits a definitive trigger signal. If the comparator interrupts the emission of trigger signals before filling the counter, it is reset, which further improves the security of the relay. If the sampling period is denoted as Δt, the operating time of the instantaneous element is given by:

$$T^{i} = k_{op}^{i} \Delta t \tag{1.12}$$

1.6.3 Inverse time relay

The dynamics of induction disk electromechanical relays are reproduced in the design of inverse time DOCRs [15]. The digitized version of the analog relay must emulate the dynamic behavior of the relay under time-varying fault currents and the operating reset state when the system current is reduced below the relay reset current.

A simplified functional diagram of a generalized version of an inverse time DOCR is shown in Fig. 1.26. The function generator receives as inputs the phasor $(I_r)_k$, which represents the fundamental component of the current, and I_{pickup}, and forms the output signals $H(I_k)$ and $J(I_k)$, where $I_k = (I_r)_k / I_{pickup}$ is the operation current.

The time curve of the relay can be expressed by the equation $T = F(I)$:

$$F(I) = \frac{J(I)}{H(I)} \tag{1.13}$$

The expression in Eq. (1.13) is a general equation that describes the operating principle of an inverse time relay. It expresses the functional relationship between the function generator ($H(I)$, $J(I)$), the integrator ($G(T)$), and the time-current characteristic $T = F(I)$.

The output signal of the integrator is a discrete approximation of the integral:

$$G_k = \Delta t \sum_{k=1}^{k} H(I_k) \tag{1.14}$$

where G_k represents the accumulated value of the integrator at the time of processing the sample k. The value of the generated function $H(I_k)$ depends on the

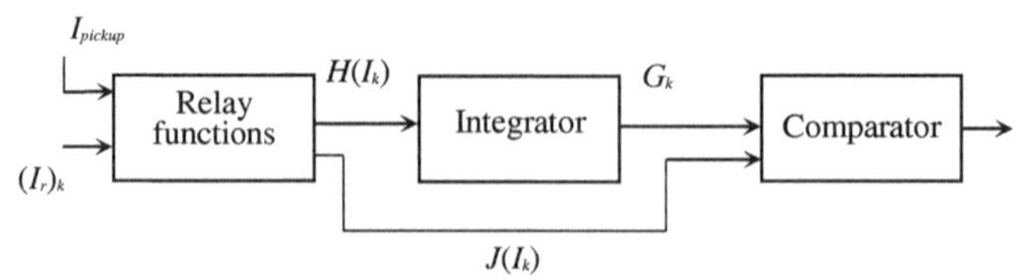

FIG. 1.26 Functional diagram of an inverse time OCR.

magnitude of the current phasor, and hence the expression in Eq. (1.14) represents the dynamic behavior of the OCR before variable fault currents. This equation forms the basis of the digital relay.

In the comparator block, the amplitude of the signal G_k is weighted with the function $J(I_k)$. The operating condition is fulfilled when:

$$\Delta t \sum_{k=1}^{k_F} H(I_k) = J(I_k) \tag{1.15}$$

The relay operates the instant k reaches a value equal to k_F and Eq. (1.15) is satisfied. The operating time is given by:

$$T = k_F \Delta t \tag{1.16}$$

Solving for Δt in Eq. (1.15) and substituting into Eq. (1.16) gives:

$$T = k_F \frac{J(I_k)}{\sum_{k=1}^{k_F} H(I_k)} \tag{1.17}$$

If for analysis purposes the current during the fault is considered constant ($I_k = I$, $H(I_k) = H(I)$, $J(I_k) = J(I)$), Eq. (1.17) is reduced to the general equation in Eq. (1.13):

$$T = F(I) = \frac{J(I)}{H(I)} \tag{1.18}$$

Within each sampling period, the current remains constant at the value calculated for the previous sample. Hence, for $I = I_k$ (a sampling period corresponding to sample k), Eq. (1.18) can also be written as follows:

$$F(I_k) = \frac{J(I_k)}{H(I_k)} \tag{1.19}$$

Solving for $H(I_k)$ in Eq. (1.19) and substituting into Eq. (1.15), we have:

$$\frac{J(I_k)}{\Delta t \sum_{k=1}^{k_F} \left[\frac{J(I_k)}{F(I_k)}\right]} = 1 \tag{1.20}$$

Eq. (1.20) is the generalized equation for the inverse time DOCR. In the same way as for analog relays, simplification of $J(I_k) = K$ is usually done for digital relays. By substituting this condition into Eq. (1.20), we obtain the form proposed in [1]:

$$\sum_{k=1}^{k_F} \left(\frac{1}{F(I_k)}\right) \Delta t = 1 \tag{1.21}$$

OCRs are devices of an electrical input signal; the representation of its operating area can be done on a numerical axis [15]. OCRs behave as maximum

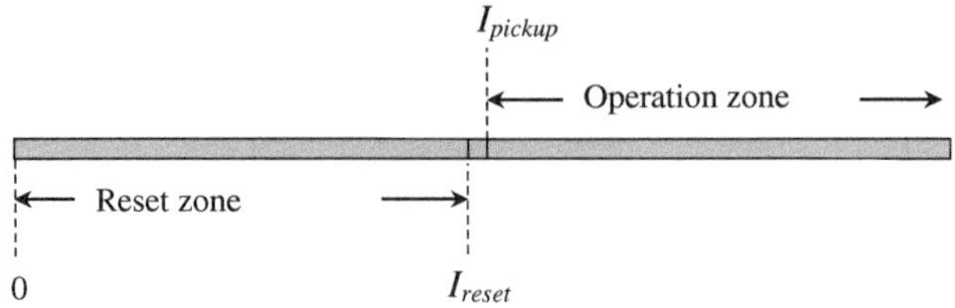

FIG. 1.27 Operation and reset zones of an OCR.

elements in their zone of operation, since they operate at signal values greater than a predefined setting, and as minimum elements within their reset zone, since they are repositioned (return of the induction disk or its electronic or digital equivalent to its initial position) for current values less than a value established by the design of the electromechanical relay or programmed in a digital relay. The operation and reset zones are shown in Fig. 1.27.

Integrator function. The functional relationship in its discrete form (Eq. 1.18) is valid both for the operating zone and for the reset zone of an inverse time relay. The function $H(I_k)$ represents the dynamic behavior of the induction disk caused by the mechanical torque of electromagnetic origin in the operation zone ($H_O\,(I_k)$) and by the retention spring of the relay in the reset zone ($H_R\,(I_k)$). The function $J(I_k)$ represents the time lever in the operating zone ($J_O(I_k)$) and the return time of the induction disk to the actuation position from the operating position ($J_R(I_k)$). The forces in each zone establish different expressions for the functions $H(I_k)$ and $J(I_k)$ for each zone of operation.

Although different expressions are defined for $H(I_k)$ and $J(I_k)$ for each zone, the integration process is the same for both zones. The integration of the function $H(I_k)$ for $I_k \geq 1$ is positive, emulating the advance of an induction disk toward the operating position in an electromechanical relay. The integration of the $H(I_k)$ function for $0 \leq I_k < 1$ is negative, representing a return of the induction disk toward its initial position. Hence, I_k determines the value and sign of the integral, and the accumulated value of the integrator G_k. Eq. (1.22) gives expressions for the integration for each zone.

$$\Delta G_k = \begin{cases} \Delta t\ H_o(I_k)\ I_k \geq 1\ \textit{Operation zone} \\ \Delta t\ H_R(I_k)\ 0 \leq I_k \leq 1\ \textit{Re position zone} \end{cases} \tag{1.22}$$

Where: $G_{k+1} = G_k + \Delta G_k$.

In an analogous way, the accumulated value of the integrator (G_k) represents the position of the induction disk of an electromechanical relay at any moment (Eq. 1.15), where the function $H(I_k)$ is defined for the operation and reposition zones. The integration of the function is positive for the operating zone, representing travel of the disk to the trip position, while the function is negative for the reset zone, representing the return of the disk to the actuation position.

The operation and reposition condition are satisfied when:

$$\sum_{k=1}^{k_{op}} H_o(I_k)\Delta t = 1, \quad \sum_{k=1}^{k_{rep}} H_R(I_k)\Delta t = 0 \tag{1.23}$$

where k_{op} and k_{rep} are the operational and repositioned samples, respectively. The integrator is bounded by values of zero and one:

$$G_k \begin{cases} \geq 0 \\ \leq 1 \end{cases} \tag{1.24}$$

When the integrator is zero, the relay is in its initial position; when the integrator equals one, the relay operates. If a new trip condition occurs during the disk reset trip (e.g., the second unsuccessful operation of a recloser), the integrator G_k will start to increase again and will indicate the relative position of the induction disk.

1.6.4 Dynamic performance

The integer function is done with Eq. (1.21), where the discrete integration emulates the disk displacement of an analog relay, and the performance depends on the dynamics of the fault current. Hence, OCRs are a unique form of protection where the operating time is unknown, and simulations are needed to check the correct operation of the relays.

A fault log was used to test the signal file. A phase-to-phase fault was logged for a 34.5-kV distribution grid, where the OCR had a moderate inverse curve [1], $TDS = 1$, and the value of $I_{pickup} = 25$ A. Using the time curve in Fig. 1.28, the operating time was obtained as 0.59 s. Fig. 1.29 shows the record of the fault current. For the sake of simplicity, the values shown in the figure are relative to the system's primary relay, and due to the dynamic evaluation the resulting time is 0.803 s (Fig. 1.29). The difference in operation time between the evaluation of the curve and the dynamic evaluation of the relay operation is an expected behavior of the overcurrent principle: this is what causes the uncertainty in the protection operation time. The code used for the dynamic evaluation of the DOCR is shown in Fig. 1.30.

1.6.5 Limitations of the OCR

An OCR uses current as a parameter to indicate the location of the short circuit, as its magnitude depends on the electrical distance to the point of fault. However, there are two fundamental problems with this approach: (a) the short-circuit current also depends on the type of fault and the operating regime of the system; and (b) the load or prefault current may be comparable to the short-circuit current, which makes it difficult to differentiate between the normal and fault regimes. The limitations of overcurrent protection are derived

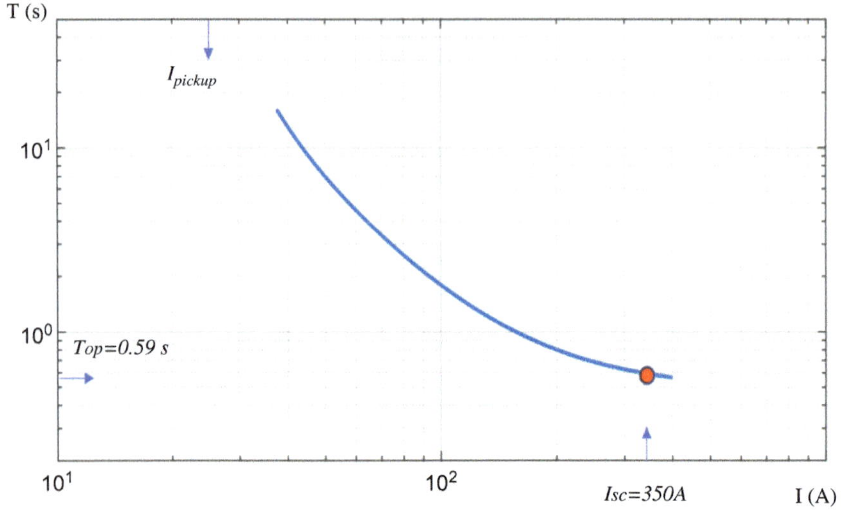

FIG. 1.28 Time curve for the OCR under test.

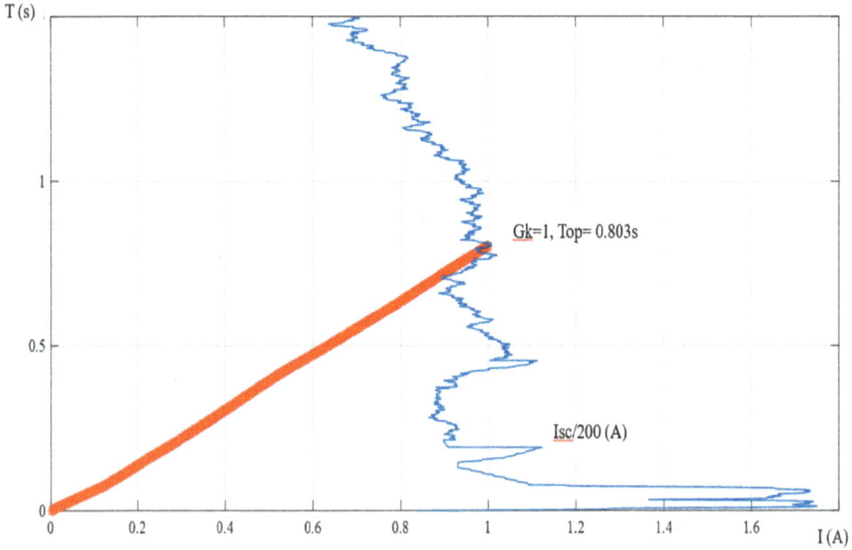

FIG. 1.29 Dynamic evaluation of the OCR during a fault current.

from these issues. Fig. 1.31 shows a radial network and a graph of the variation in the short-circuit current as a function of the electrical distance (l) to the point of fault, where curve 1 corresponds to the maximum level of the short-circuit current and curve 2 to the minimum level. The figure also shows a graph of $T=f\ (l)$ that reflects the relationship between the operating time T and the

```
A=19.61; B=0.491; P=2;              %Time curve coefficients
Ipickup=25;                         %Pickup Current
load Ifault;                        %Phasor of fault signal
At=(1/60)/16;                       %inter-sample time
for i=1:length(Ifault),
    T(i)=(A)/((Ifault (i)/Ipickup)^P-1)+B;  %Time curve IEEE
    H(i)=1/T(i);Cont(i)=i;
    P=At*H(i);                      %integral of the time curve
    Gk=Gk+P;AGk(i)=Gk;              % integrator
    if Gk >= 1,                     %Operating condition
       break
    end
 end
Time=Cont/(16*60);                   %Calculation of operating time
```

FIG. 1.30 Code for the dynamic evaluation of the DOCR.

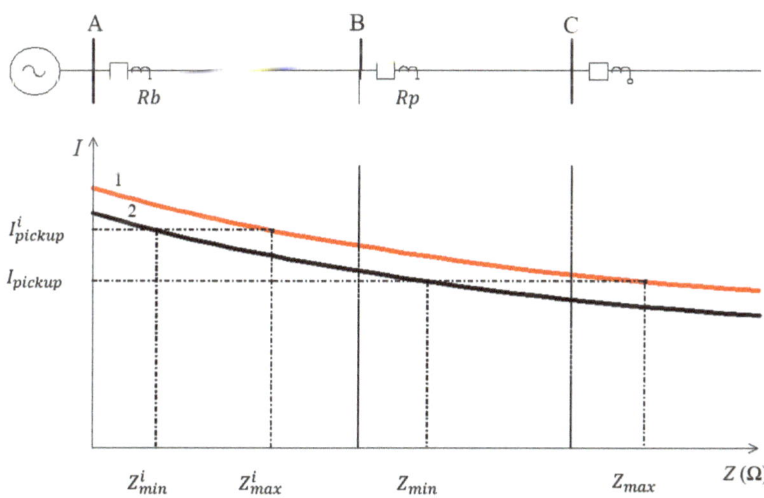

FIG. 1.31 Short-circuit current as a function of electrical distance.

distance to the fault for relays *Rp* and *Rb*. In relay *Rb*, the pickup currents are I^i_{pickup} for the instantaneous relay and I_{pickup} for the inverse time element.

A primary limitation of overcurrent protection is that its reach depends on the type of short circuit and the operating conditions of the electrical system. From Fig. 1.32, it can be observed that the reach of the instantaneous relay *Rp* is between the limits I^i_{min} and I^i_{max}, which introduces uncertainty into the reach of the relay; it may even be the case that the instantaneous relay zone disappears completely under minimal generation conditions. The reach of the inverse time element also shifts between the limits l_{min} and l_{max} and may eventually unprotect the adjacent line for minimum generation, as shown in Fig. 1.32. This is particularly likely in the case of phase protection, where

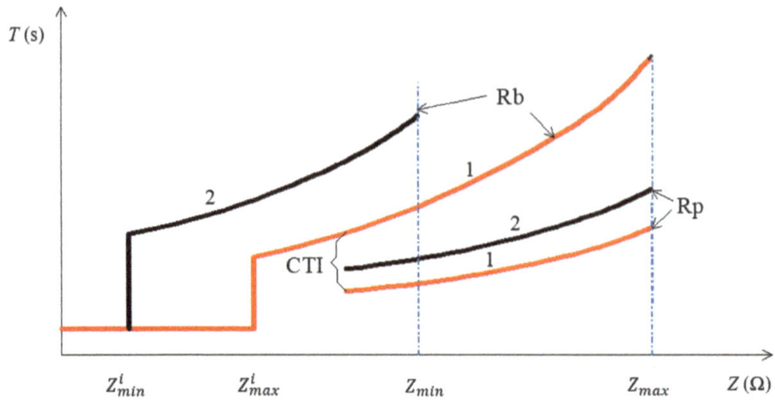

FIG. 1.32 Time curves for maximum/minimal fault current.

the maximum I_{load} is used for setting I_{pickup}, thus limiting its sensitivity for fault detection under minimum generation conditions.

A secondary limitation of overcurrent protection is that the operating time depends on the type of fault and the operating conditions of the system. This is illustrated in the graph of $T=f(Z)$ in Fig. 1.32, in which the curves for minimum generation (marked 2) represent longer operating times than for maximum generation (marked 1). Although coordination between relays *Rb* and *Rp* is guaranteed for peak generation conditions by the *CTI*, the coordination margin is much higher for other operating regimes.

The voltage control can improve the sensitivity of the phase OCR. This solution is used in backup protection for external faults in generators [4], but it is not adequate for the protection of lines. In this case, a phase-to-phase short circuit in one of the lines emerging from a bus reduces the voltage of all the lines. On the nonfault lines, the current during the fault may be greater than the nominal value, since the negative sequence impedance of the motors is less than the positive sequence impedance. Under these conditions, the overcurrent protections of the nonfailed lines may not operate correctly, even when they have voltage retention. In the case of overcurrent protections against ground faults, there are no sensitivity problems associated with I_{load} since they respond only to the zero-sequence component of the current.

In the following chapters, we will introduce alternative solutions to improve the performance of the relays.

1.7 Practical applications

Overcurrent protection devices for the protection of distribution networks include relays, fuses, reclosers, and sectionalizers. These devices lack directionality, meaning that their application is limited to radial systems, in which the power flow is unidirectional. Many distribution systems are radial, and this constitutes the main field of application for overcurrent protection. In these systems, the fault current decreases with the distance between the substation and the fault point, which allows the protection zones to be defined using a current level criterion.

Time curves are specific to the type of overcurrent device (relay, fuse, or recloser) and reflect its principle of operation. In a fuse, the inverse time curve is obtained by melting a tin, copper, or silver, where the heat generated in the fuse link is proportional to the square of the current multiplied by its resistance. The melting time of the fuse link and the extinction of the electric arc determine the operating time of the fuse. In a recloser, the time curve is obtained based on the movement of a piston affected by an electromagnetic field, which causes a displacement proportional to the magnitude of the current.

1.7.1 Coordination of overcurrent devices

In a protection coordination study, a suitable selection of devices must be made. The protection system must distinguish between normal and abnormal conditions, thus allowing for selective and coordinated operation of each device. The main elements that must be protected in a distribution substation are transformers, subtransmission lines, and feeders. The protection system for a transformer depends on its capacity and importance. Two basic arrangements are used, depending on whether the transformer is greater or less than 10MVA [16]: for transformers greater than 10MVA, differential protection and OCRs are used as backup protection, while for transformers less than 10MVA, fuses are used on the primary side and phase and ground relays on the low-voltage side.

A phase protection system for transmission and subtransmission lines is defined according to their length, where short lines use differential line protection (87L) and long lines use pilot protection. In both cases, ground fault protection with DOCRs (67N) is used. In some situations, a ground distance relay is used, verifying the reach in lines with mutual coupling. Protection for distribution circuit feeders depends on the size and type of the substation: for distribution circuit feeders in area and rural substations with installed capacities of less than 5 MVA, automatic reclosers are used with overcurrent protection for phase and ground protection, whereas for feeders in substations with installed capacities greater than 5 MVA, phase and ground OCRs are used, and an automatic reclosing function (79) is enabled.

The coordination criterion for overcurrent devices [17,18] has been used to establish recommended ranges or margins of application, which may be modified based on accumulated experience and knowledge of the particular electrical system.

Relay-relay coordination. The coordination of relays fundamentally involves devices located in the same substation (protection of transformers in high-low voltage and protection of transformer-protection of feeder) but may also be applied to radial subtransmission lines or lines that supply stepped substations. The coordination of ground and phase devices is done separately.

The criterion used to define the degree of inversion of the time curves for an OCR depends on the type of device to be backed up or protected, the proximity of the generation source, and the operations carried out in the electrical circuit. Extremely inverse curves are typically used to back fuses, whereas inverse curves are generally used to back reclosers. In systems where the magnitude of the short-circuit current is strongly dependent on the capacity of the generation system, the use of very inverse curves is recommended. This very inverse characteristic is also applicable to commissioning in which currents comparable to the fault currents are generated (cold load).

A minimum coordination margin of 0.2–0.3 s is applied between the curves of the time-current characteristics of the protection devices, for the maximum short-circuit current common to both relays. Verification of the coordination must be done at both ends of the line in a situation involving meshed nets, since it is possible that the curves may cross.

Relay-fuse coordination. Fuses are the most basic and ineffective overcurrent device. The basic functions of fuses are to sense current system conditions, to interrupt a fault quickly, and to coordinate with other protection devices. The process of interrupting the current in the fuse, after heating and melting of the fuse link when an overcurrent circulates, consists of three basic stages: (i) initiation of the electric arc, in which the fuse link melts and an electric arc is generated due to the potential difference between the fuse terminals; (ii) elimination of the arc by means of a process of confinement and cooling inside the fuse; and (iii) current interruption at the next zero crossing, given the earlier mentioned conditions.

The most common feeder protection scheme is the relay-fuse. This arrangement is applied between devices located in different locations, where the relay in a substation acts to protect a distribution circuit, and the fuse acts to protect a branch on the distribution line [18,19]. The fuse will only operate for permanent faults on its branch, and the relay as a backup device can be adjusted to give fast operation in such a way that it operates before the fuse (save fuse scheme), a reclose is set. If the fault has not been released, operation of the fuse is allowed. The coordination criterion establishes a minimum margin of 0.3 s between the maximum trip time curve and the relay characteristic for the maximum short-circuit current common to both devices.

The 13.8-kV radial distribution system shown in Fig. 1.33 serves as an example for carrying out relay-fuse coordination. If the B-C branch is very important, a recloser can be used; if the use of a recloser is not justified, operation of the instantaneous relay is allowed for any fault in that branch. The circuit is then reenergized, and the instantaneous relay is blocked. If the fault persists, the fuse will blow. In this scheme, there is the possibility of maintaining service on the branch for temporary faults, while sacrificing the power supply to the remaining branches via the operation of the instantaneous relay.

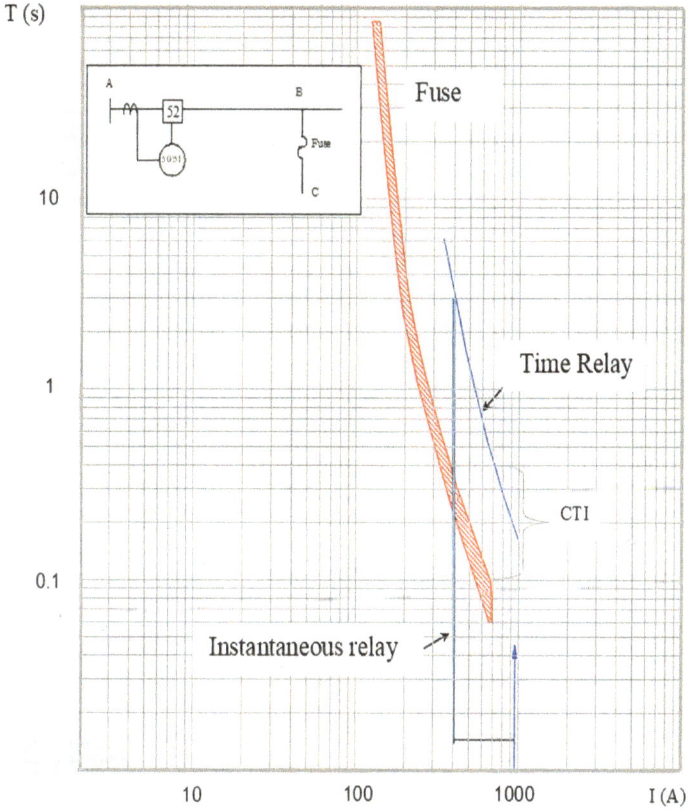

FIG. 1.33 Relay-fuse coordination.

Relay-recloser coordination. Studies of electrical distribution systems have established that approximately 80%–95% of the total faults in a system are of a transitory nature and short duration [18,19]. The automatic recloser provides trip and reclose functions, thus eliminating prolonged shutdowns from electrical distribution systems due to temporary faults or transient overcurrent conditions. If a fault is permanent, the recloser will remain open after a predetermined number of operations (trip and reclose) in order to isolate the faulty section from the rest of the system. The principle of operation of a recloser is very similar to that of a switch and a relay, with the main difference between the two devices being that the recloser has a lower current interrupting capacity and its cost is considerably lower.

The timing of the relays must be long enough to allow the entire sequence of recloser operations. An important factor in obtaining this selectivity is the reset time of the relays, since if the relay does not have time to fully reset, the next sequence will start with some travel on the disk. As a rule, selectivity can be obtained if the relay run time for any current is less than twice the time delay characteristic of the recloser.

The recloser time curves are adjustable, and a recloser can undergo more than one instantaneous operation with the intention of eliminating a temporary fault condition. The most common sequence of a restorer is two fast operations followed by two slow 2A–2B operations. An example is shown in Fig. 1.34, where curves A and B correspond to the instantaneous and time-delayed sequences of the restorer, respectively. It is recommended that the first sequence of the restorer is not instantaneous, as studies have shown that the possibility of eliminating temporary faults in the first dead time of the restorer is higher if a certain time (2 s) is allowed to allow for deionization of the air. The coordination criterion establishes a minimum margin (0.3–0.4 s) between the time-current characteristics for the maximum short-circuit current common to both overcurrent devices.

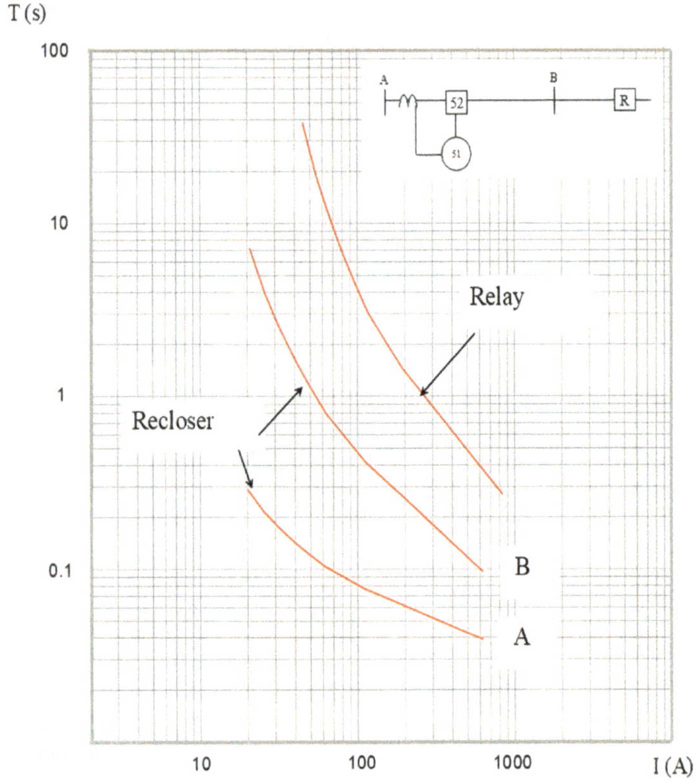

FIG. 1.34 Example of relay-recloser coordination.

1.8 Summary

The main drawback of overcurrent protection is that its protection zone and operating time depend largely on the type of short circuit and the operating regime of the system. This is manifested as sensitivity limitations on phase protection and at high operating times, especially for low generation levels. The generalized expression for an inverse time OCR for both the operating and reset zones of the relay allows us to carry out a relay analysis and solves design problems in a straightforward way, even in the presence of nonlinear functions.

References

[1] IEEE C37.112, IEEE Standard Inverse-Time Characteristic Equations for Overcurrent Relays, 2018.

[2] IEC Standard 255-4, Single Input Energizing Measuring Relays with Dependent Specified Time, first ed., IEC Publication 255-4, 1976.

[3] IEEE recommended practice for protection and coordination of industrial and commercial power systems (IEEE buff book), IEEE Std 242-2001 (Revision of IEEE Std 242-1986) [IEEE Buff Book], 17 Dec 2001, pp. 1–710, https://doi.org/10.1109/IEEESTD.2001.93369.

[4] J. Lewis Blackburn, T.J. Domin, Protective Relaying, Principles and Applications, third ed., CRC Press Taylor & Francis Group, 2006.

[5] H. Zayandehroodi, A. Mohamed, H. Shareef, M. Mohammadjafari, Impact of distributed generations on power system protection performance, Int. J. Phys. Sci. 6 (16) (2011) 3873–3881.

[6] J.C. Gómez, M.M. Morcos, Coordination of voltage sag and overcurrent protection in DG system, IEEE Trans. Power Deliv. 20 (1) (January 2005) 214–218.

[7] P.G. McLaren, K. Mustaphi, G. Benmouyal, S. Chano, A. Girgis, C. Henville, M. Kezunovic, L. Kojovic, R. Marttila, M. Meisinger, G. Michel, M.S. Sachdev, V. Skendzic, T.S. Sidhu, D. Tziouvaras, Software models for relays, IEEE Trans. Power Deliv. 16 (2) (2001) 238–245.

[8] IEEE Committee Report, Computer representation of overcurrent relay characteristics, IEEE Trans. Power Deliv. 4 (3) (July 1989) 1659–1667.

[9] S.E. Zocholl, Testing dynamic characteristics of overcurrent relays, in: 20th Annual Western Protective Relay Conference, Spokane Washington, October 19-21, 1993.

[10] J.G. Proakis, D.G. Manolakis, Digital Signal Processing, fourth ed., Prentice Hall, 2006.

[11] S.M. Brahma, P.L. De Leon, R.G. Kavasseri, Investigating the option of removing the antialiasing filter from digital relays, IEEE Trans. Power Deliv. 24 (4) (2009) 1864–1868.

[12] IEC 61000-4-7, Testing and Measurement Techniques—General Guide on Harmonics and Interharmonics Measurements and Instrumentation, second ed., 2002-08.

[13] Q.I. Li, L. Qian, S. Woodruff, D. Cartes, Prony analysis for power system transient harmonics, EURASIP J. Adv. Signal Process. 2007 (2007), 48406.

[14] IEEE Guide for Protective Relay Applications to Distribution Lines, IEEE Std. C37.230-2007.

[15] M. Titarenko, I. Noskov-Dukelsky, Protective Relaying in Electric Power Systems, Peace Publishers, Moscow, U.R.S.S, 1963.

[16] IEEE Std C37.91-2000, IEEE guide for protective relay applications to power transformers, in: Power Systems Relay Committee of the IEEE Power Engineering Society, March 2000.

[17] IEEE Std C37.113-1999, IEEE guide for protective relay applications to transmission lines, in: Power Systems Relaying Committee (PSRC) of the IEEE Power Engineering Society, September 1999.
[18] A.F. Naiem, A.Y. Almoataz, Y.G. Hegazy, Distribution Systems with Distributed Generation Protection Strategy for Fuse-Recloser Coordination, LAMBERT Academic Publishing, 2013.
[19] J.J. Burke, Power Distribution Engineering: Fundamentals and Applications, CRC Press, 2017.

Chapter 2

Adaptive operating time for overcurrent relays

2.1 Introduction

The coordination criterion is established for maximum fault currents since the time curves converge; these curves diverge at lower current values, giving longer operating times. This behavior is typical of OCRs and has been found to be very useful where the device frequently operates under overload conditions. However, backup protection can result in long backup times due to this divergence of the time curves, subjecting the electrical system to high currents that can increase the thermal and mechanical stresses on the primary equipment. Another problem that affects backup times is the difference in load currents at each location of the relays, which produces greater divergence for minimum fault currents. Lastly, the difference in the time curves for the primary and backup protection devices makes the coordination process difficult, resulting in long delays in the operation of backup protection.

The presence of cogeneration sources, DGs and RESs, can modify the fault current by increasing or decreasing the contributions to the fault [1,2]. The OCRs must be adjusted and coordinated based on the changes due to these additional contributions; however, their intermittent nature can make it difficult to weigh these properly [3]. Since the initial contribution of an RES is reduced after several cycles, the real impact on coordination must be evaluated [4]. The operational and topological diversity of active networks can result in a degradation of the protection functions, which may have low sensitivity and high operating times. A diverse range of very interesting applications of OCRs in medium- and low-voltage systems have been presented [5–10]. However, the performance improvement in overcurrent protection is limited under the operating conditions of many active networks. Adaptive protection schemes, protection functions supported by communication links, and online adjustment systems should therefore be evaluated.

The introduction of communication channels is a well-known and frequently used alternative to protection schemes and is generally used in transmission networks. However, the feasibility of its application in medium- and low-voltage networks is feasible, due to the reduction in costs and the increase in infrastructure in substations, in this way, the adaptation of the relay settings

Overcurrent Relay Advances for Modern Electricity Networks. https://doi.org/10.1016/B978-0-323-91777-3.00002-7

is feasible today [11,12]. It is also necessary to consider its application on a large scale, due to the high number of distribution networks associated with the electrical network. However, the modernization trend of smart grids currently allows the implementation of protection systems through communication links and with the feasibility of phasor measurements. These aspects need to be applied in protection systems to solve the problem of sensitivity for the detection of faults and to reduce fault release times. Thus the implementation of adaptive protection techniques is feasible.

2.2 Adaptive time

The time coordination criterion is established to satisfy Eq. (2.1) for any value of the current [5].

$$t_b \geq t_p + CTI \tag{2.1}$$

The main purpose is to ensure that the coordination criterion is met. To achieve this, the backup relay t_b must operate in a similar way to the primary device t_p plus a *CTI*. From Fig. 2.1, it can be observed that the coordination criterion in Eq. (2.1) is established for the maximum fault current seen by both relays. However, for minor fault currents or fault locations along the line, the time interval is greater than the *CTI*. Another important factor is the effect of the load current (I_{load}), when the pickup current (I_{pickup}) of the backup relay R_2 is greater than the value of I_{pickup} for the primary relay R_1. In this case, which occurs very frequently, the time curve is asymptotic on its vertical axis, which results in divergence of the curves and higher *CTI*s for minimal faults.

The aim of using adaptive time is to obtain a constant *CTI* between the primary and backup relays over the full range of fault current. In this way, the

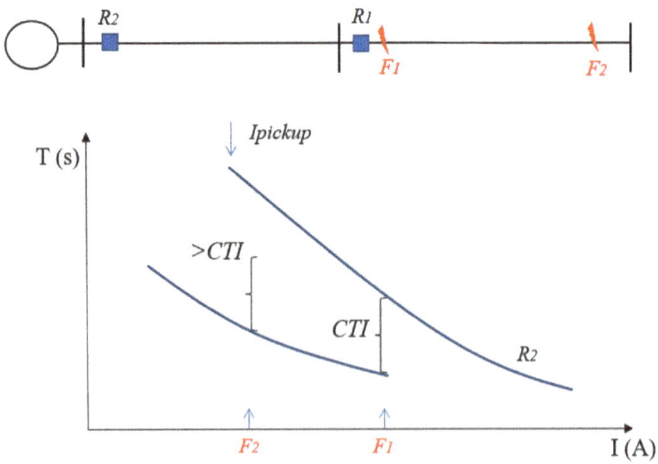

FIG. 2.1 The coordination criterion.

effect of I_{load} and the divergence of time curves for coordination with the backup relay can be avoided.

The backup relay *Relay A* and the primary relay *Relay B* are shown in Fig. 2.2. Due to the I_{load} effect, the backup relay will be slower than the primary relay. Although the time curve determines the time operation, I_{pickup} will have a greater impact on the operation time. There are two ways to achieve a constant backup time delay for any value of the fault current. The first involves modifying the *TDS* for each value of the fault current, which represents actualization for different values of the fault current; this compensates for the divergence of the curves and maintains a constant *CTI*.

If t_b is defined as in functions *J(I)* and *H(I)* in Chapter 1, the coordination criterion becomes:

$$\frac{J_b(I)}{H_b(I)} = t_p(I) + CTI \tag{2.2}$$

For each sample *k*, this gives:

$$J_b(I_k) = H_b(I_k)\left[t_p(I_k) + CTI\right] \tag{2.3}$$

A better solution is to compute the curve marked 5 in Fig. 2.2. Due to the effect of I_{load}, this curve is not obtained from the *TDS* (curve 1), and in order to find curve 5, the shape of the time curve needs to change. If curve 5 is similar to curve 1, the *CTI* must be included, and hence the time expression for the backup relay must be computed based on the I_{pickup} of the primary device. This means that curve 5 has the same horizontal asymptote as curve 1 (Fig. 2.2). The

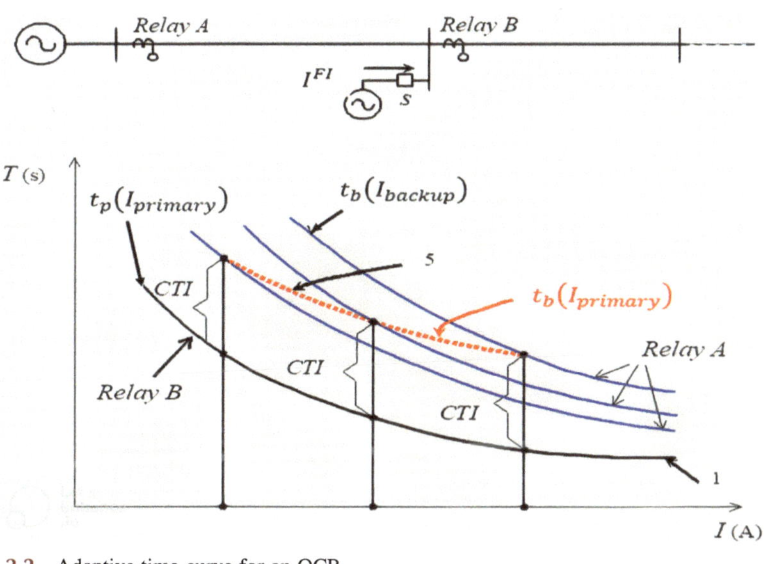

FIG. 2.2 Adaptive time curve for an OCR.

dynamics of the primary device is then replicated by the proposed relay, giving a faster operating time. The difference between these approaches is that the first dynamically modifies the time curve for each fault current value (Fig. 2.2), while the second uses only curve 5 to give a constant *CTI* over the entire current range of coordination.

When switch *S* is closed (Fig. 2.2), the source contribution I^{FI} causes an infeed current effect, in which the primary relay (Relay B) accelerates but the fault current of Relay A is the same, while the primary relay reduces its time, the backup relay remains the same. Under these conditions, coordination is not lost, but the *CTI* will be higher. However, in the proposed coordination method, the performance is different, as both relays have a similar time curve and the infeed effect is minor. Furthermore, the operating condition described before also applies to feeder main relays, where the effect of the I_{load} causes an increase in the backup times, as shown in Fig. 2.3.

The backup relay reproduces the dynamic behavior of the primary relay, and the time function of the proposed relay must include the time function of the primary relay, as follows:

$$t_{proposed} = t_p\left(I^{primary}\right) + CTI \tag{2.4}$$

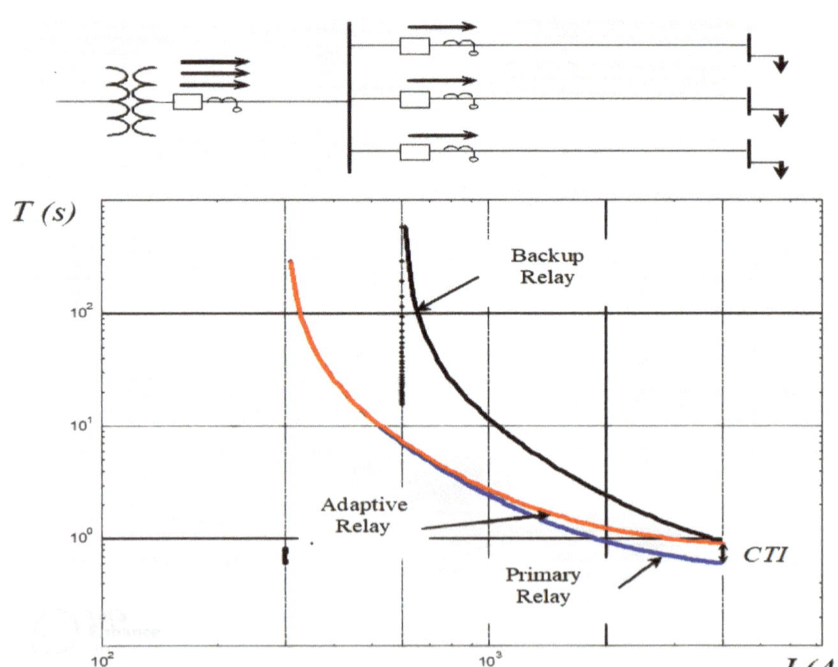

FIG. 2.3 Coordination of the main relay and feeder devices.

where $I^{primary}$ is a multiple of the current of the primary protection, calculated using I_{pickup} and the fault current (I_{sc}) as $I^{primary}=I_{sc}/I_{pickup}$. In this way, the proposed relay emulates the operation dynamics of the primary relay, including the function t_p ($I^{primary}$) and maintaining the CTI for any fault current value. The analytical expression for the time curve of the primary protection can be applied to any overcurrent device such as a fuse or recloser. Thus the proposed adaptive criterion can be applied to the protection of both radial networks with OCR and mesh networks with DOCRs.

The dynamic equation describing the proposed relay is obtained as follows:

$$G_k = \Delta t \sum_{k=1} H_{proposed} \tag{2.5}$$

where:

$$H_{proposed} = \frac{1}{t_{proposed}} = \frac{1}{t_p\left(I^{primary}\right) + CTI} \tag{2.6}$$

The proposed method involves determining the function $H_{proposed}$ that allows the OCR to operate with a CTI with respect to the primary protection, for all coordination intervals. This means a change in the form of the relay time curve, which in an analog relay would represent a change in the relay design.

The operating condition is obtained when:

$$G_k = \Delta t \sum_{k=1}^{k_{op}} H_{proposed}\left(I_k^{primary}\right) = 1 \tag{2.7}$$

Correct operation of the relay is achieved when $k = k_{op}$ and Eq. (2.7) is fulfilled.

2.3 Evaluation of the adaptive operating time

The alternative approaches to adapting the operating time described in the previous section were evaluated for constant and variable fault currents. For constant fault currents, the time curves $T = F(I)$ are presented for both relays, and the adaptation process is shown to comply with the coordination criterion for each current value. For variable fault currents, the behavior of the relay was evaluated during the integration process and at the time of operation of both relays.

Constant fault current. In Fig. 2.4, the proposed function $H(I)$ is calculated using Eq. (2.6), and a characteristic curve is obtained for each value of I. The function $H(I)$ generates a unique characteristic curve that guarantees compliance for any current value. This is a different type of curve than for the primary relay.

Variable fault current. In the tests, the adaptive relay integration process was evaluated with variable fault currents. Fig. 2.5 illustrates the test I_{sc}

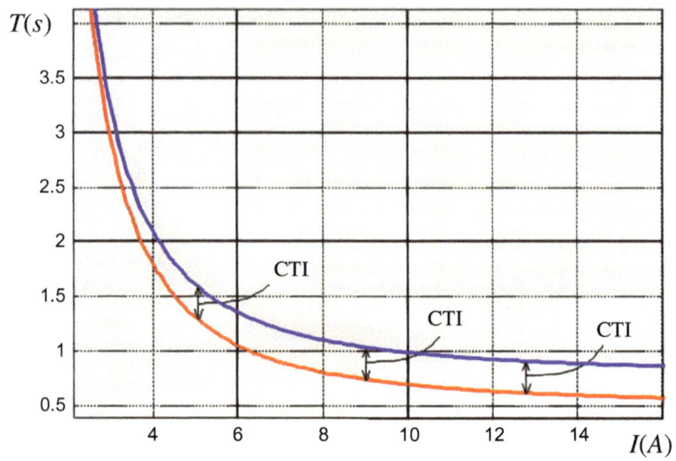

FIG. 2.4 Time curve adaptation for constant current: function $H(I)$.

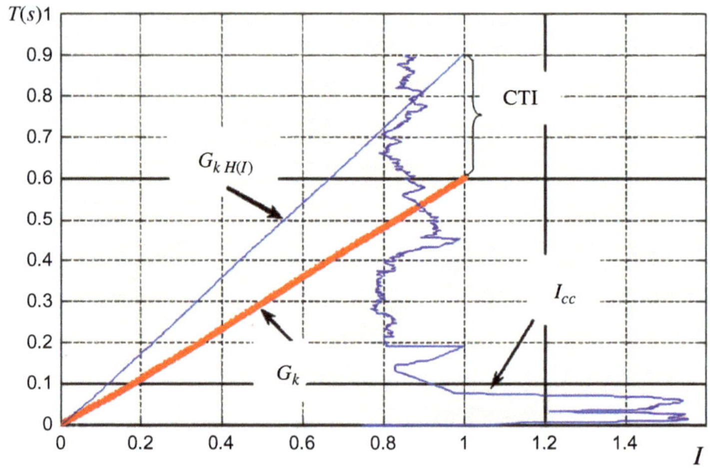

FIG. 2.5 Integration of the function $H(I)$ for dynamic fault current.

(esc: 1:15) used in the evaluation. The accumulated value behavior of the primary relay integrator is shown by the thick line marked G_k. It can be seen that the adaptive backup relay integrator $G_{k\ H(I)}$ differs from the conventional one, and hence the integration process is also different. The time difference at the instant of operation is 866–576 = 288 samples at a rate of 16 samples per cycle, which represents a *CTI* of 0.3 s.

The operation time of the adaptive relay depends on the time curve of the primary protection. In Fig. 2.6, the shaded area represents the reduction in operating time between a conventional relay and the adaptive relay. Through the

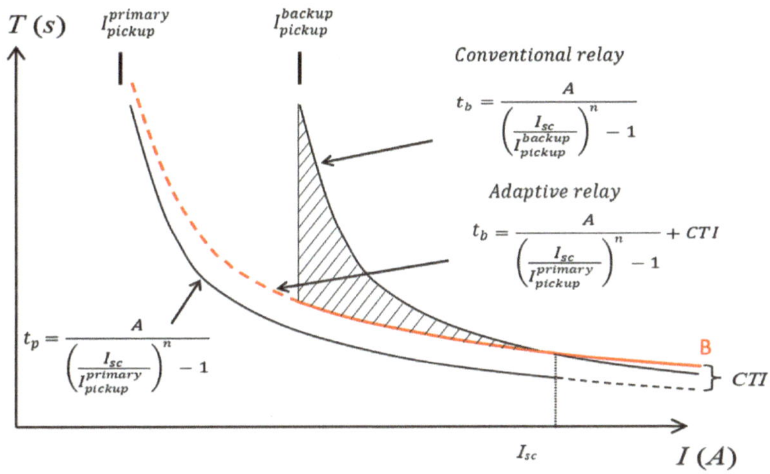

FIG. 2.6 Time curve for the proposed adaptive relay.

process of adapting the operating time, automatic coordination of the adaptive relay with the primary protection device is achieved. It is not necessary to apply any coordination procedure, even when there are changes in the operating conditions of the electrical system, such as changes in topology, generation, and load, that modify the magnitude of the short-circuit current. If the value of the short-circuit current is increased, the *CTI* is maintained.

Coordination between a series of adaptive relays is carried out based on the principle described before, where the analytical expression for each relay is the same as for the primary relay plus the *CTI*. This process should be limited to a few relays, as a lack of convergence of the curves will result in longer operating times, as indicated by point *B* in Fig. 2.6. However, in the coordination process, reducing the operating time of one relay allows for improvements in the times of neighboring relays, with a strong impact on coordination.

Since the proposed relay uses $I^{primary}$, its time curve converges to the value of I_{pickup} for the primary device; the time curve is therefore the one marked curve 5 in Fig. 2.2. This means that the proposed relay can operate at currents lower than its value of I_{pickup}, and coordination is maintained by the function in Eq. (2.6). To increase the sensitivity of the adaptive relay, the value of I_{pickup} for the adaptive relay can be the same as for the primary relay, as shown by the broken line in Fig. 2.6. However, it is possible to limit the sensitivity of the relay by modifying its structure.

2.4 Functional structure of the proposed relay

A functional diagram of the proposed adaptive OCR is shown in Fig. 2.7. A constant delay in the operation of the relay is guaranteed, regardless of the

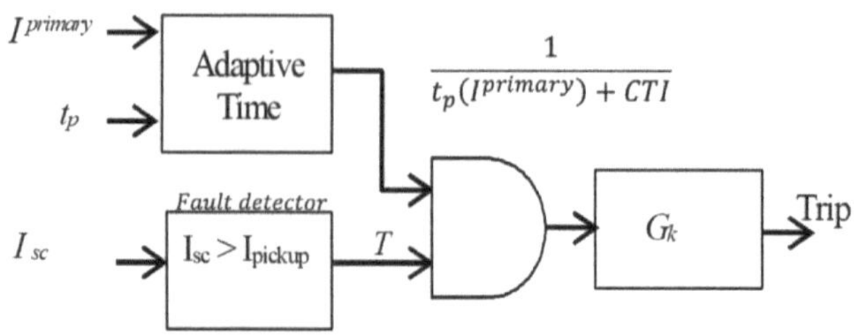

FIG. 2.7 Diagram of the proposed adaptive relay.

magnitude of the fault current. The adaptive operating time block is used to calculate the value of the function $H(I_{primary})$ in Eq. (2.6), and the integration G_k in Eq. (2.7) emulates the dynamics of the primary device. The *CTI* is added to ensure coordination between the adaptive relay and the primary device, and the output signal of the adaptive time block is integrated in the presence of variable fault currents by the time element if a fault condition is detected (signal *T*).

As can be seen from Fig. 2.2, to obtain the proposed curve, the backup relay must be adjusted using the value of I_{pickup} for the primary relay. It is therefore necessary to decouple the functions of fault detection and determination of the operation time for a conventional relay. Hence, the operating time of the adaptive relay, which depends on the operating time of the primary protection, is independent of I_{pickup}, which acts only as a fault detector [13]. When I_{pickup} for the backup relay is used as a fault detector, the relay time curve is as shown by the solid red line in Fig. 2.6. Finally, at the output element, the trip condition is verified, and the output signal is sent to the trip circuit of the breaker.

To achieve safe operation, the fault detector must be able to discriminate between I_{load} and I_{sc}, using the value of I_{pickup} for the primary or backup relay (Fig. 2.7). In another approach, the value of I_{pickup} for the adaptive relay does not have a lower limit, meaning that the operation of the adaptive relay will be feasible in conditions where the primary device has no sensitivity. This condition does not represent a loss of coordination, since the fault will then be released by the adaptive relay.

2.4.1 Implementation

The structure of the inverse time OCR (Fig. 2.8) shows the relationship between the main stages of the relay, which are described later. In Stage 1, which involves the phasor estimation of the current signal, two filter stages are used: an analog filter and a digital filter.

In Stage 2, the operation time of the relay is determined based on the estimate of the operation time for the primary protection device. To do this, an analytical expression for the operating curve of the primary device is needed, and the *CTI* is added to ensure that the correct coordination time between the relay

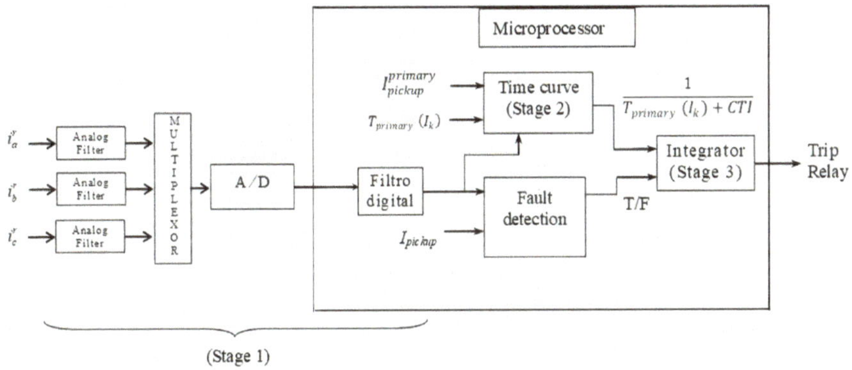

FIG. 2.8 Functional structure of the proposed OCR.

and the primary protection is obtained. This process results in a shorter operation time for the proposed OCR compared to a conventional relay (Fig. 2.9).

The time control determines the time curve of the relay (Fig. 2.10). This integration process is carried out in Stage 3 to establish the operational dynamics of the relay under variations in the short-circuit current.

To carry out the coordination process, an analytical expression for the time curve of the primary protection device is required, which may be a fuse, electromechanical relay, or recloser. The process of adaptation is carried out by

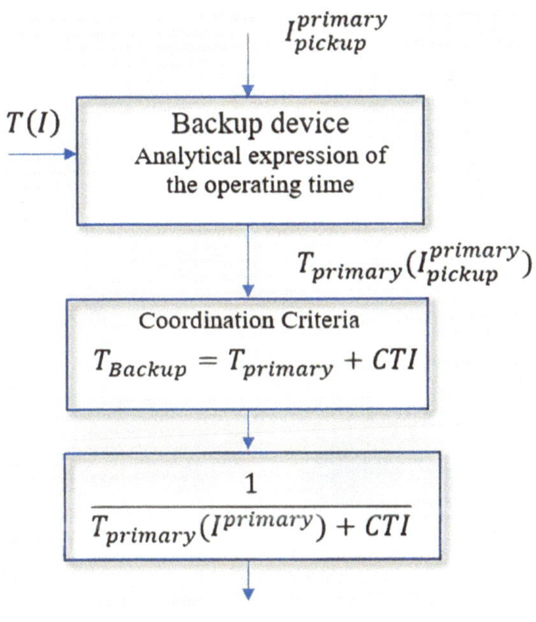

FIG. 2.9 Calculation of the time curve by Stage 2.

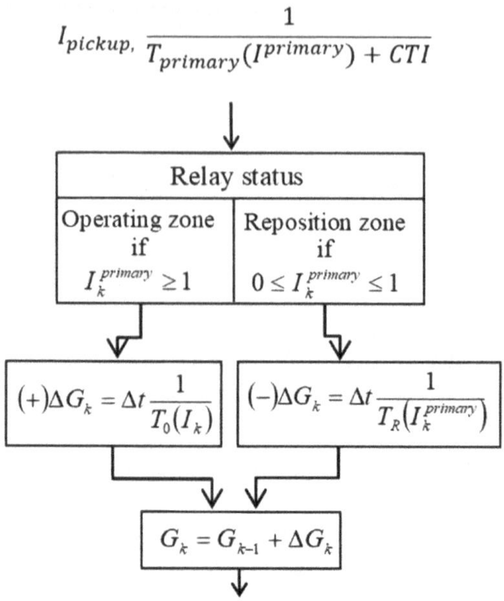

FIG. 2.10 Integration stage of the time curve.

obtaining analytical expressions that represent the operating dynamics of the relay. To achieve this, an analytical representation for the operating characteristic of the backup device is required.

When an analytical expression for the operating characteristic of the primary element is available (as is the case with many digital relays), the adaptation process is simplified. However, when the equation for the time behavior is unknown, pairs of values can be obtained from the time-current graph for use with an offline curve fitting program. The best possible fit can be automatically selected from a set of expressions proposed in the literature. Thus an analytical representation of the time curve is obtained to be included in the formulation of adaptation of the operating time.

2.5 Time curve fitting algorithms

The time curve characteristics for inverse time electromechanical OCRs are given in the form of families of curves published by the manufacturers. These time curves generally reflect the average values obtained by testing the relays. To comply with the standardized curves, it is necessary that the operation of the relay is within a given time interval [14].

Historically, the time curve characteristic of an electromechanical relay was determined by the displacement dynamics of the induction disk [15–18]. The position of the disk is calculated based on the integral of the angular velocity, which depends on the difference between the electromechanical torque and the spring torque. The electromagnetic torque is proportional to the square of the magnetic flux caused by the current. If the magnetic circuit is linear, the operating time depends inversely on the square of the current. In the design of a relay, saturation of the magnetic circuit (induction disk) was used to calculate the time-current characteristic. In analog electronic relays the process of charging a capacitor toward the integration function, the input voltage to the integrator has a nonlinear functional relationship with the current, which serves to shape the time-current characteristic. Digital relays numerically emulate integration and allow for great flexibility in terms of shaping the time-current characteristics.

In the design of a DOCR, mathematical modeling of the time-current characteristics is also required [19–23]. From the equation for the desired characteristic, the nonlinear function of the current is determined. Historically, several mathematical models have been proposed for the time characteristics of inverse time relays. For example, Wylie [24] used tables containing pairs of time and current values to represent the characteristics. Knable [25] proposed a model based on determining the slope of the characteristic at different points. Warrington [26,27] and Hieber [28] proposed exponential equations for the characteristics, while Radke [29], Albrecht [30], Sachdev [31], Singh [32], and Damborg [33] suggested the use of polynomial equations. The IEC 255-4 standard [34] and, more recently, the IEEE C37.112-1996 standard [14] propose exponential equations for inverse time-current characteristics.

Two basic approaches have been used to model the family of curves based on *TDS* as a parameter. In the first, one of the characteristics of the family of curves is modeled and an independent equation (a function of the *TDS*) is used to calculate the operating time corresponding to another characteristic [31]. The second possibility is to model the entire family of curves using a single polynomial, which expresses the operating time as a function of the current and the *TDS* [30,33].

In this section, we report the results of a comparative evaluation of the main mathematical models of the time curve characteristics of an inverse time OCR. A curve fitting program was developed in Matlab and used to study the characteristics of a set of electromechanical and digital relays. This program applied the fit models proposed in the literature and was used to perform a statistical comparison between them to determine the best fit. The study focused on the modeling of a single characteristic of the family, and polynomial models that included the *TDS* were not considered. The results indicated that there is no single equation that offers the best adjustment to the characteristics of all relays.

2.5.1 Equations of the time-current characteristics

It has been established [16,31] that the equations used to set the time curve characteristics for an OCR must meet the following requirements: (a) they must have a reduced setting error, (b) the resulting time curves must be asymptotic with regard to the pickup current, (c) the time curves must be asymptotic at a certain value of the operating time (which implies closely reproducing the shape of the relay curve over a wide range of currents), and (d) the equations should be as simple as possible. The most widely accepted equations for fitting a single time-current characteristic are presented later. The mathematical treatment of the generation of the family of curves is not the object of study in this work.

A. *Exponential equations*

The IEC 255-4 [34] standard proposes the following equation:

$$T = \frac{A}{I^n - 1} \tag{2.8}$$

where A and n are constants. This rule can be used to establish various different types of time curves. The IEEE C37.112-1996 standard [14] proposes the equation:

$$T = \frac{A}{I^n - 1} + B \tag{2.9}$$

Eq. (2.8) is a particular case of Eq. (2.9), and the curves resulting from both equations are asymptotic with regard to I_{pickup}. The time curve in Eq. (2.9) is asymptotic at B, which facilitates coordination with expulsion-type fuses since the arc extinction time gives a definite time character for large multiples of the time curve. In contrast, the time curve in Eq. (2.8) is asymptotic with regard to the current axis; this characteristic is typical in Europe and Japan, where distribution networks are made up of short three-wire lines in which expulsion fuses are not used [18]. The IEEE C37.112-1996 standard [14] establishes bands that allow the type of curve to be classified by multiplying the basic curves by factors of 1.15 and 0.85.

Hieber [28] proposed an experimental time curve based on a more complex exponential fit equation, as follows:

$$T = C + \frac{K}{\left(I - h + wI^{-2I}\right)^q} - b\left(\frac{I}{50}\right)^n \tag{2.10}$$

where C, K, h, w, q, b, and n are design constants.

B. *Polynomial equations*

Radke [29] proposed the following logarithmic polynomial:

$$\log(T - B) = A_0 + A_1(\log I) + A_2(\log I)^2 + \cdots \tag{2.11}$$

where B, A_0, A_1, A_2, ... are constants. Eq. (2.11) is not asymptotic with regard to the pickup current, and the operating time does not decrease monotonically as the current increases. Sachdev et al. [31] therefore added a term to Eq. (2.11) to solve the first of these problems, as follows:

$$\log(T-B) = \frac{A_{-1}}{\log I} + A_0 + A_1(\log I) + A_2(\log I)^2 + \cdots \tag{2.12}$$

where A_{-1} is another constant. However, Eq. (2.12) still has the drawback that the time does not decrease monotonically with current, which has limited its application.

Sachdev [31] and Singh [32] proposed three polynomial equations to solve both of the problems described before, as follows:

$$\log T = A_0 + \frac{A_1}{\log I} + \frac{A_2}{(\log I)^2} + \cdots \tag{2.13}$$

$$T = A_0 + \frac{A_1}{I-1} + \frac{A_2}{(I-1)^2} + \cdots \tag{2.14}$$

$$T = A_0 + \frac{A_1}{I-1} + \frac{A_2}{I^2-1} + \cdots \tag{2.15}$$

It is also interesting to evaluate a polynomial in which the time T is a function of the logarithm of the current:

$$T = A_0 + \frac{A_1}{\log I} + \frac{A_2}{(\log I)^2} + \cdots \tag{2.16}$$

Equations of digital overcurrent relays. Historically, manufacturers of digital relays published families of time curves for their devices, and in the early days of digital relays analytical equations were also given. When designing a relay, a family of curves is generated from the equation for the time characteristic, and a multiplication coefficient is used to obtain the remaining time curves in the family. This coefficient, denoted as D in the following equations, was the digital equivalent of the *TDS*. Following, we present the equations for eight digital relays that were obtained from the information given by the manufacturers at the time.

$$T = \frac{5.67D}{I^2-1} + 0.352D \tag{2.17}$$

$$T = 0.2D\left(\frac{28.2}{I^2-1} + 0.1217\right) \tag{2.18}$$

$$T = 0.2D\left(\frac{0.0515}{I^{0.02}-1} + 0.114\right) \tag{2.19}$$

$$T = \frac{0.4797D}{I^{1.5625}-1} + 0.21359D + 0.028 \tag{2.20}$$

$$T = \left(\frac{2.93}{I^2-1.026} + 0.0112\right)(2.244)(14D-5)/9 \tag{2.21}$$

$$T = \frac{120D}{I - 1} \tag{2.22}$$

$$T = \frac{91.2011D}{I^{1.96}} \tag{2.23}$$

$$T = D\left(A_0 + \frac{A_1}{I + B} + \frac{A_2}{I^2} + \frac{A_3}{I^3}\right) \tag{2.24}$$

where A_0, A_1, A_2, A_3, and B are constants.

Eqs. (2.17)–(2.20) are versions of Eq. (2.9). Eq. (2.21) represents a slight modification of Eq. (2.9), which makes it asymptotic at $I = 1.026$. Eq. (2.22) is a version of Eq. (2.8), while Eq. (2.23) is based on a modification of Eq. (2.8) to make it asymptotic with regard to the time axis rather than I_{pickup}. Finally, Eq. (2.24) is a polynomial equation that has certain differences from Eq. (2.15).

2.5.2 Time curve fitting program

There are two methods of representing time curves digitally [35]. The first is to store pairs of time and current values in memory and to determine the operating time corresponding to a current value by direct interpolation. This method can be used for devices with relatively fixed characteristics and few adjustments, although for devices with many settings, such as relays, a great deal of memory is required for data storage. The second method is to apply time curve fitting, which involves selecting an appropriate equation and determining and storing the coefficients that produce a satisfactory fit. This is the most widely used method, and it is the one that is applied in this work.

There are two approaches that can be used for curve fitting, the application of which depends on the error associated with the data [35], as follows:

Least squares regression. This is suitable for input data with noise. A curve is obtained that represents the general trend of the data. Since individual data points may be incorrect, the algorithm does not attempt to intersect each point.

Interpolation. This is suitable when the input data are accurately known. The process consists of fitting one or more curves that pass directly through each of the points. An estimate of values is made between well-determined discrete points.

The block diagram in Fig. 2.11 illustrates the classification of the time curve fitting algorithms. The goal of least squares regression algorithms is to reduce the error between the estimated values and the data using least squares optimization techniques. In linear regression, the model depends linearly on the parameters to be determined, and the fit equation may be a straight line (simple linear regression) or a polynomial (polynomial regression). There is also multiple linear regression, in which the function contains two or more independent variables. In nonlinear regression, the model is a nonlinear function of the parameters or coefficients.

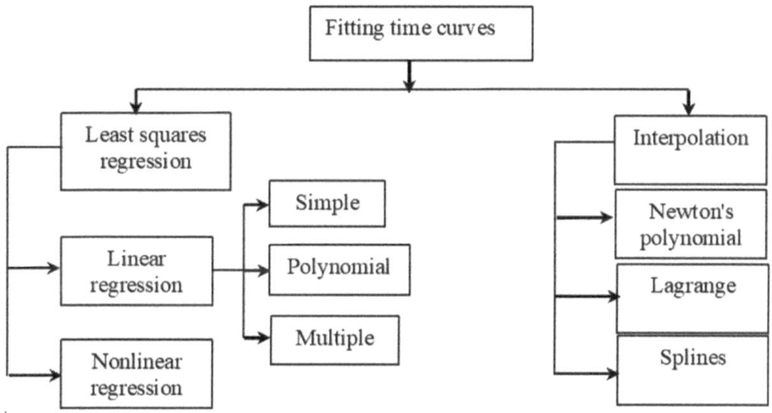

FIG. 2.11 Classification of time curve fitting algorithms.

Interpolation algorithms can be used to fit the values based on exact data. To achieve this purpose, polynomials are used. Newton's algorithm is preferable when the order of the polynomial is unknown, but when the order is known, the Lagrange algorithm is the most suitable. An algorithm called Splines uses low-order polynomials to interpolate subsets of data and is particularly suitable when the data contains abrupt changes.

The time curves for OCRs published by the manufacturers contain small errors. One source of error is that these characteristics are average curves resulting from relay tests. Errors are also introduced in the process of reading the values from the curves, which are generally logarithmic. To reduce the effect of the latter source of error, the IEEE Protection Committee has recommended that manufacturers provide tables of time-current values, in addition to the curves for the relays [16]. Many digital relay manufacturers publish equations for the time-current characteristics, which solve this problem.

The presence of these errors in the data makes it advisable to use least squares regression techniques to fit relay time curves. In this work, nonlinear regression is used to adjust the exponential equations and polynomial linear regression to adjust the polynomial equations. Interpolation techniques such as the Splines algorithm have also been used in commercial relay coordination programs.

2.5.2.1 *Nonlinear regression*

Nonlinear data fitting is used with models in which there is a nonlinear dependence on their parameters. Nonlinear regression minimizes the residuals obtained based on the sum of the squares. A solution is found in an iterative way, and the success of the solution is highly independent of good starting values for the parameters.

The Gauss-Newton method is an optimization algorithm that aims to minimize the sum of the squares of the residuals between the data and a nonlinear equation. A Taylor series expansion is used to express the original nonlinear equation in an approximately linear form and the least squares technique can then be used to obtain new estimates for the parameters, moving in the direction of minimization of the residual. The relationship between the nonlinear equation and the data can be expressed in a general way as follows:

$$y_i = f(x_i; a_0, a_1, \cdots, a_m) + e_i \tag{2.25}$$

where y_i is the measured value of the dependent variable, $f(x_i; a_0, a_1, \ldots, a_m)$ is a function of the independent variable x_i and a nonlinear function of $a_0, a_1, \ldots, a_m$, and e_i is the error. In abbreviated form, we have:

$$y_i = f(x_i) + e_i \tag{2.26}$$

The nonlinear model can be expanded in a Taylor series around the parameters and truncated after its first derivative. For example, for the case of two parameters, we have:

$$f(x_i)_{j+1} = f(x_i)_j + \frac{\partial f(x_i)_j}{\partial a_0}\Delta a_0 + \frac{\partial f(x_i)_j}{\partial a_1}\Delta a_1 + e_i \tag{2.27}$$

where j is the initial value and $j+1$ is the subsequent value and Δa_0 and Δa_1 are the differences between the initial and preceded values for a_0 and a_1, respectively. The linearized model with respect to the parameters is as follows:

$$y_i - f(x_i)_j = \frac{\partial f(x_i)_j}{\partial a_0}\Delta a_0 + \frac{\partial f(x_i)_j}{\partial a_1}\Delta a_1 + e_i \tag{2.28}$$

In matrix form, this is:

$$\{D\} = [Z_j]\{\Delta A\} + \{E\} \tag{2.29}$$

More explicitly, we have:

$$\begin{Bmatrix} y_1 - f(x_1) \\ y_2 - f(x_2) \\ \vdots \\ y_n - f(x_n) \end{Bmatrix} = \begin{bmatrix} \frac{\delta f_1}{\delta a_0} & \frac{\delta f_1}{\delta a_1} \\ \frac{\delta f_2}{\delta a_0} & \frac{\delta f_2}{\delta a_1} \\ & \\ \frac{\delta f_n}{\delta a_0} & \frac{\delta f_n}{\delta a_1} \end{bmatrix} \begin{Bmatrix} \Delta a_0 \\ \Delta a_1 \end{Bmatrix} + \begin{Bmatrix} e_0 \\ e_1 \end{Bmatrix} \tag{2.30}$$

The objective is to solve Eq. (2.34) for $\{\Delta A\}$, which is used to recalculate the values toward the minimum:

$$\Delta a_0 = a_{0,j+1} - a_{0,j}, \Delta a_1 = a_{1,j+1} - a_{1,j} \tag{2.31}$$

The convergence criterion of the iterative process is defined as the result of Eq. (2.31) falling below a predefined tolerance.

2.5.2.2 Polynomial regression

For nonlinear data patterns, a fit can be obtained with curved line models. An alternative in which data integrity is preserved involves fitting with polynomial models [35]. Least squares regression can be extended to fit data using a polynomial of degree m:

$$y = a_0 + a_1x + a_2x^2 + \cdots + a_mx^m + e \tag{2.32}$$

The sum from $i = 1$ to the total number of data points n of the square of the residual is as follows:

$$Sr = \sum_{i=1}^{n} \left(y_i - \left(a_0 + a_1x + a_2x^2 + \cdots + a_mx^m\right)\right)^2 \tag{2.33}$$

The partial derivative with respect to each of the coefficients of the polynomial is calculated as follows:

$$\frac{\partial Sr}{\partial a_0} = -2\sum \left(y_i - \left(a_0 + a_1x + a_2x^2 + \cdots + a_mx^m\right)\right)$$

$$\frac{\partial Sr}{\partial a_1} = -2\sum x_i \left(y_i - \left(a_0 + a_1x + a_2x^2 + \cdots + a_mx^m\right)\right)$$

$$\vdots$$

$$\frac{\partial Sr}{\partial a_m} = -2\sum x_i^m \left(y_i - \left(a_0 + a_1x + a_2x^2 + \cdots + a_mx^m\right)\right)$$

We can then develop the following set of normal equations:

$$a_0n + a_1\sum x_i + a_2\sum x_i^2 + \cdots + a_m\sum x_i^m = \sum y_i$$

$$a_0\sum x_i + a_1\sum x_i^2 + a_2\sum x_i^2 + \cdots + a_m\sum x_i^{m+1} = \sum x_iy_i$$

$$a_0\sum x_i^m + a_1\sum x_i^{m+1} + a_2\sum x_i^{m+2} + \cdots + a_m\sum x_i^{2m} = \sum x_i^my_i$$

The summations are carried out from $i = 1$ to n. The equations are linear and have $m + 1$ unknown quantities, $a_0, a_1, \ldots, a_m$. The problem of determining least squares polynomials of degree m is equivalent to solving a system of $m + 1$ simultaneous linear equations.

The error in the polynomial regression can be quantified as a standard error of the estimate:

$$S_{y/x} = \sqrt{\frac{Sr}{n - (m + 1)}} \tag{2.34}$$

2.6 Curve fitting program

The flowchart in Fig. 2.12 illustrates the implementation of the time curve fitting program, which was developed in Matlab. Subroutines for nonlinear regression and polynomial regression are used to give different alternatives for fitting

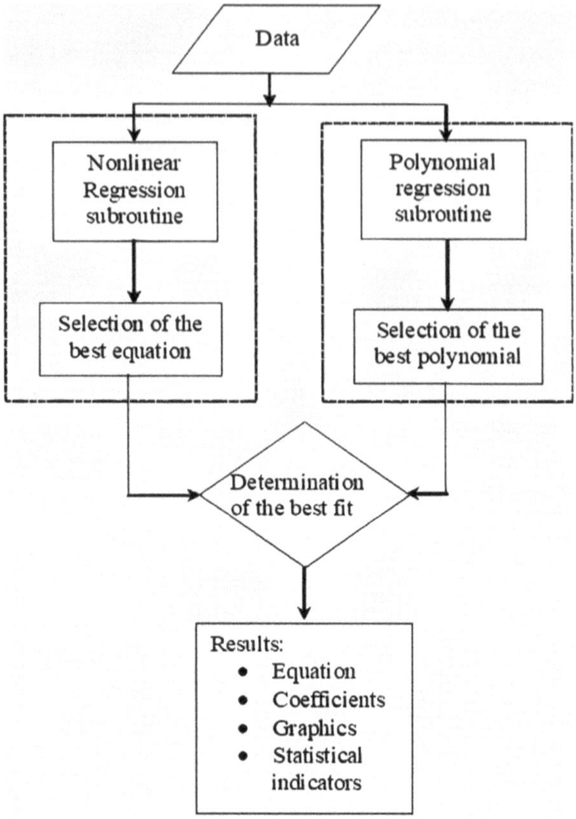

FIG. 2.12 Flowchart for the curve fitting program.

the time curves. The selection of the order of the polynomial is done automatically by the program, which finds the equation that results in the best fit. Since the polynomial equations depend on the choice of terms, the program defines the best structure of the polynomial and calculates the best fit of all the equations. The process with the nonlinear regression equations is defined by selecting the best fit, and hence the best polynomial or nonlinear regression expression that gives the best fit to the time curve is obtained. The best adjustment equation is not known in advance, since it depends on the time curve to be adjusted, meaning that this option is of great value. This option is also important for setting the characteristics of other overcurrent protection devices, such as fuses or automatic reclosers.

The program presents the results in both tabular and graphical form. The tables include the coefficients of the fit equation and some statistical indicators that can be used as a measure of the error. The graphs include the actual and fitted time-current and error-current curves. These statistical indicators are

(a) the sum of the squares of the errors, S; (b) the mean of the error, $EMED$; (c) the maximum error deviation, $EMAX$; and (d) the standard deviation of the error, σ.

Relay time curve fitting. For electromechanical relays, values are obtained from the curves published by the manufacturers, which can introduce an error of appreciation. According to the recommendations of a report by the IEEE Protection Committee [16], 10 points should be used, corresponding to current values with uniform spacing on a logarithmic scale. There are no definitions of accuracy criteria, but it is accepted that a good fit [16,29,31,36] is one where for any current value, the time calculated by the equation has an error of no more than three cycles with respect to the exact value (taken from the published curve, for example). For currents that represent high multiples of I_{pickup}, the error may exceed this limit, but should not be greater than 5%. These accuracy requirements are generally applied starting at a multiple of two, although manufacturers typically give time-current characteristics starting at a multiple of 1.5. Curves based on multiples of between 1 and 1.5 are considered to have some uncertainty and are not useful for coordination [31,33].

2.6.1 Fuse fitting

In many protection systems for distribution circuits, a relay must be coordinated with other types of overcurrent protection, such as fuses and reclosers. In this section, we evaluate the curve fitting process for the time curves of fuses and reclosers.

The characteristics of fuses are not defined analytically, so the resulting settings are validated with respect to the time-current values. These values were obtained from tables given by commercial power system analysis programs, from which 10 uniformly distributed points were taken [16]. Since there are no accuracy criteria for fuses, the criteria used for relays were taken as a reference. When setting fuses, there is no critical region in which other accuracy criteria can be defined, unlike for relays; in a fuse, the entire current range is critical, and the shape of the characteristic operating curve is more variable than for relays. In this work, we expand the accuracy criterion used for relays to the case of setting fuses. A setting of 10 cycles is proposed as acceptable, representing a maximum uncertainty of 166 ms, and the entire range of values used by manufacturers is analyzed.

The data for the time curves of the studied fuses are presented in Table 2.1.

Fig. 2.13 shows the graphical results obtained from the time curve fitting program. For each case, the input and adjusted time-current curves, the error deviation (ΔE), and the error in cycles (Er) are presented. It can be observed that the settings for all fuses are poor. Tables 2.2 and 2.3 present the statistical indicators of the errors resulting from the application of the adjustment algorithms, where the best fit results are highlighted in bold. The following conclusions can be drawn:

TABLE 2.1 Time curve for fuses.

Fuse 1		Fuse 2		Fuse 3		Fuse 4	
I	*T*	*I*	*T*	*I*	*T*	*I*	*T*
9.611	2545.03	29.7	550	741.3	59.40	7.794	1124.41
10.348	855.824	29.7	367.63	771.8	22.45	8.462	250.12
12.862	92.132	30.1	144.82	792	7.71	9.378	69.597
17.285	9.837	31	56.99	869.3	3.12	10.182	26.102
21.662	2.902	34	16.16	905	2.5	11.284	10.547
27.825	0.992	35.5	11.01	954.1	1.75	13.718	2.582
50.662	0.105	39.5	7.6	1089.1	0.9	16.849	0.799
67.524	0.043	54.7	3.3	1163.7	0.69	22.238	0.215
109.148	0.011	64.4	2.25	1346.4	0.44	35.318	0.039
155.99	0.004	79.9	1.3	1497.7	0.32	59.048	0.009
		126.6	0.52	1953.6	0.16		
		249.3	0.14	2032	0.15		
		569.4	0.03	2260.3	0.12		

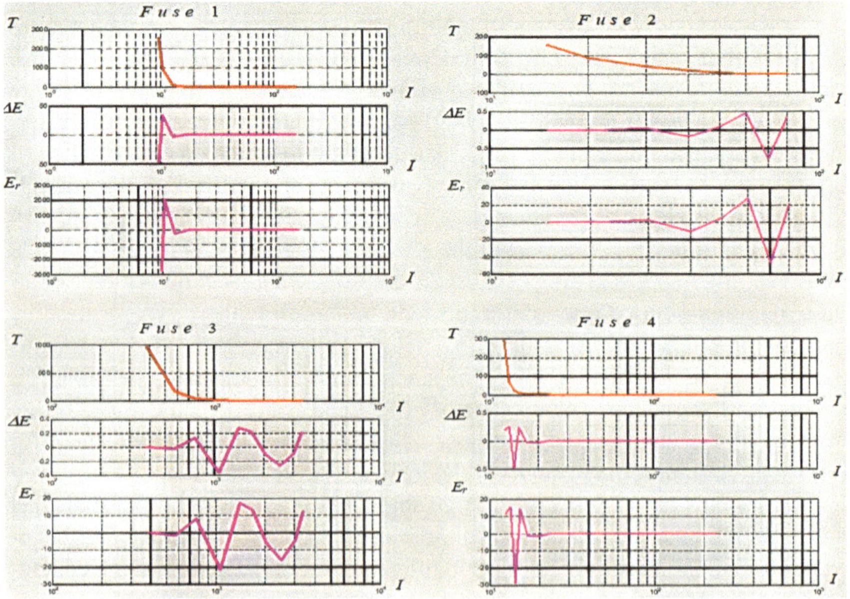

FIG. 2.13 Graphical results from the curve fitting program for commercial fuses.

TABLE 2.2 Results of curve fitting with exponential equations for fuses.

Fuse	Eq. (2.8)				Eq. (2.9)				Eq. (2.10)			
	S	*EMED*	*EMAX*	σ	*S*	*EMED*	*EMAX*	σ	*S*	*EMED*	*EMAX*	σ
1	8.1e5	62.101	562.32	342.06	5.9e6	000	2.1e3	1.0e3	3.5e6	−0.71	1.5e3	1.3e3
2	148.48	1.679	7.039	5.449	3.0e3	7.967	30.99	31.99	19.79	000	2.58	Inf
3	9.6e4	44.37	180.9	117.32	8.3e5	000	850.32	407.93	808.8	000	18.93	20.11
4	2.6e4	12.57	117.43	61.27	7.2e4	000	248.33	120.33	0.43	000	0.382	0.46

TABLE 2.3 Results of curve fitting with polynomial equations for fuses.

Fuse	Eq. (2.13)				Eq. (2.14)				Eq. (2.16)			
	S	*EMED*	*EMAX*	σ	*S*	*EMED*	*EMAX*	σ	*S*	*EMED*	*EMAX*	σ
1	**3.2e3**	**−1.576**	**45.78**	**28.51**	1.4e3	000	69.89	60.92	9.7e3	000	187.48	155.87
2	74.55	0.283	7.33	6.105	**0.95**	**000**	**0.767**	**0.691**	1.776	−0.001	0.946	0.942
3	17.15	0.362	3.84	2.071	**0.416**	**000**	**0.363**	**0.322**	1.14	−0.015	0.761	0.534
4	**0.399**	**0.017**	**0.482**	**0.315**	1.0e3	000	21.11	15.97	2.1e3	000	31.88	23.19

- Exponential equations generally give poor fitting results, because they were developed to fit relay curves. They are not suitable for setting curves with degrees of inversion lower than relays.
- Polynomial expressions are more suitable than exponential expressions for curve fitting for fuses. The polynomial expressions in Eqs. (2.12) and (2.13) generally give the best fits for fuses.
- Acceptable setting errors for fuses cannot be achieved with this algorithm. It is therefore advisable to explore the variant of fit based on the $H(I)$ function.

2.6.2 Recloser fitting

This section presents the fitting results for commercial reclosers. The operating characteristics of three reclosers were analyzed, and the time-current values were obtained from tables given in commercial power system analysis programs. Ten uniformly distributed points were used [16], and the accuracy criteria were the same as for relays.

The data for the time curves of the studied reclosers are presented in Table 2.4.

Fig. 2.14 shows the graphical output, including the actual and adjusted operating time curves and the error percentage (E_r) per cycle. It can be seen that generally good fitting results are obtained. Recloser 3 has an operating

TABLE 2.4 Input recloser data.

Recloser 1			
Digital T = (4.211/(M^1.8−0.35) + 0.035)*(14*TD−5)/9			
Recloser 2		Recloser 3	
I(A)	*T(s)*	*I(A)*	*T(s)*
98.996	6.194	100.447	9.75
186.163	1.449	140.036	4.905
336.164	0.416	209.697	2.453
619.587	0.145	294.851	1.208
948.331	0.083	397.51	0.429
1477.73	0.055	457.828	0.207
1704.11	0.05	518.627	0.069
1888.66	0.047	531.727	0.052
2359.91	0.043	593.661	0.045
2956.81	0.043	3027.33	0.044

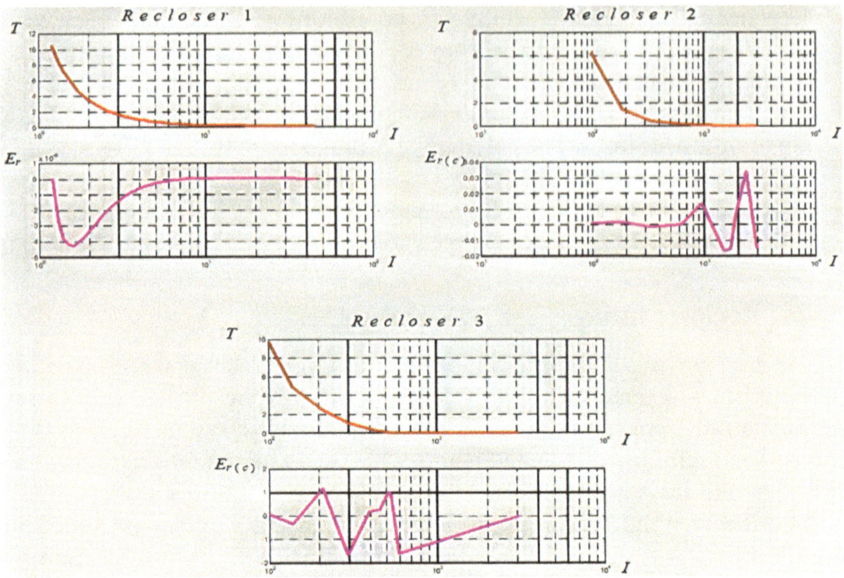

FIG. 2.14 Graphical results from the curve fitting program for commercial reclosers.

characteristic in which part of the curve behaves as a definite time for large current values. Tables 2.5 and 2.6 present the statistical indicators of the resulting errors. The following observations can be made:

- The curve fitting for recloser 1 is acceptable only for the exponential expression in Eq. (2.9), while recloser 2 shows good results for both the exponential and polynomial expressions. The exponential function in Eq. (2.10) gives the best fit for recloser 1, while the polynomial expression in Eq. (2.16) offers the best fit for reclosers 2 and 3.
- The time curves for the reclosers are very similar to those for the relays, showing acceptable settings for these types of overcurrent devices.

2.7 Curve fitting using the *H*(*I*) function

Another alternative for fitting the time curves of overcurrent protection devices is to use the *H(I)* function. The curve fitting process was applied to the set of devices described in Sections 2.7.1–2.7.2. Fuses 1 and 2 and reclosers 1 and 3 were considered.

Fig. 2.15 shows the fitting results for the fuses. In general, the errors obtained were much smaller than for the fitting based on *T(I)*. For fuse 1, the error was reduced from more than 2000 cycles to 2 cycles based on *H(I)*, while for fuse 2, the error was reduced from 40 cycles to 8. Tables 2.7 and 2.8 present the fitting results for the fuse curves. The following can be concluded:

TABLE 2.5 Results of curve fitting with exponential equations for reclosers.

Recloser	Eq. (2.8)				Eq. (2.9)				Eq. (2.10)			
	S	EMED	EMAX	σ	S	EMED	EMAX	σ	S	EMED	EMAX	σ
1	33.32	−2.609	2.063	2.182	**000**	**000**	**000**	**000**	0.129	−0.131	0.149	0.254
2	0.014	−0.029	0.047	0.044	000	000	0.005	0.003	000	000	0.002	0.002
3	0.369	0.065	0.311	0.229	90.11	000	7.83	4.245	0.031	000	0.091	0.124

TABLE 2.6 Results of curve fitting with polynomial equations for reclosers.

Recloser	Eq. (2.13)				Eq. (2.14)				Eq. (2.16)			
	S	EMED	EMAX	σ	S	EMED	EMAX	σ	S	EMED	EMAX	σ
1	336.98	−7.98	8.673	8.209	25.60	−2.236	1.791	1.91	5.671	−3.28	2.736	2.846
2	000	000	0.001	0.001	000	000	0.001	0.001	**000**	**000**	**000**	**000**
3	0.01	000	0.05	0.051	0.008	000	0.06	0.044	**0.002**	**000**	**0.028**	**0.024**

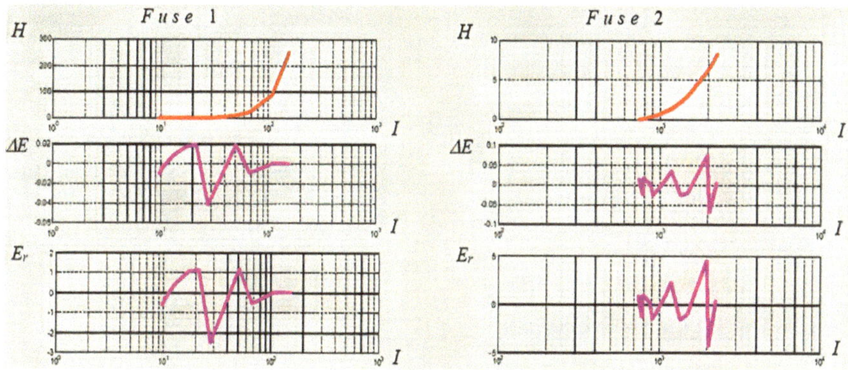

FIG. 2.15 Graphical results from the curve fitting program for fuses based on $H(I)$.

- Exponential expressions give large errors for the fuse curves. Similar behavior was obtained as for the adjustment based on the $T(I)$ function.
- The polynomial expression in Eq. (2.14) gave the best fit for the cases studied here.
- The polynomial expression in Eq. (2.13) gave the worst results.

The fitting results for reclosers are shown in Fig. 2.16. For recloser 1, good fits are achieved with the exponential expressions in Eqs. (2.9) and (2.10), whereas for recloser 3, no acceptable fits are achieved with either expression. Tables 2.9 and 2.10 present the fitting results for the reclosers.

2.7.1 Comparison between curve fits using the functions *H*(*I*) and *T*(*I*)

The curve fits obtained for the operating characteristic based on $T(I)$ and $H(I)$ were acceptable for all the cases studied. An adequate analytical representation was achieved for each device analyzed.

However, for the fuses, there was a marked difference between the settings based on $T(I)$ and $H(I)$. The $T(I)$ function gave very large adjustment errors of up to 300 s, while the $H(I)$ function gave smaller errors. For the reclosers, the curve fits obtained were acceptable for both variants.

Uniform behavior of the proposed adjustment models was not found, since the results strongly depended on the type of time curve to be analyzed and the number of data points entered. For example, the polynomial expression in Eq. (2.13) yielded acceptable adjustments for the characteristics of relays with $T(I)$ but gave poor results for relays with $H(I)$. For fuses, the expression based on $H(I)$ gave much better results than $T(I)$. For recloser 3, an acceptable fit was obtained with $T(I)$ but the fit was poor with $H(I)$.

An adequate analytical representation of the operating characteristic of fuses and reclosers was achieved using both adjustment variants (i.e., $T(I)$

TABLE 2.7 Results of curve fitting for fuses with exponential equations based on $H(I)$.

Fuse	Eq. (2.8)				Eq. (2.9)				Eq. (2.10)			
	S	EMED	EMAX	σ	S	EMED	EMAX	σ	S	EMED	EMAX	σ
1	1.848	0.254	0.851	0.513	0.687	000	0.471	0.370	7.0e4	−34.07	249.99	188.10
2	2.982	0.318	0.669	0.546	0.176	000	0.251	0.148	171.82	−2.36	8.33	5.862

TABLE 2.8 Results of curve fitting for fuses with polynomial equations based on $H(I)$.

Fuse	Eq. (2.13)				Eq. (2.14)				Eq. (2.16)			
	S	EMED	EMAX	σ	S	EMED	EMAX	σ	S	EMED	EMAX	σ
1	7.0e4	36.54	248.99	88.598	**0.003**	**000**	**0.041**	**0.018**	5.96	000	1.46	0.813
2	130.66	1.583	6.66	3.299	**0.015**	**000**	**0.076**	**0.035**	0.019	−0.002	0.082	0.039

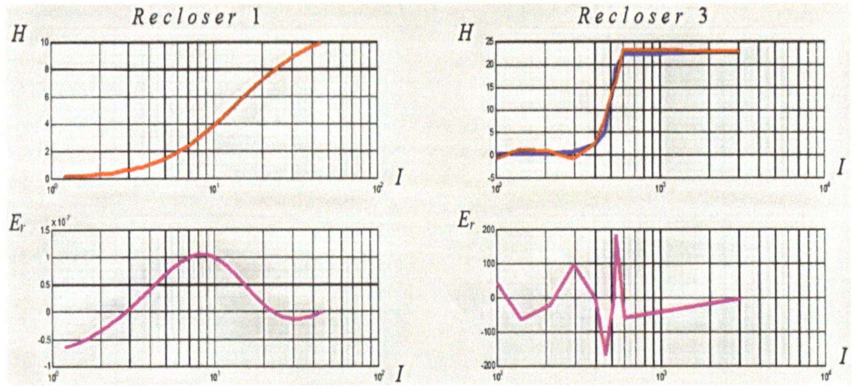

FIG. 2.16 Graphical results from the curve fitting program for reclosers with *H*(*I*).

and *H*(*I*)). From the results, it was observed that there was no function (*H*(*I*) or *J*(*I*)) that could achieve better results for all cases, and that the responses of the algorithms and equation models were highly dependent on the input time curve.

The best option for obtaining a suitable analytical representation for an overcurrent device was to use both variants within the time curve fitting program; this process was done offline, so the computational load was not important in this case.

2.8 Coordination of overcurrent protection

Coordination of the adaptive relay with the primary protection device is done automatically, meaning that the criterion must be met when the adaptive relay is used to back up a relay, fuse, or recloser. It is also necessary to define the criterion for the adjustment of the relay that acts as backup for the adaptive relay. Depending on the tools available to perform coordination, the curve for the adaptive relay may be graphically represented. The coordination process is performed directly with the adaptive relay operating time curve based on the conventional coordination criterion. If a graphical representation of the adaptive relay is not possible, coordination can be performed with the primary device to the level of the short-circuit current at the location of the adaptive relay. For relays and reclosers, this criterion is met by prolonging the operating characteristic of the device supported by the adaptive relay to the short-circuit level indicated for coordination between the adaptive relay and its backup. For fuses, the criterion is fulfilled by shifting their characteristic down to the short-circuit level of the adaptive relay. In both cases, the coordination interval used is 2**CTI*.

TABLE 2.9 Results of curve fitting for reclosers with exponential equations based on $H(I)$.

Recloser	Eq. (2.8)				Eq. (2.9)				Eq. (2.10)			
	S	EMED	EMAX	σ	S	EMED	EMAX	σ	S	EMED	EMAX	σ
1	121.25	1.99	1.113	4.16	**000**	**000**	**000**	**000**	000	0.011	0.005	0.0152
3	413.39	0.566	11.91	7.684	200.14	0.180	7.411	6.326	1.6e3	−8.73	22.72	28.45

TABLE 2.10 Results of curve fitting for reclosers with polynomial equations based on $H(I)$.

Recloser	Eq. (2.13)				Eq. (2.14)				Eq. (2.16)			
	S	*EMED*	*EMAX*	σ	*S*	*EMED*	*EMAX*	σ	*S*	*EMED*	*EMAX*	σ
1	1.1e9	−3.2e3	3.4e4	1.1e4	0.862	−0.372	0.107	0.309	2.62	0.616	0.312	0.539
3	1.4e3	7.758	21.68	12.66	**22.503**	**000**	**3.04**	**1.581**	24.97	000	3.609	1.665

2.8.1 Coordination of an adaptive relay and conventional relay

The 13.8-kV radial system shown in Fig. 2.17 is used as an example to illustrate coordination between two conventional relays (C and A) and relay (B), which for comparison purposes is defined as conventional and adaptive. The maximum levels of short-circuit current are shown for each coordination point between the protection devices.

Fig. 2.17 shows the results of relays coordination, which was carried out using commercial software for the analysis of electrical systems [19]. Relay C has built-in instantaneous and inverse time elements. For comparison purposes, two time curves are shown for the protection located on bus 1: curve B_1 like a conventional relay and curve B_2 like an adaptive relay. The curve for the adaptive relay was obtained by editing the operating characteristic of

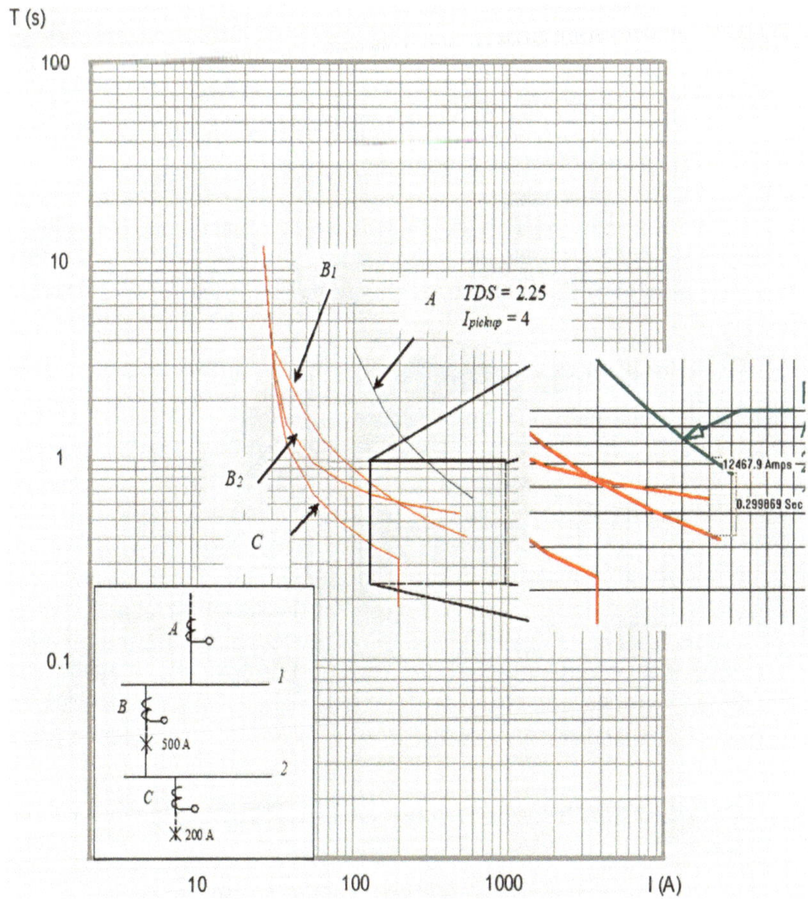

FIG. 2.17 Example of coordination with conventional relays.

relay C with the relay editor provided by the same commercial program. The process was as follows: the operating time curve of the primary relay C was copied; the adaptive relay was edited by adding the CTI (0.2–0.3 s) to the time values (10 points) describing this time curve, while the current values remained unchanged; and finally, the resulting curve was stored under another name. The results of coordination between relay A and the conventional relay B_1 are shown, and the values of TDS and I_{pickup} resulting from this coordination are indicated.

The reduction in the operating time of device B_2 can be observed in the backup zone for C, with a maximum decrease of 1.0 s. It can also be observed that for the coordination point with C (instantaneous element setting = 200 A), an interval of 0.3 s is maintained between both relays throughout the coordination interval. From the coordination examples shown before, it can be seen that introducing the adaptive relay causes a slight increase in the operating time of the B_2 relay on bus 1. However, the advantages of increased sensitivity and the reduction in the operating time of the adaptive relay justify their use and the slight increase in the operating time of the relay.

If the slight increase in the operating time of the adaptive relay is disadvantageous, a hybrid time curve can be designed based on a combination of the adaptive curve B_2 up to the coordination point of Bus 2, and the conventional curve B_1 up to Bus 1. This process is very simple in terms of the protection logic, since it will depend only on activating the curve as a function of the fault current. For dynamic faults, the accumulated value of the integrator G_k will depend on the integration of both functions, as shown in Fig. 2.18.

2.8.2 Coordination of a conventional relay, adaptive relay, and fuse

This section presents an example of the coordination between a fuse C, an adaptive relay B, and a conventional relay A. Fig. 2.19 shows a radial system of 13.8 kV, where the maximum levels of short-circuit current are indicated and it is assumed that the feeder cable between buses 1 and 2 is short enough that the short-circuit levels are comparable. As indicated before, the adaptive relay represents a replica in analytical form of the maximum clearing curve of the fuse plus the coordination CTI, and the curve for this adaptive relay is obtained by means of the analysis software via the same process that was used for relays. Fig. 2.19 shows the coordination results. The conventional relay A can be coordinated directly with the adaptive relay if there is a graphical representation of it or can be coordinated directly with fuse B using a coordination range of 2*CTI.

2.8.3 Coordination between a conventional relay, adaptive relay, and recloser

The diagram in Fig. 2.20 shows a 13.8-kV radial network containing a recloser C, an adaptive relay B, and a conventional relay A, and the results of

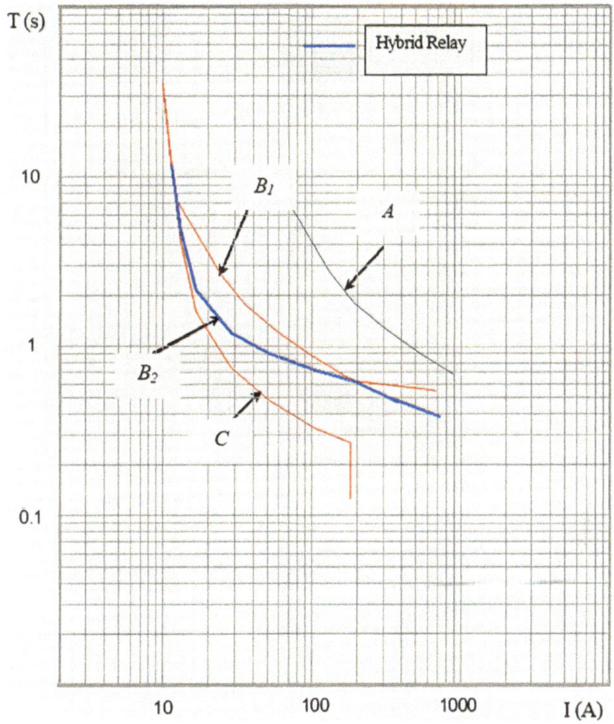

FIG. 2.18 Time curve for a hybrid relay.

coordination are shown in the same figure. The cumulative recloser curve is considered for coordination purposes, meaning that it is not necessary to consider the restoration time and the effect of over travel. The time curve for the adaptive relay is the same as for the accumulated recloser curve plus the coordination *CTI* and is obtained using the analysis software based on the same process that was used for relays and fuses. Coordination of the conventional relay can be done directly with the adaptive relay curve or directly with the recloser C.

2.9 Comparison between overcurrent relays

This section presents a comparative study of the sensitivity and operating times of three types of OCR: conventional, negative sequence, and adaptive.

A. *Sensitivity analysis*

This analysis was performed on the electrical system shown in Fig. 2.21A. The minimum short circuit is located on bus 4. The variation in the impedances Z_3 and Z_4 simulates a multiterminal configuration and a variation in I_{load}, respectively. For this analysis, the phase relays were located on bus 2 (Relay B).

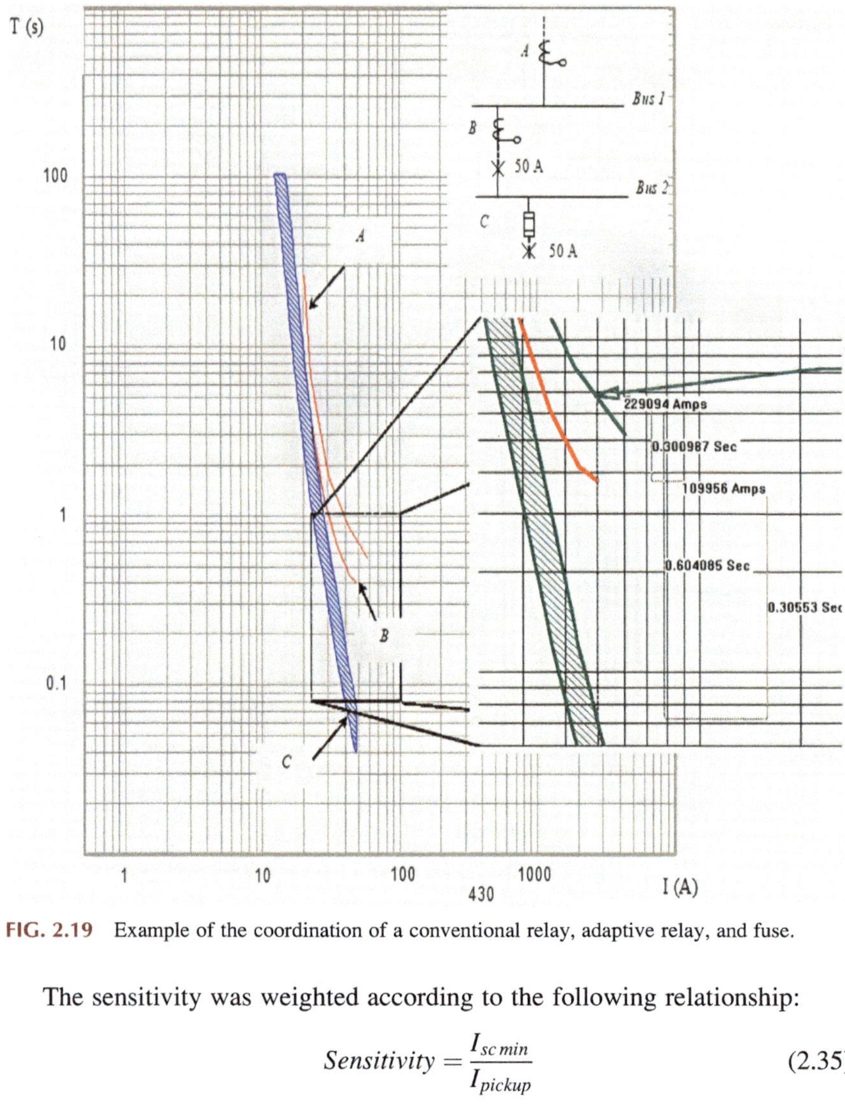

FIG. 2.19 Example of the coordination of a conventional relay, adaptive relay, and fuse.

The sensitivity was weighted according to the following relationship:

$$Sensitivity = \frac{I_{sc\,min}}{I_{pickup}} \tag{2.35}$$

The methods used to determine the sensitivity values for the conventional and negative sequence relays are described in [37,38], respectively. For the adaptive relay, it is assumed that the value of I_{pickup} (1) is not limited by the primary device. This study focused on a particular system, and although the study conditions may change due to topological dependence, operational and protection schemes, it was possible to obtain a qualitative overview of the sensitivity of the three types of OCR.

The minimum fault current was determined as a phase-to-phase fault under minimum generation conditions, located on bus 4, with an adequate sensitivity

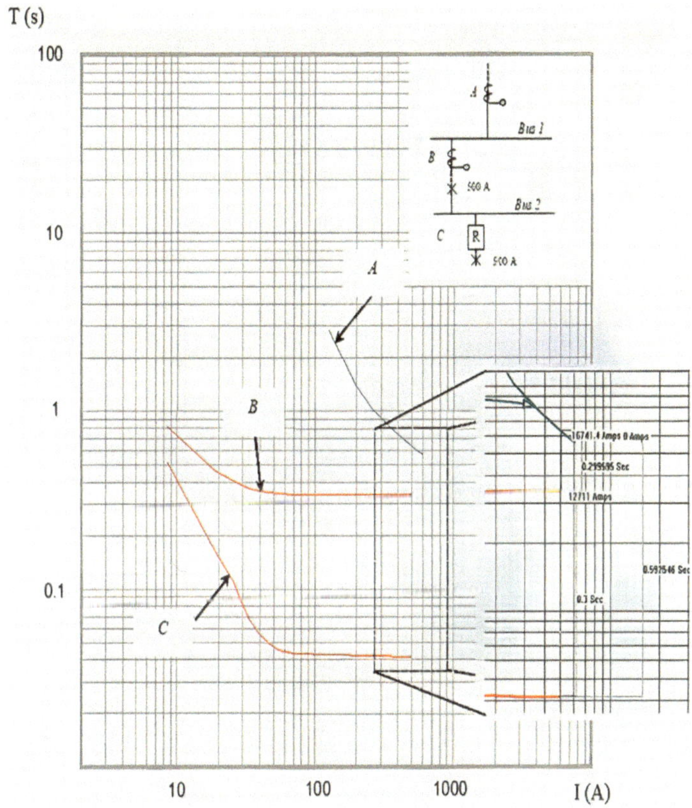

FIG. 2.20 Example of coordination between a conventional relay, adaptive relay, and restorer.

of 1.5 [38]. Fig. 2.21B shows the sensitivity values of the relays for different values of I_{load} and varying contributions from the generation source (P_{Gen}). It can be observed that the conventional relay (shown as plane 1) and the negative sequence relay (plane 2) are insensitive to changes in demand and maintain the same level of sensitivity, since the value of I_{pickup} is independent of the dynamics of the load in these cases. In contrast, the adaptive relay (plane 3) has a setting that is dependent on the dynamics of the load, which results in an increase in the sensitivity of the relay during minimum demand conditions compared to the other relays.

During conditions of higher power contribution (P_{Gen}) in the network, the sensitivity of these types of relays increases (higher fault current). There is a greater increase in the sensitivity of the adaptive relay compared to the other relays, and it can be seen that overall, the adaptive relay has greater sensitivity than the conventional and negative sequence relays under conditions of minimum demand and greater power contribution, due to its greater dependence on the dynamics of I_{load} and the variation in the operating state (power

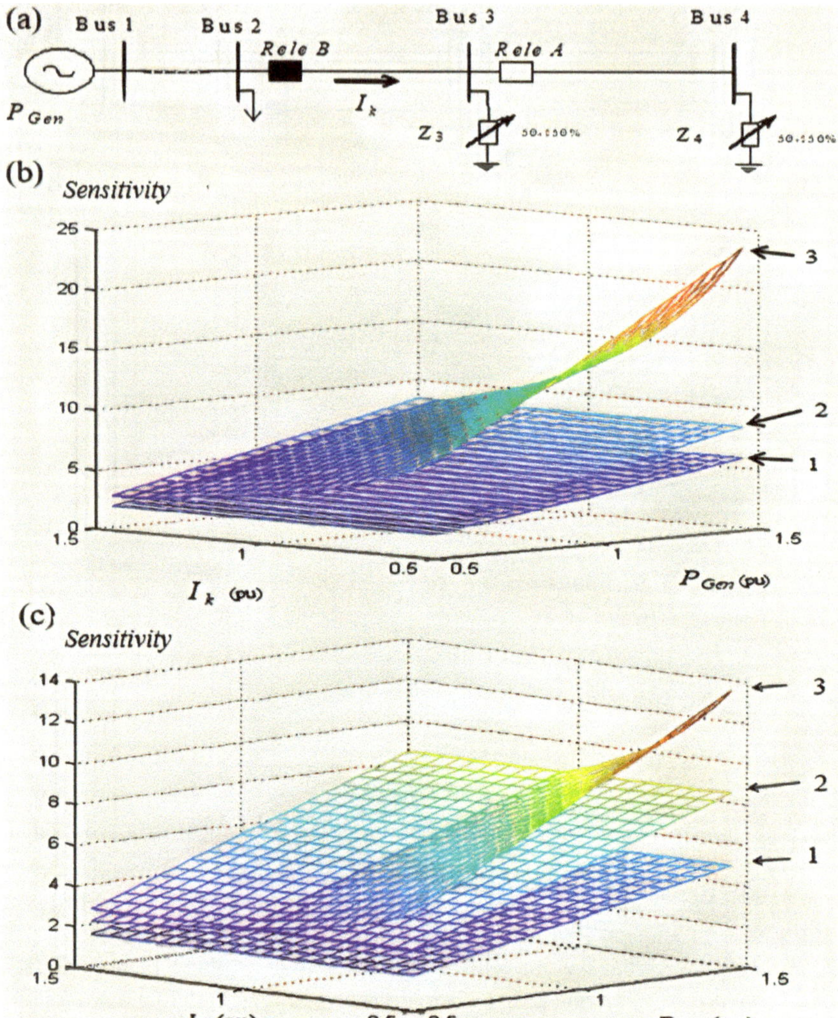

FIG. 2.21 Sensitivity of phase OCRs.

contribution). The sensitivity of the negative sequence relay is similar to that of the adaptive relay only under conditions of maximum demand and for two-phase faults. The sensitivity of the adaptive relay is always higher than that of the conventional relay.

In multiterminal lines (Fig. 2.21C), the negative sequence relay has greater sensitivity than the conventional relay, as it is not affected by the value of I_{load}, and its I_{pickup} is greater than the slowest protection of the feeders of the multi-terminal configuration. Under conditions of maximum demand, the sensitivity of the negative sequence relays is the best, whereas under conditions of minimum demand, the sensitivity of the adaptive relay is the highest.

From these results, we can conclude that the conventional relay has the lowest sensitivity, and the negative sequence relay is more sensitive when used in multiterminal lines (but only for two-phase faults). The adaptive relay, on the other hand, has a higher sensitivity when used in radial networks, and in multiterminal networks it has higher sensitivity than the negative sequence relay under minimum demand conditions, which is when the relay needs it most.

B. *Operation time analysis*

An operation time analysis was applied to the radial system in Fig. 2.21A. The value of I_{pickup} was determined based on Z_3 and Z_4, which were both set to 100%. Fig. 2.22 shows the results for the phase relays. It can be observed that the backup time for the negative sequence relay is lower than for the conventional relay. This is because the adaptive relay has a constant delay of *CTI*, which results in shorter backup times than the other relays, regardless of the magnitude of the fault.

If the short-circuit current is lower than I_{pickup} for the primary device, the adaptive relay will be able to detect the fault (based on fault detection logic);

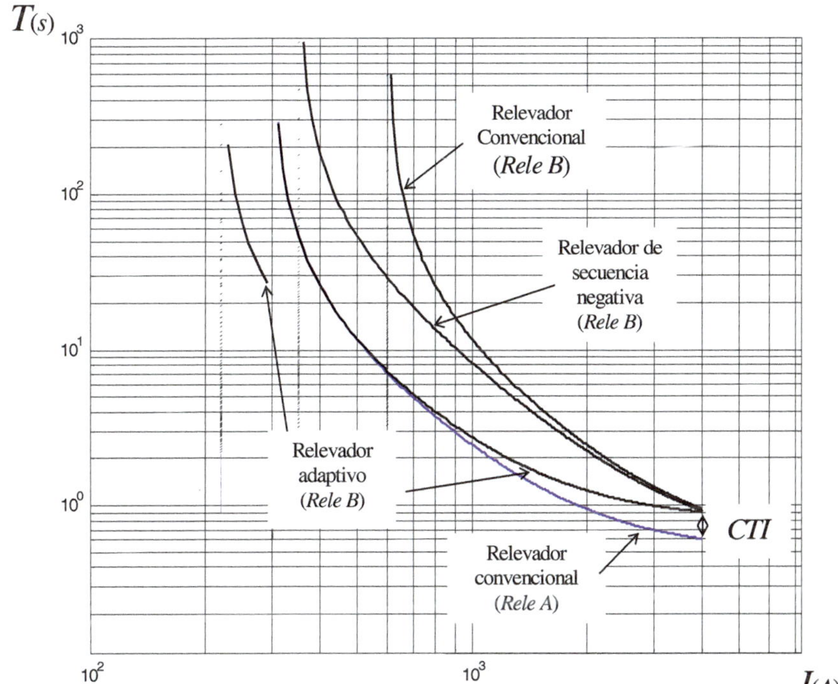

FIG. 2.22 Comparison of time curves for OCRs.

therefore, the I_{pickup} of the adaptive relay that performs only fault detection functions, it can be activated and define the operating time, as in conventional relays. This criterion can be optional to the user, and the application of this criterion offers a solution to the loss of sensitivity of the primary protection, especially under conditions of minimum demand. The adaptive relay time curve will be the same as for the primary protection without the *CTI*. To activate this logic, the fault detection conditions must be met (signal T in Fig. 2.21) and the condition $I_{sc} < I_{pickup}$ must be fulfilled. The minimum settings of I_{pickup} for minimum demand conditions are 300 A for the primary relay and 220 A for the adaptive relay (Fig. 2.22). For any fault between 220 and 300 A, the adaptive relay will have higher sensitivity than the primary relay, meaning that the fault will be detected through high impedance.

2.9.1.1 Location of adaptive overcurrent protection

When applying the proposed adaptive relay, protection should be located at those points where conventional protection faces difficulties, particularly in terms of sensitivity. The increased sensitivity of the adaptive relay can allow it to operate for faults that are outside the scope of conventional protection. This is an incidental backup of adaptive protection for faults that are beyond the sensitivity of the endorsed protection. The adaptive relay is coordinated automatically with conventional coordination devices, which represents an advantage in terms of the protection operation time and a considerable simplification in the application of the OCR.

2.10 Summary

The proposed adaptive relay has a backup time delay of *CTI* for any short-circuit current. The adaptive relay uses the time curve of the primary protection device to obtain a similar shape, meaning that coordination is automatic between the proposed adaptive relay and the overcurrent primary protection (fuse, relay, or recloser). The use of the proposed adaptive relay offers the possibility of solving problems related to sensitivity and coordination between protection devices.

The overcurrent relay can be modeled using functions: the time curve can be visualized by means of two functions, where $J(I)$ represents the coefficient A, and the function $H(I)$ represents the denominator of the standardized curves. To modify the operation time, the function $H(I)$ that represents the operation dynamics of the primary protection is determined.

Coordination of the adaptive relay requires an analytical expression for the time curve of the primary protection device. Exponential or polynomial expressions have been proposed in the literature for the adjustment of the time curves for electromechanical relays, and these expressions can also be applied to fuses and reclosers.

Different results are obtained from the adjustment of the time curves based on *T(I)* and *H(I)*. When setting fuses, there was a marked difference between the settings based on *T(I)* and *H(I)*: the *T(I)* function yielded very large adjustment errors, while *H(I)* gave smaller errors. For reclosers, the adjustments obtained were acceptable for both variants.

From the results presented here, it can be concluded that there is no single equation that provides the best fit for all the protection devices studied. This confirms the importance of the proposed curve fitting program having access to several fitting models and selecting the most suitable model for each characteristic analyzed. Uniform behavior of the proposed adjustment models was not observed, since the results strongly depend on the type of curve to be analyzed and the number of data points entered. It was determined that an acceptable fit could be obtained with 10 points on the curve.

References

[1] A. Girgis, S. Brahma, Effect of distributed generation on protective device coordination in distribution system, in: Power Engineering, 2001. LESCOPE '01. 11–13 July 2001, 2001, pp. 115–119.

[2] C.W. So, K.K. Li, Protection relay coordination on ring-fed distribution network with distributed generations, in: TENCON'02. Proceedings. 2002 IEEE Region 10 Conference on Computers, Communications, Control and Power Engineering, vol. 3, 28-31 Oct. 2002, 2002, pp. 1885–1888.

[3] IEEE Std 242, IEEE recommended practice for protection and coordination of industrial and commercial power systems, in: IEEE Buff Book, IEEE, Inc., New York, NY, U.S.A, May 1990.

[4] IEEE Std C37.95-2002 IEEE guide for protective relaying of utility-consumer interconnections.

[5] ANSI/IEEE Std 141-1986, IEEE Recommended Practice for Electric Power Distribution for Industrial Plants.

[6] C. Yanxia, Y. Xianggen, Z. Zhe, C. Deshu, The research of the overcurrent relays based on phase-to-phase differential current -adaptive setting and coordination, in: *Transmission and Distribution Conference and Exposition*, 2003 IEEE PES, Volume 1, 7–12 Sept. 2003, 2003, pp. 250–255.

[7] T. Tunyagul, P. Crossley, P. Gale, J. Zhao, Design of a protection relay for use with a measuring CT, in: Power Engineering Society Summer Meeting, 2000. IEEE, Volume 3, 16–20 July 2000, 2000, pp. 1390–1395.

[8] D.N. Vishwakarma, Z. Moravej, ANN based directional overcurrent relay, in: Transmission and Distribution Conference and Exposition, 2001 IEEE/PES, Volume 1, 28 Oct.-2 Nov. 2001, 2001, pp. 59–64.

[9] S. Lotfi-fard, J. Faiz, R. Iravani, Improved overcurrent protection using symmetrical components, IEEE Trans. Power Deliv. 22 (2) (2007) 843–850.

[10] I. Zamora, A.J. Mazon, K.J. Sagastabeitia, J.J. Zamora, New method for detecting low current faults in electrical distribution systems, IEEE Trans. Power Deliv. 22 (4) (2007) 2072–2079.

[11] K.R. Shah, E.D. Detjen, A.G. Phadke, Feasibility of adaptive distribution protection system using computer overcurrent relaying concept, IEEE Trans. Ind. Appl. 24 (5) (1988) 792–797.

[12] M.S. Sachdev, T.S. Sidhu, B. Chattopadhyay, B. Takuldar, G. Mc-Donald, R. Chan, Design and evaluation of an adaptive protection system for a distribution network, in: Cigré Paper 34–202, París, 1995.

[13] Patent. A. Conde, Relevador de sobrecorriente de tiempo inverso con funciones independientes de deteccion de falla y tiempo de operación. MX/a/2013/011384, 2017-06-23, 2017.

[14] IEEE Std C37.112-1996, IEEE Standard Inverse-Time Characteristic Equations for Overcurrent Relays, September 1996.

[15] W.E. Glassburn, W.K. Sonnermann, Principles of induction-type relay design, AIEE Trans., Part III, Power Appar. Syst. 72 (4) (1953) 23–27.

[16] IEEE Committee Report, Computer representation of overcurrent relay characteristics, IEEE Trans. Power Deliv. 4 (3) (1989) 1659–1667.

[17] S.E. Zocholl, Testing dynamic characteristics of overcurrent relays, in: 20th Annual Western Protective Relay Conference, Spokane Washington, October 19-21,1993, 1993.

[18] G. Benmouyal, S.E. Zocholl, Time-current coordination concepts, in: 21st Annual Western Protective Relay Conference, Spokane, Washington, October 18-20,1994, 1994.

[19] E.O. Schweitzer, A. Aliaga, Digital programmable time-parameter relay offers versatility and accuracy, IEEE Trans. Power Appar. Syst. PAS-99 (1) (1980) 152–157.

[20] G. Benmouyal, A log-table based algorithm for implementing microprocessor time-overcurrent relays, IEEE Trans. Power Appar. Syst. PAS-101 (9) (1982) 3563–3567.

[21] G. Benmouyal, Design of a digital multi-curve time overcurrent relay, IEEE Trans. Power Appar. Syst. 5 (4) (1990) 1725–1731.

[22] G. Benmouyal, Some aspects of the digital implementation of protection time functions, IEEE Trans. Power Deliv. 5 (4) (1990) 1705–1713.

[23] C.A. Kramer, W.A. Elmore, Flexible inverse overcurrent relaying using a microprocessor, IEEE Trans. Power Deliv. 5 (2) (1990) 915–923.

[24] R.A. Wylie, G.G. Auer, R.C. Ender, Automatic selection of distribution system protective devices, in: Proceedings of the IEEE 1965 PICA Conference, May 1965, pp. 683–705.

[25] A.H. Knable, Overcurrent relays characteristics in a simple form for use in digital computers, in: AIEE Summer General Meeting, Denver, Colorado, June 1961, Paper No. CP 62-1090, 1961, pp. 1–7.

[26] A.R.V.C. Warrington, Protective Relays: Their Theory and Practice, vol. I, John Wiley and Sons, Inc, New York, 1962.

[27] A.R.V.C. Warrington, Protective Relays: Their Theory and Practice, vol. II, John Wiley and Sons, Inc, New York, 1969.

[28] J.E. Hieber, Empirical equations of overcurrent relay curves for computer application, in: IEEE Winter Power Meeting, New York, N.Y., January/February 1965, Paper No. 31 CP 65-91, 1965, pp. 1–11.

[29] G.E. Radke, A method for calculating time-overcurrent relay settings by digital computer, IEEE Trans. Power Appar. Syst. 82 (Special Supplement) (1963) 189–205.

[30] R.E. Albrecht, M.J. Nisja, W.E. Feero, G.D. Rockefeller, C.L. Wagner, Digital computer protective device coordination program I- general program description, IEEE Trans. Power Appar. Syst. 83 (4) (1964) 402–411.

[31] M.S. Sachdev, J. Singh, R.J. Fleming, Mathematical models representing time-current characteristics of overcurrent relays for computer application, in: IEEE Paper A78 131-5, January 1978.

[32] J. Singh, M.S. Sachdev, R.J. Fleming, A.E. Krause, Digital IDMT directional overcurrent relays, in: Second International Conference on Developments in Power System Protection, IEE Publication No. 185, 1980, pp. 84–87.

[33] M.J. Damborg, R. Ramaswami, S.S. Venkata, J.M. Postforoosh, Computer aided transmission protection systems design, IEEE Trans. Power Appar. Syst. PAS-103 (1) (1984) 51–59.

[34] IEC Standard 255-4, Single Input Energizing Measuring Relays with Dependent Specified Time, first ed., IEC Publication 255-4, 1976.

[35] S.C. Chapra, R.P. Canale, Numerical Methods for Engineers, McGraw-Hill, New York, 1988.

[36] S.S. Begian, R.O. Bigelow, D.M. Lyle, G.E. Radke, W.F. Reinke, R.F. Silva, J.R. Truitt, H.Y. Tsien, E.L. Worthington, A computer approach to setting overcurrent relays in a network, in: Proceedings of the IEEE 1967 PICA Conference, May 1967, pp. 447–457.

[37] S.H. Horowitz, A.G. Phadke, J.S. Thorp, Adaptive transmission system relaying, IEEE Trans. Power Deliv. 3 (1988) 1436–1445.

[38] G.D. Rockefeller, C.L. Wagner, J.R. Linders, Adaptive transmission relaying concepts for improved performance, IEEE Trans. Power Deliv. 3 (1988) 1446–1458.

Chapter 3

Overcurrent coordination via metaheuristic algorithms

3.1 Introduction

The coordination of DOCRs involves some degree of uncertainty, due to its complexity and nonlinearity, and it is therefore difficult to claim that global optimal results have been found for different electrical networks and operational conditions. In the same way as for optimization algorithms, there is no particular formulation that is considered to be the best. This has given impetus to a very interesting algorithmic search to find coordination solutions by means of different formulations. There are various criteria that can be used to select the most suitable formulation for a specific application, for example, its robustness in terms of finding solutions under a higher number of restrictions in large electrical networks, its search engine to find complex solutions as it happens in active networks, and whether it offers results within statistical indicators of central tendency for possible online applications.

In 1988 the first technique introduced to solve the problem of protection coordination used a deterministic approach. The coordination problem was initially considered as a nonlinear parameter optimization problem and was tackled using linear programming [1]. Deterministic methods have been extensively improved in order to obtain adequate solutions for the coordination of DOCRs. These include nonlinear programming (NLP), interval linear programming (ILP), mixed nonlinear programming (MINLP), and quadratically constrained programming (QCQP) [2–4]. Although these deterministic methods have undergone vast improvements, they require a significant number of iterations and high computational resources. At the same time, they cannot guarantee an optimal solution, as the algorithm can easily become trapped in a local minimum, an effect that is strongly dependent on the initial starting point.

Following the success of approaches to solving the coordination problem using a deterministic approach, heuristic and metaheuristic optimization methods were introduced based on the concept of evolution. The first of these methods to be introduced was the genetic algorithm (GA) [5]; since then, various techniques in this category have been explored to solve the problem and have been shown to be capable of overcoming most of the limitations of deterministic methods. These algorithms use a population technique to carry out a

Overcurrent Relay Advances for Modern Electricity Networks. https://doi.org/10.1016/B978-0-323-91777-3.00003-9

global search, which improves the solution. Other techniques have included particle swarm optimization (PSO), the harmony search algorithm (HSA), self-adaptive differential evolution (SADE), the covariance matrix evolution strategy (CMA-ES), the seeker algorithm (SA), the cuckoo search algorithm (CSA), ant colony optimization (ACO), and biogeography-based optimization (BBO) [6–12].

The possibility of combining deterministic and metaheuristic approaches has recently been explored by researchers with the aim of finding the best in each category to solve the issue of DOCR coordination. The main factors that determine which algorithms will be combined to obtain the best solution for a problem are the capabilities of the algorithm, the efficiency improvement, parameter control, convergence speed, and the quality of the solutions. Approaches are combined to exploit the individual performance of each algorithm, since it has been observed in the literature that it is not practical to expect that a hybrid algorithm will perform well for all optimization problems. The application of hybrid algorithms to DOCR coordination has yielded very good results [13–17].

The DOCR coordination model commonly relies on single-objective functions, with weightings assigned to the terms that make up the function. The main characteristic of many optimization problems is that these are multiobjective problems. One recent approach to carrying out coordination in interconnected networks is to formulate this as multiobjective problem [17,18].

3.2 DOCR coordination via metaheuristic algorithms

The use of metaheuristic algorithms cannot guarantee a global solution for DOCR coordination, and the results are considered to be good when compliance with the equality and inequality restrictions is reached. It is important to specify the population size and the number of generations to be evaluated, since the number of feasible solutions and the solution time depend on these aspects. The nature of an evolutionary algorithm is that it achieves the best solution within the search space; however, its performance is strongly affected by the diversity of the problem. The problem of coordination of protection devices in electrical networks involves a high level of topological diversity, and the operating conditions are very diverse. Hence, the need for new optimization algorithms is constant, and researchers are always in search of better solutions.

Another alternative is to view the optimization problem as a multiobjective problem, in which minimization of the relay times is carried out independently. The selection of the solution is made based on a dominance criterion, and a combination of all the OFs will be evaluated. In the specific case of DOCR coordination, the functions are not convex because the OFs are obtained with the time curve expression evaluated for the primary as the backup relays, that is, they are similar functions. This is why the use of a multiobjective formulation does not represent a great advantage over weighted functions for DOCR coordination.

The optimization problem becomes even more challenging when constraints are considered. If the evaluated decision vector satisfies the constraint at any given point, it is said to be a feasible solution. There are various types of constraints that can be added to the coordination problem, of which the two main types are equality and inequality constraints, also known as hard constraints. These types of constraints place bounds on the OF so that it does not try to search beyond these points, even though there is a high possibility that better results will be obtained. Flexible or soft constraints such as the penalty method can also be included in the OF; however, unlike hard constraints, it is not obligatory that these are satisfied.

3.2.1 Data preparation for relay coordination formulation

The complexity of the problem increases exponentially as the size of the power system grows. In meshed systems, the coordination of DOCRs is formulated as an optimization problem, with the aim of obtaining the minimum operation time. It is not possible to guarantee the global optimum with metaheuristic methods, but from the point of view of optimization, the objective function is minimized to obtain acceptable operating times for the protection of the electrical network.

The operation time interval (*CTI*) between the backup relay (*Tb*) and primary relay (*Tp*) must guarantee the sequence of operation in case of nonoperation of the primary relay. It is recommended that the same type of time curve be used for the OCRs to avoid crossing of the curves, which is why in radial lines it is only necessary to verify the coordination for the maximum fault current (F_1). However, for meshed networks, in which it is possible for the curves of the DOCRs to cross, it is recommended that the coordination check be performed at both ends of the line (Fig. 3.1).

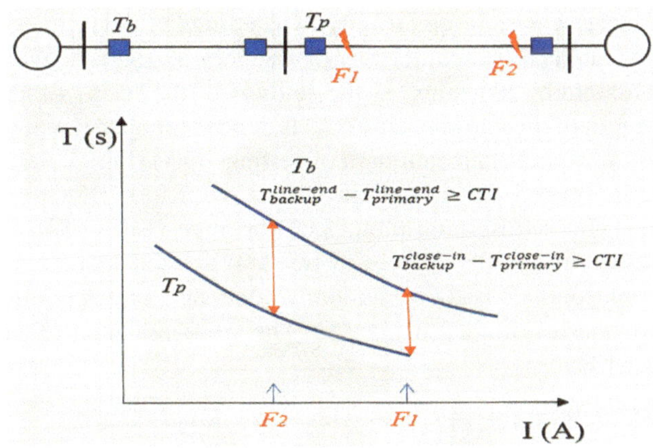

FIG. 3.1 OCR coordination.

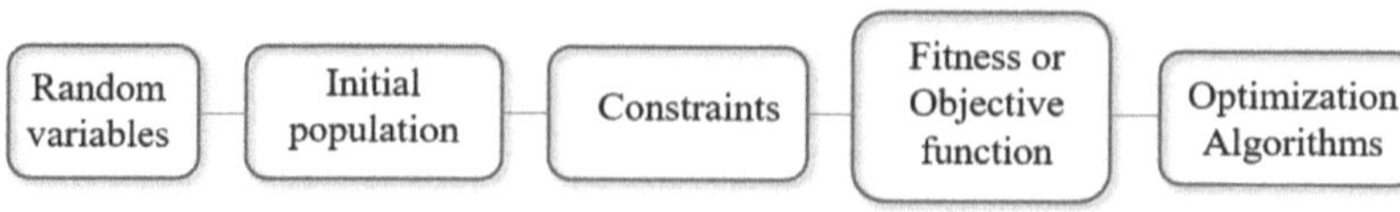

FIG. 3.2 Data preparation for relay coordination via optimization algorithms.

The diagram in Fig. 3.2 shows the data preparation and the generation of vectors and matrices for the coordination formula. The random variables form the initial population that will be evaluated by the relay time curve functions, where the range of each variable must fulfill the coordination requirements to avoid constraint violations if the ranges are very limited. Optimization algorithms are more successful when the solution space is larger. The initial population size represents the number of relays, and the equality constraints are imposed by the time curve expression for OCRs, while the inequality restrictions are determined by the ranges of the random variables and, more importantly, compliance with the coordination criterion established for each coordination pair.

The OF for DOCR coordination is a minimization problem that establishes criteria for searching for the solution. This must take into consideration the coordination criterion and other metrics for weighting the distance of the solution. A diverse range of OFs have been proposed, and some have been compared with each other [19]. The optimization problem can be solved by means of various optimization algorithms; however, the main data preparation steps for relay coordination shown in Fig. 3.1 must be followed. The details of each data structure are given as follows.

3.2.1.1 Random variables

A DOCR provides current-dependent time operation, and the relay adjustable settings (ASs) determine the degree of freedom of the analytical expression for the time curve. For example, if only two settings of the relay are adjusted, it is referred to as having two degrees of freedom. A total of five degrees of freedom is possible by adjusting all the settings of the relay, which results in NCCs. The length of the chromosome (C_x) is dependent on the number of relays (*NR*) in the system and their numbers of settings (*NS*). Fig. 3.3 shows how a chromosome is created for a network system with four DOCRs, for which two (*TDS* and plug setting multiple (*PSM*)) or three (*TDS*, *PSM*, and type of characteristic curves (*TCC*)) degrees of freedom are considered.

Each chromosome consists of randomly generated values within the upper and lower boundaries of each of its *ASs*. The value assigned to each gene is given by Eq. (3.1):

$$AS = AS_{lower} + \left[AS_{upper} - AS_{lower}\right] \times Rand, Rand \in [0, 1] \tag{3.1}$$

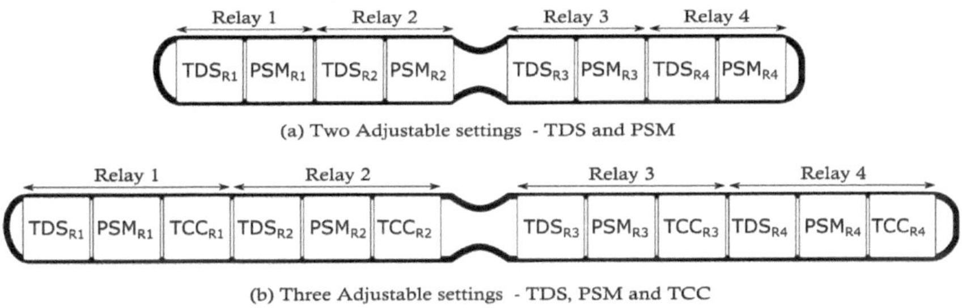

FIG. 3.3 Length of a chromosome.

For example, if the coordination problem involves different types of time curve, such as the use of different degrees of freedom to search for better solutions, the *TCC* random generated values are obtained using Eq. (3.2):

$$TCC = \lceil Randi \rceil, \; Randi \in [1, NC] \tag{3.2}$$

where *NC* is the number of characteristic curves to choose from. The relay characteristic curve will depend on the parameters of the chosen *TCC*.

3.2.1.2 *Initial population*

The population is formed of individuals known as chromosomes, which represent different traits providing them with a unique capacity to minimize the OF. The chromosomes that make up the population consist of one or more genes, as shown in Fig. 3.3, which are mapped uniquely to a setting or decision variable of the problem. The length of each chromosome depends on the total number of decision variables of the problem, which are formed of alleles, integers, or real values.

The algorithm does not depend on the initial starting points in the same way as deterministic methods. Instead, the initial population is created with random genes for each chromosome in the population. When creating these random genes, the lower and upper boundaries required by the problem should be respected. This initial population allows the algorithm to start the exploration process within a larger search space and to identify a specific area for the algorithm to exploit to obtain a solution in successive generations.

In nature, chromosomes do not contain similar genes, which is why all organisms are different. In the GA, this is one of the many requirements that must be considered and is why each chromosome and population differ from one generation to the next. All the chromosomes in the population are evaluated using an OF to determine a fitness value for the given problem. These fitness values determine the degree of approximation of each chromosome with regard to the solution at each generation; the closer these are, the higher the fitness of the solution.

The population consists of different traits, which provide them with a unique capacity to minimize the OF. The population size (*PS*) can remain constant or can be modified throughout the iterations to improve the convergence rate of the algorithm. In this research, the population size is kept constant and depends solely on the number of individuals and their length. Eq. (3.3) shows the general matrix arrangement of the initial population, where each row represents an individual in the population and each column represents a relay setting, *PN* is the population number and *NR* is the number of relays.

$$PS = \begin{bmatrix} (TDS_1 \cdots PSM_{NS})_{1,r1} & (TDS_1 \cdots PSM_{NS})_{1,r2} & \cdots & (TDS_1 \cdots PSM_{NS})_{1,NR} \\ \vdots & \vdots & \ddots & \vdots \\ (TDS_1 \cdots PSM_{NS})_{PN,r1} & (TDS_1 \cdots PSM_{NS})_{PN,r2} & \cdots & (TDS_1 \cdots PSM_{NS})_{PN,NR} \end{bmatrix} \tag{3.3}$$

3.2.1.3 *Constraints*

The coordination of DOCRs in large interconnected systems is a highly constrained optimization problem that has been solved using various techniques. There are restrictions that limit the effect of the decision variables on its cost function, making it even more difficult to solve using conventional methods. When viewed as an optimization problem, coordination requires that both the hard and soft constraints on the formulation of the population and the OF should be considered. Hard constraints apply to the ranges within which the decision variables should operate, while soft constraints are used to assist in making the final decision on the best individual in a population. For example, the decision variables that determine the characteristic curves for DOCRs are the *PSM* and *TDS* for two degrees of freedom.

Equality constraint

The operating time of a relay is established based on the time curve equation [20,21], depending on the number of ASs considered, as shown in Eq. (3.4):

$$T = F(PSM, TDS, A, p, B) \tag{3.4}$$

Inequality constraint

The variables used for the optimization algorithms should be bounded within ranges defined by minimum and maximum values. The use of limited maximum intervals is desirable to force the algorithm to give results with shorter relay times. For example, ranges for the *TDS* and *PSM* are considered as follows:

$$(PSM_{min} \leq PSM \leq PSM_{\max})_\beta \tag{3.5}$$

$$(TDS_{min} \leq TDS \leq TDS_{max})_\beta \tag{3.6}$$

The most important inequality constraint is the **coordination criterion**, since this results in the desired time sequence for the operation of the relays:

$$(Tb - Tp \geq CTI)_{\vartheta} \tag{3.7}$$

where ϑ is the total number of coordination pairs and β is the total number of relays. The 14-bus IEEE test system has 50 coordination pairs and 40 relays, while the 30-bus IEEE test system has 120 coordination pairs and 82 relays. In the coordination process, a sequence of operation is established in which the primary relay is the first option to clear the fault, and the backup relay operates only when the primary relay fails to operate. In terms of protection, both functions are extremely important and must be properly coordinated. However, the functional dependency between relays causes the resetting of one relay to affect the others.

3.2.1.4 Objective function

The main purpose of this is to define the importance of an individual in the population by obtaining its overall performance within the problem domain. The individual from a set of relay settings that performs best after each iteration is the fittest of the population. In this research, DOCR coordination is formulated as a minimization optimization problem that aims to obtain optimal relay settings that minimize the operation time of each relay on the network, while maintaining selectivity and dependability throughout the interconnected system in regard to any abnormal condition that may occur. The OF used to address the problem of coordination has been modeled using several different approaches, which are not discussed in detail here but can be found in various references. Two different approaches to formulating the problem are based on weighted and multiobjective formulations.

Weighted objective function

This is an a priori approach that combines two or more single objectives to create a weighted objective function. The weights represent the importance of each single objective and are obtained by evaluating different sets of weights to determine the problem's unknown Pareto front that accurately represent the importance of each function. For all optimization problems that use an a priori approach, it is important to determine beforehand which function has a higher impact on the OF. A single objective formulation is very helpful in terms of providing a straightforward solution to any optimization problem and is the most widely used approach to solving the problem of DOCR coordination.

The OF is formulated to minimize the operation time of the relays, while the restrictions are formulated to meet the coordination criteria for the primary and backup relays. In addition, limits on variables such as *TDS* and *PMS* are considered. Due to the diversity of topologies and operating conditions of electrical networks, each coordination problem is different, and hence a range of different OFs and optimization algorithms have been proposed; although there is no dominant OF or algorithm, certain algorithms are more widely used than others.

In this chapter, a comparison between different OFs or optimization algorithms is not made. The function used in this chapter is the one presented in Ref. [22]:

$$OF = \min\left\{\frac{NV}{NCP} + \omega_1\left(\frac{\sum_{p=1}^{NCP} Tp}{NCP}\right) + \omega_2\left(\frac{\sum_{b=1}^{NCP} Tb}{NCP}\right) + \omega_3\left(\sum_{pb=1}^{NCP} CTI\right)\right\} \quad (3.8)$$

where *NCP* is used for normalized and represents the number of coordination pairs, and the number of violations is *NV*. The factors ω_1, ω_2, and ω_3 are used to weight the terms of the OF thought Pareto frontier. Depending on the function with the most dominant weighting, different results will be obtained; for example, if the weighting is greater for *Tp*, the results will give preference to minimization of the primary operation times.

For an optimal solution, there should be zero miscoordination among the coordination pairs of the tested network. As the interconnected system becomes larger with the addition of lines and equipment, there is unlikely to be a reduction in miscoordination without expanding the range of each decision variable. The reduction of miscoordination is a dilemma in the optimal coordination of DOCRs, as decreasing the number of miscoordinated relays sometimes increases the operation time, and vice versa.

The weights are obtained by evaluating various sets to obtain a Pareto front that accurately represents the effect and behavior of each objective on the weighted OFs. For example, the factors used for the coordination problem that result in a minimal OF are shown in Fig. 3.4. The data on the entire set of weights are utilized, and the results obtained for each can be found. In the search for a Pareto front for relay coordination, it is important to evaluate different operating conditions, load demand, and short-circuit contributions (Zth). In the process of coordinating relays, the time functions that make up the *OF* are similar, which results in a proportional variation of the functions. Hence, the best solution is the one obtained through the lowest value obtained from the OF functions.

Multiobjective function

The coordination of DOCRs depends on various single objectives that are simultaneously optimized to obtain the lowest operation time for the network relays. A multiobjective (a posteriori) approach allows each objective to be optimized without using any decision preferences to guide the search for the best solution. During the optimization process, a compromise is found between the objectives, creating multiple Pareto-optimal solutions. This multiobjective approach computes a set of Pareto-optimal solutions (or all of them) to obtain

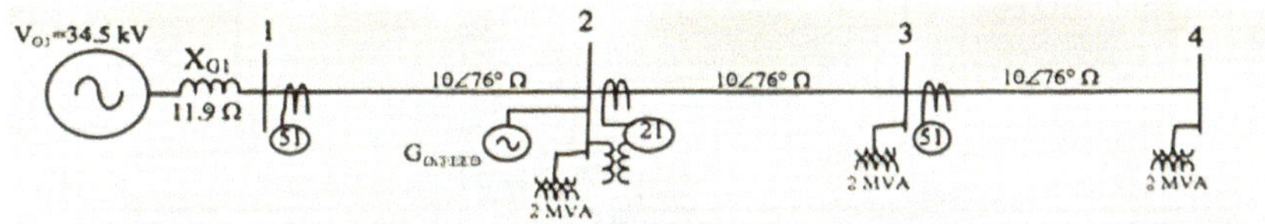

Weights				A-op (s)			A-tb (s)			A-CTI (s)			Overall (s)		
Set	w1	w2	w3	Max	Med	Min	Max	Med	Min	Max	Med	Min	Max	Med	Min
1	0.25	0.25	0.25	0.7554	0.8385	0.7960	1.0786	1.1719	1.1247	0.4571	0.4771	0.4819	2.2911	2.4875	2.4028
2	0.25	0.25	0.5	0.7497	0.7925	0.7971	1.0756	1.0983	1.1249	0.4524	0.4618	0.4775	2.2777	2.3526	2.3995
3	0.25	0.25	0.75	0.7574	0.7908	0.8597	1.0645	1.0875	1.2024	0.4576	0.4569	0.4867	2.2795	2.3352	2.5488
4	0.25	0.25	1	0.7571	0.8246	0.8556	1.0862	1.1364	1.1825	0.4696	0.4663	0.4774	2.3129	2.4273	2.5156
5	0.25	0.5	0.25	0.7707	0.7378	0.7754	1.0862	1.0381	1.1070	0.4555	0.4563	0.4601	2.3123	2.2322	2.3425
6	0.25	0.5	0.5	0.7922	0.7692	0.8488	1.1200	1.0733	1.1825	0.4651	0.4532	0.4824	2.3773	2.2957	2.5137
7	0.25	0.5	0.75	0.7900	0.8172	0.8065	1.1095	1.1429	1.1207	0.4619	0.4691	0.4598	2.3614	2.4292	2.3870
8	0.25	0.5	1	0.7225	0.8267	0.8115	1.0163	1.1471	1.1157	0.4309	0.4594	0.4644	2.1697	2.4332	2.3916
9	0.25	0.75	0.25	0.7239	0.8103	0.8232	1.0342	1.1292	1.1481	0.4425	0.4611	0.4608	2.2006	2.4006	2.4321
10	0.25	0.75	0.5	0.7342	0.8278	0.8132	1.0272	1.1508	1.1593	0.4379	0.4696	0.4769	2.1993	2.4482	2.4494
11	0.25	0.75	0.75	0.7274	0.8401	0.8226	1.0333	1.1665	1.1381	0.4383	0.4644	0.4647	2.1990	2.4710	2.4254
12	0.25	0.75	1	0.7820	0.7856	0.8120	1.0821	1.1312	1.1477	0.4516	0.4826	0.4657	2.3157	2.3994	2.4254
13	0.25	1	0.25	0.7572	0.8034	0.8010	1.0359	1.0961	1.1100	0.4355	0.4412	0.4626	2.2286	2.3407	2.3736
14	0.25	1	0.5	0.7498	0.7860	0.8360	1.0401	1.0926	1.1357	0.4323	0.4511	0.4625	2.2222	2.3297	2.4342
15	0.25	1	0.75	0.7460	0.8177	0.8262	1.0492	1.1275	1.1290	0.4413	0.4604	0.4623	2.2365	2.4056	2.4178
16	0.25	1	1	0.7462	0.8307	0.8248	1.0540	1.1369	1.1305	0.4452	0.4635	0.4643	2.2454	2.4311	2.4196
17	0.5	0.25	0.25	0.7308	0.7815	0.7943	1.0715	1.0765	1.1438	0.4584	0.4538	0.4933	2.2607	2.3118	2.4314
18	0.5	0.25	0.5	0.7647	0.8214	0.8573	1.0882	1.1348	1.1747	0.4736	0.4623	0.4747	[illegible]	2.4185	2.5067
19	0.5	0.25	0.75	0.7923	0.8100	0.8234	1.1062	1.1411	[illegible]	[illegible]	0.4669	0.4734	2.3593	2.4200	2.4350
20	0.5	0.25	1	0.7933	0.8115	0.8146	1.0855	1.1491	1.1691	0.4543	0.4771	0.4725	2.3231	2.4377	2.4362
21	0.5	0.5	0.25	0.7321	0.7659	0.7984	1.0496	1.0586	1.1265	0.4554	0.4524	0.4653	2.2371	2.2769	2.3902
22	0.5	0.5	0.5	0.7828	0.8260	0.8088	1.0850	1.1679	1.1370	0.4490	0.4785	0.4693	2.3168	2.4724	2.4151
23	0.5	0.5	0.75	0.7646	0.8090	0.7744	1.0786	1.1272	1.0973	0.4555	0.4604	0.4577	2.2986	2.3966	2.3294
24	0.5	0.5	1	0.7691	0.7834	0.7856	1.0941	1.0822	1.1071	0.4571	0.4507	[illegible]	[illegible]	2.3163	2.3561

	Zth = 15				Zth = 8				Zth = 13		
F1	F2	F3	OF	F1	F2	F3	OF	F1	F2	F3	OF
0.099168	0.4318	0.0002	0.53122	0.07674	0.395531	5.04E-05	0.47232	0.09246	0.4274	0.00057	0.52042
0.099168	0.3701	0.0007	0.47002	0.07674	0.339027	0.000151	0.41591	0.09246	0.3663	0.0005	0.4593
0.099168	0.3393	0.001	0.43942	0.07674	0.310774	0.000202	0.38771	0.09246	0.3358	0.00047	0.42873
0.099168	0.3084	0.0012	0.40882	0.07674	0.282522	0.000252	0.35951	0.09246	0.3053	0.00043	0.39817
0.099168	0.2776	0.0015	0.37822	0.07674	0.25427	0.000302	0.33131	0.09246	0.2748	0.0004	0.36761
0.099168	0.2467	0.0017	0.34762	0.07674	0.226018	0.000353	0.30311	0.09246	0.2442	0.00036	0.33705
0.099168	0.2159	0.002	0.31702	0.07674	0.197766	0.000403	0.27491	0.09246	0.2137	0.00033	0.30648
0.099168	0.1851	0.0022	0.28642	0.07674	0.169513	0.000454	0.2467	0.09246	0.1832	0.0003	0.27592
0.099168	0.1542	0.0024	0.25582	0.07674	0.141261	0.000504	0.2185	0.09246	0.1526	0.00026	0.24536
0.119001	0.401	0.0002	0.52021	0.09208	0.367279	5.04E-05	0.45941	0.11095	0.3969	0.00055	0.50837
0.119001	0.3701	0.0005	0.48961	0.09208	0.339027	0.000101	0.43121	0.11095	0.3663	0.00052	0.47781
0.119001	0.3393	0.0007	0.45901	0.09208	0.310774	0.000151	0.40301	0.11095	0.3358	0.00049	0.44724
0.119001	0.3084	0.001	0.42841	0.09208	0.282522	0.000202	0.37481	0.11095	0.3053	0.00045	0.41668
0.119001	0.2776	0.0012	0.39781	0.09208	0.25427	0.000252	0.34661	0.11095	0.2748	0.00042	0.38612
0.119001	0.2467	0.0015	0.36721	0.09208	0.226018	0.000302	0.3184	0.11095	0.2442	0.00038	0.35556
0.119001	0.1851	0.002	0.30601	0.09208	0.169513	0.000403	0.262	0.11095	0.1832	0.00031	0.29443

FIG. 3.4 Weighting factors.

the best compromise solution, as there may be more than one solution to the problem at any given iteration.

Unlike in the single objective approach, there is no decision maker to determine the most important objective function. The main advantage of this approach is the approximation of the true Pareto front using multiple solutions, whereas the disadvantages are an increase in the execution time of the algorithm and the fact that Pareto optimality is not guaranteed. A multiobjective approach uses a paradigm based on novelty to search for nondominated solutions within unexplored areas, which helps to boost the exploration process. The objectives used to solve the problem of DOCR coordination are given by the following minimization functions:

(a) Primary relays

$$f_1\left(\vec{x}\right) = \min\left\{\sum_{p=1}^{NCP} Tp\right\} \tag{3.9}$$

(b) Backup relays

$$f_2\left(\vec{x}\right) = \min\left\{\sum_{b=1}^{NCP} Tb\right\} \tag{3.10}$$

(c) Coordination time interval

$$f_3\left(\vec{x}\right) = \min\left\{\sum_{pb=1}^{NCP} CTI_{pb}\right\} \tag{3.11}$$

The purpose of solving the problem of DOCR coordination using a multiobjective approach is to determine whether this method provides a better solution to the problem. Nondominated solutions to the Pareto-optimal front obtained for DOCR coordination with a multiobjective approach are shown in Fig. 3.5.

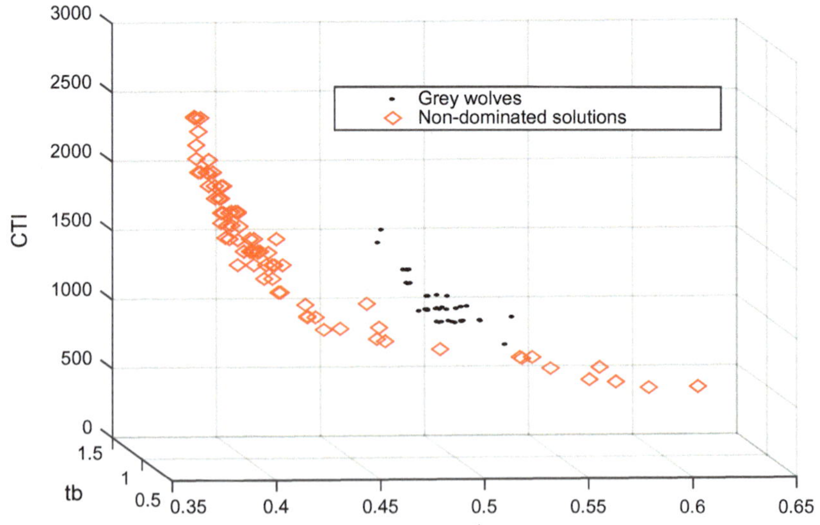

FIG. 3.5 Multiobjective Pareto front evaluation.

3.3 Genetic operators

Evolutionary algorithms mimic the process of biological evolution to solve simple and complex optimization problems using the Darwinian principle of the survival of the fittest. This method starts with a population containing a predefined number of individuals (referred to as chromosomes in this research) with a unique associated fitness value that represents an objective value to the problem. Chromosomes with better unique fitness values are selected from each generation and reproduced using operators inspired by genetics to form a new population that will act as the next generation.

According to the Schema Theorem, this new population should have a better fitness value, and hence provide a better approximation to a solution to the problem. To prevent a premature solution (local minimum), the newly formed chromosomes, referred to as child chromosomes, undergo a mutation process in which genes on selected chromosomes are randomly modified using an appropriate mutation factor, with the aim of widening the search space to obtain a global solution instead. Like many optimization solutions, the GA algorithm provides more than one solution to a given problem, and the result is therefore dependent on a stopping criterion set by the user.

3.3.1 Population (P_x)

These algorithms use a population consisting of certain individuals that have a unique fitness value for a corresponding optimization problem. Each chromosome has decision variables that, when combined, represent a point in the entire search space and a potential solution to the problem. The size of the population strongly depends on the simplicity or complexity of the problem and how accurate the solution needs to be, since a larger population increases the search space and affects the ability of the algorithm to locate a global solution. The size of the population has a direct effect on the execution time of the algorithm; the size should be smaller for simple problems and larger for complex problems.

During the crossover process, two or more individuals are reproduced, resulting in new chromosomes that are expected to have certain traits from each of these "parent" chromosomes rather than all the genes of only one parent. Another nature-inspired operator, known as mutation, is used to improve the diversity of the population after each generation.

3.3.2 Selection

After the evaluation process, each C_x in the population is assigned a fitness value depending on its performance in the problem domain. The fittest chromosomes will be weighted based on this value and will have a higher probability of being selected to produce one or more child chromosome to form the new population. Based on the concept of elitism, chromosomes that perform poorly in this evaluation will have little probability of being selected and are replaced by newly generated chromosomes to maintain the size of the population at a constant value. The selection process is done by evaluating the whole population or a portion of it; this choice depends on the execution time, as evaluating all possible solutions in a population can be somewhat time consuming for complex problems.

In the evaluation process, a fitness value is assigned to each C_x in the population to reflect its performance in the problem domain. The possibility of selecting a given C_x as a parent to produce new offspring for the next generation depends on its fitness: the better the fitness, the higher the chance for the C_x to be selected. A ranking mechanism is used to sort all the chromosomes in the population based on the evaluated fitness values, as shown in Fig. 3.6.

Each chromosome fitness represents a portion of the population's selection area, as illustrated in Fig. 3.7. The percentage of the area covered depends directly on the fitness value of each chromosome after the evaluation process. In Fig. 3.8, chromosome C_{X1} has the largest percentage of the entire selection area, covering 50% of the entire area. Since it has a larger selection area, it is likely to be selected more than once to form the new generation, unlike chromosome C_{X_4}, which covers a lower percentage of the selection area. The high probability of the fittest chromosome being selected ensures that the next generation contains the best chromosomes. At the same time, however, this poses a risk to the diversity of the population, as these will dominate the selection process after a couple of generations and will probably lead to a premature solution, referred to as "stagnation" of the algorithm.

The percentage of the selection area for each chromosome can be obtained using a fitness-based approach. The raw fitness value of each chromosome is used to determine the percentage of the total sum of all the fitness values of the current population to which it corresponds. Each chromosome is then assigned a percentage of the selection area using Eq. (3.12):

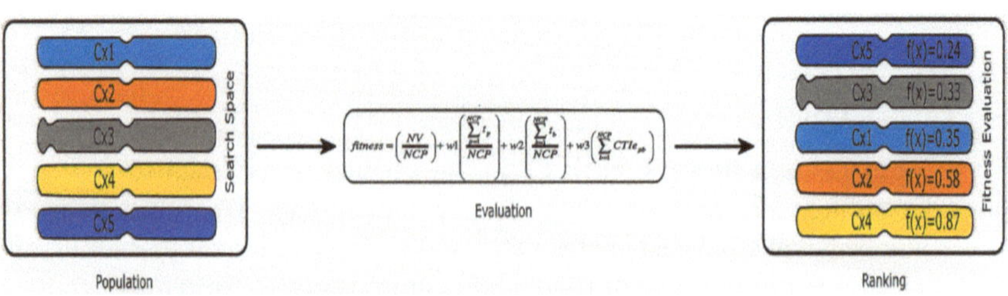

FIG. 3.6 Ranking mechanism based on fitness evaluation.

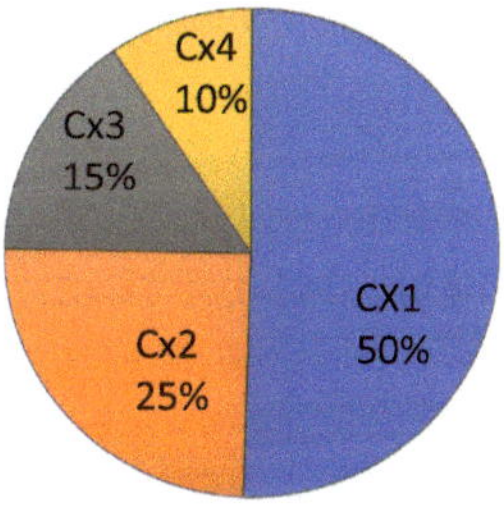

FIG. 3.7 Selection area of the population.

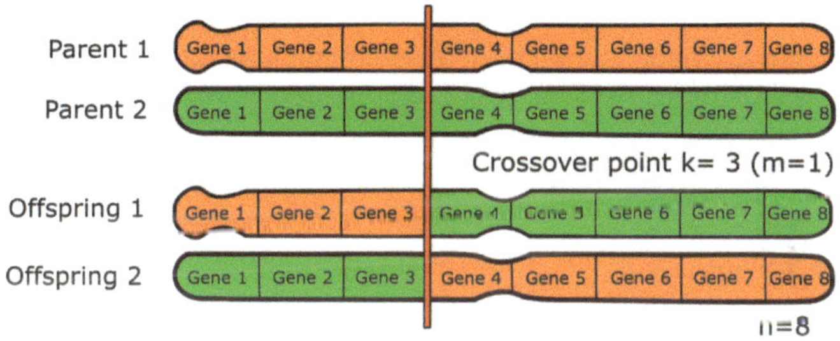

FIG. 3.8 Single-point crossover, $k=1$.

$$P_i = \frac{x_i}{\sum_{j=1}^{N} x_j} \tag{3.12}$$

where:

P_i = Percentage value assigned to the chromosome on the ith position.
x_i = Fitness or ranking value assigned to the chromosome on the ith position.
x_j = Fitness or ranking value assigned to the chromosome on the jth position.
N = Total number of chromosomes on the population.

After the chromosomes C_x are sorted based on their performance, they are each assigned a value based on their ranking, calculated as the inverse of the ranking position. This value is used to obtain the selection range and the percentage of the selection area occupied by each C_x.

3.3.2.1 Elitism

Darwin's theory of evolution states that the stronger individuals of a population are more capable of surviving and reproducing to create the next generation. Since most selection methods are somewhat stochastic, it is important to ensure the inclusion of fitter chromosomes or to pass these untouched to the newly generated population. When applying operators imitating natural selection, such as crossover and mutation, it is entirely possible that the best chromosome will be changed, resulting in a different chromosome and degrading the convergence of the algorithm. In this way, mutation and crossover can cause the loss of the best chromosome after several generations. The elitism method ensures that the best (or some of the best) chromosomes are included unchanged in successive generations. However, it is very likely that the optimization algorithm will become trapped in a particular area and will cease to explore other areas of the solution space. In the relay coordination problem, several mechanisms can be used to avoid this situation, such as crossover and mutation.

3.3.3 Crossover

The chromosomes in the mating pool that were created during the selection process are used in the crossover process to generate offspring chromosomes. Crossover is the main genetically inspired operator that is used to manipulate the genes of the chromosomes. This is the process that determines whether the newly created population is better than its predecessor. The Schema Theorem assumes that on average, a newly created chromosome has better fitness than the previous population. The chromosomes from the mating pool are randomly chosen and genes are interchanged to create new chromosomes using one of the available crossover methods. The aim of these methods is to produce new chromosomes made up of genes from both parents, with the expectation that these will provide better performance during the evaluation process. The higher the fitness of these child chromosomes in comparison to the parents, the better the convergence curve of the GA. Crossover can be done using two or more parent chromosomes. There are three main variations of the crossover process: single point, multipoint, and uniform crossover.

3.3.3.1 Single-point crossover

This is the simplest of the crossover methods and relies on the random generation of a single point to determine the exchange of genes between the parent chromosomes. This single point k is determined uniformly at random between positions 1 and $n-1$ on the chromosome, where n is the number of genes in the chromosome.

The selected parent chromosomes are selected randomly from the mating pool. The single point is randomly generated for each pair of parents and is used to determine the point at which each chromosome will be cut and divided. The genes located after the crossover point are exchanged between the parents in order to exchange genetic information, resulting in two offspring, as illustrated in Fig. 3.8.

The simple crossover method is very effective when it is necessary to reduce the algorithm execution time, as it does not require many operations. The population disruption in this approach is dependent on the length of the chromosomes and the number of times a chromosome is selected as a parent.

Multipoint crossover

This method is similar to single-point crossover with the exception that two or more crossover points k_i are randomly generated rather than one, where $k_i \in \{1,2, \ldots, n-1\}$, n is the number of genes on each chromosome. Once generated, the cut points should not be repeated and the chromosomes should be sorted before the reproduction process so that the cut points appear in ascending order. As shown in Fig. 3.9, all genes from cut point k_i to cut point (k_i+1) will be exchanged between the parents. This exchange is done for each cut-point interval.

The possible number of cut points m increases if the chromosomes have a greater number of genes, which in turn increases the execution time of the algorithm. The population disruption is dependent on the number of cut points and the length of the chromosomes.

3.3.4 Mutation

This is another basic operator based on natural selection, which promotes diversity within a population to prevent stagnation of the algorithm. It injects random alleles into selected chromosomes to explore the neighborhood of the search space, in a process that is somewhat similar to that of the crossover operator within the population search space. The mutation operator is important in terms of the convergence of the algorithm, as it permits the algorithm to continue searching over several generations to obtain a global solution rather than generating a possible solution using only the crossover operator.

In 1975 De Jong recommended using a mutation probability of n^{-1}, where n is the length of the chromosome, to determine the percentage of the chromosome that is mutated during this process. Other researchers proposed different schemes that included a mutation mask *mut* like the crossover mask, which is generated randomly but instead of swapping the selected chromosome gene, it replaces it with a new gene, as shown in Fig. 3.10.

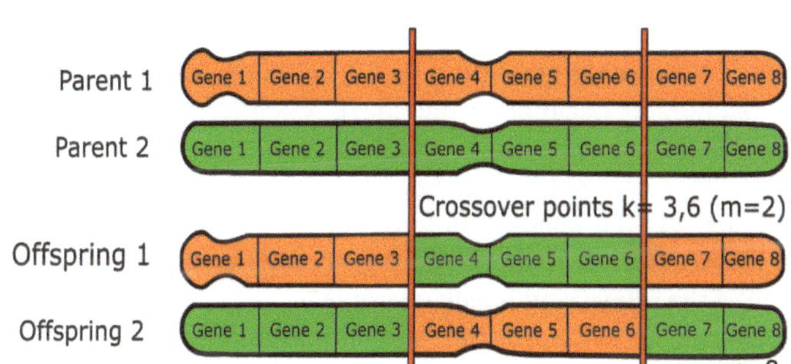

FIG. 3.9 Multipoint crossover, $k_i = 3{,}6$.

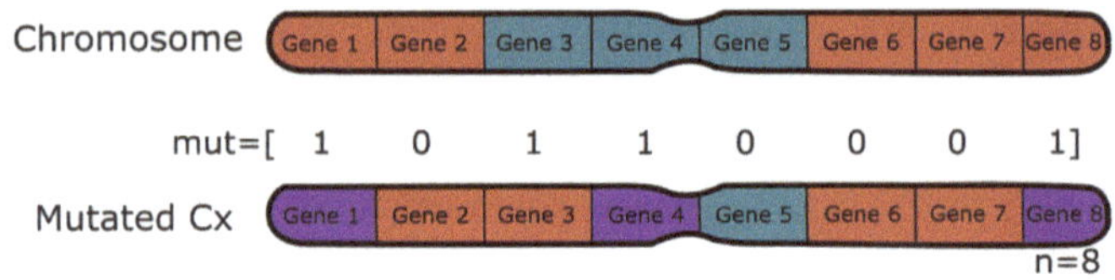

FIG. 3.10 Mutation of a selected chromosome.

The number of chromosomes to be mutated should be specified before the population is initialized and should remain constant over the generations. Some researchers have shown that it is very effective to increase this number, especially if the population disruption increases. Specific mutation is another way of implementing the mutation process whenever there is a compromise of diversity, in which specific chromosomes from the population undergo mutation. A reduction in diversity in the population over successive generations can occur during the crossover process, which results in a faster convergence of the algorithm, limiting the algorithm's ability to continue to searching its surroundings. The chromosomes that do not defer from others in the population should be determined and mutated to prevent this from occurring.

3.4 Metaheuristic optimization algorithms

In the literature, various metaheuristic algorithms have been explored, with adequate results for coordination purposes; however, due to the diversity in the problem variables and the configurations and operating states of the electrical network, no formulation has been shown to give robust results that can improve the other algorithms. If it is seen as an endless search, it is necessary to evaluate the results not only from the point of view of optimization of the function, but also based on the quality of the results and their impact and feasibility in terms of the setting and coordination of relays. For example, with the aim of reducing operating times, some researchers have proposed reducing the *TDS* to 0.1 or even 0.01; this does not necessarily reduce the operation times, but it can compromise the safety of the relay operation, since some relays operate very fast and can be fooled by transient conditions in the network, for example, the energization of capacitors or reactors, or may have less tolerance to temporary overload conditions.

Likewise, the search for the global optimum can often be algorithmically very exhausting and the benefits in terms of protection may be insignificant. In many problems, it is sufficient to find a good solution that allows for adequate performance of the relays, and which guarantees the protection and safety of the electrical network. In addition, since the time asymptote of the time curve for many coordination problems may include a flat region, variations in the short-circuit current due to changes in topological or generation conditions, for example, contributions from DG, may not give rise to significant variations in operation time.

Some hybrid methods that combine metaheuristic and deterministic methods must be weighed based on computational effort in order to determine their benefit in terms of coordinating relays. A specific formulation for the coordination solution of the most widely used metaheuristic methods for DOCR coordination is presented as follows.

3.4.1 Genetic algorithms

The GA imitates and uses various natural concepts from evolution. Evolutionary algorithms were first proposed in the early 1950s in a publication by A.M. Turing called "Computing Machinery and Intelligence," which included some principles of evolutionary theory and aimed to allow a machine to think. GA mimics the process of biological evolution to solve simple and complex optimization problems using the Darwinian principle better known as survival of the fittest [23]. This method starts with a population containing a predefined number of individuals, referred to here as chromosomes, that have a unique associated fitness value which represents an objective value for the problem. The chromosomes with best fitness value in each generation are selected and reproduced with genetically inspired operators to form the next population. According to the Schema Theorem, this new population should give better performance and result in a better solution.

To prevent the algorithm reaching a premature solution (local minimum), the newly formed chromosomes, referred to as child C_x, undergo a mutation process which randomly modifies some of the genes using an appropriate mutation factor, with the aim of expanding the search space to obtain a global solution instead. Like many optimization solutions, the GA provides more than one solution to a given problem, and the result is therefore dependent on the stopping criterion set by the user (Fig. 3.11).

All the chromosomes in the population are evaluated using an objective function to determine a fitness value for the given problem. These fitness values determine the approximation each chromosome has with regard to the solution for each generation; the closer these are, the better fitness of the chromosome. Chromosomes in nature do not contain similar genes, which is why all organisms are different. In the GA, this is one of many requirements that needs to be met, which is why each chromosome and population are different from one generation to the next. During the crossover process, two or more individuals are reproduced, resulting in new chromosomes that are expected to have traits from each of these "parent" chromosomes rather than all the genes of only one parent. Another naturally inspired operator known as mutation is used to improve the diversity of the population in each generation.

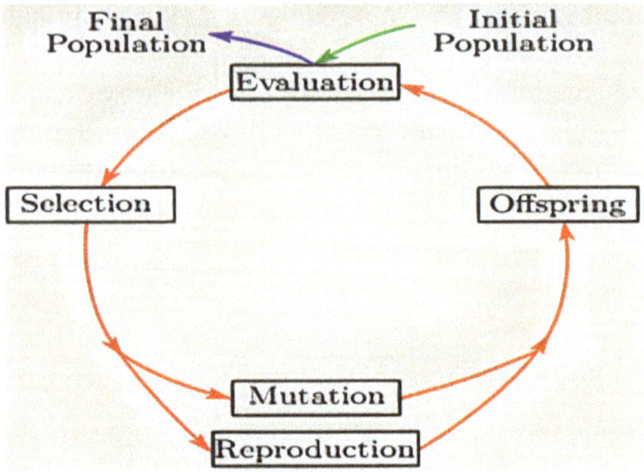

FIG. 3.11 Principle of operation of a genetic algorithm.

3.4.1.1 Use of GA for DOCR coordination

The GA has been extensively used to solve the problem of coordination of DOCRs and has been proven to provide competitive results compared to other similar techniques. This Section describes the basic concepts and the operators utilized in this research in order to adapt the algorithm to give a solution.

The initial population of the GA is created using uniformly random values within the range of the operational adjustable relay settings. Each chromosome is evaluated to determine its fitness value using the weighted objective function. Selection, crossover, and mutation are applied to create a new population that is expected to outperform its predecessor. Each relay may be part of several different coordination pairs, as a primary relay may have several backup relays. The selection process is carried out using a roulette method to ensure that diversity is maintained throughout the execution of the algorithm, and the ranking is assigned a value using the ranking-based approach to promote exploration of the neighboring search area.

Crossover is the main genetic operator used in the GA to exchange genes among the selected parent chromosomes to create the next generation. The method employed in this research to perform the crossover is single-point crossover, as shown in Fig. 3.12. The single-point method is used due to its simplicity and to reduce the computational effort of the algorithm. The computational time of the algorithm increases as more crossover points are considered.

As described before, mutation is another of the basic operators of a GA that involves the random swapping of relay settings on a selected C_x to force the algorithm to search outside the current search area. In the initial stages of the algorithm, 3% of the newly generated population is selected to undergo this process. In the single-point crossover method, there is a high possibility that the same parents will be selected to generate offspring for the new population, which can cause stagnation in early stages of the process and lead to a local minimum solution.

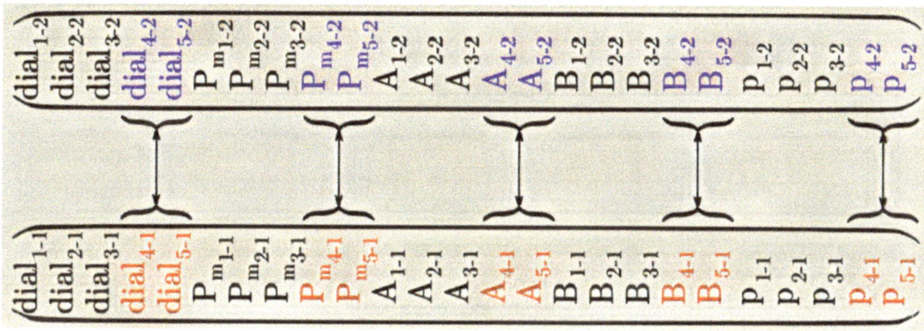

FIG. 3.12 Exchange of relay settings via a crossover operator.

It is therefore necessary to use crossover in a GA. Depending on how this is applied and which measures are taken, it can result in increased or decreased diversity in the new population, which again may lead to stagnation of the algorithm. A specific form of mutation is applied to prevent a diversity reduction by identifying identical chromosomes C_x in the new population. Mutation is applied to random genes to assist in keeping the population as diverse as possible.

Relay coordination using GA

Various steps and naturally inspired operators were incorporated into the GA to find a solution to the problem of relay coordination. The steps used in this implementation of the GA are shown in Fig. 3.13.

3.4.2 Invasive weed optimization

Weeds are plants with invasive habits, which absorb resources and nutrients to continue growing and reproducing [24]. From the point of view of optimization, this is a description of a robust system that can adapt to search spaces, allowing solutions to be found. Invasive weed optimization (IWO) was proposed by Mehrabian and Lucas [25] and is a numerical stochastic algorithm that imitates colonization by weeds (Fig. 3.14).

The initial implementation is similar to a GA in which a chromosome represents a weed, and the weed population is randomly generated and then weighted for ranking. Each weed generates seeds (S), which are weighted and classified depending on their aptitude using Eq. (3.13):

$$S_i = S_{min}\left[(F_i - F_{min})\frac{S_{max} - S_{min}}{F_{max} - F_{min}}\right] \tag{3.13}$$

where:

S_i is the seeds of weed i

S_{max} and S_{min} are the range of allowed seeds

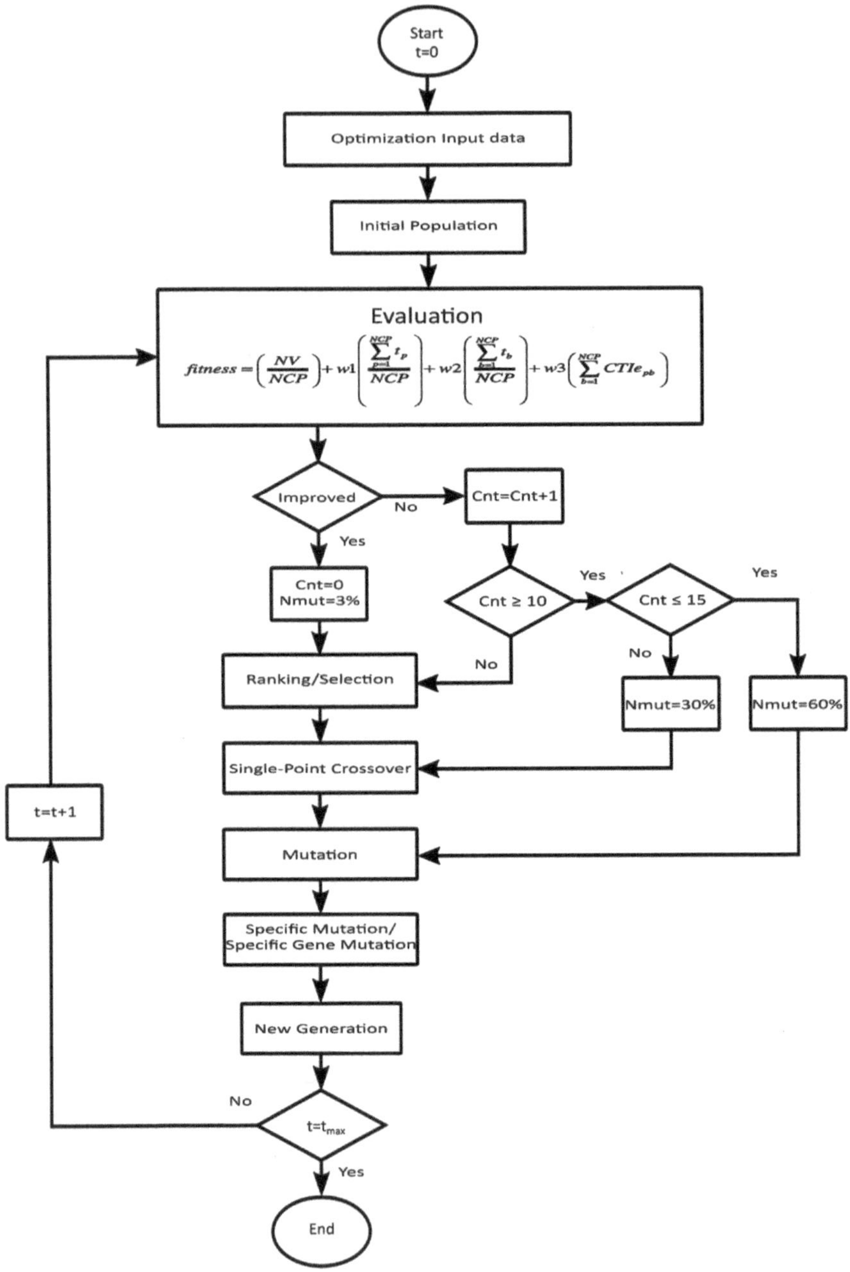

FIG. 3.13 Flowchart for the genetic algorithm.

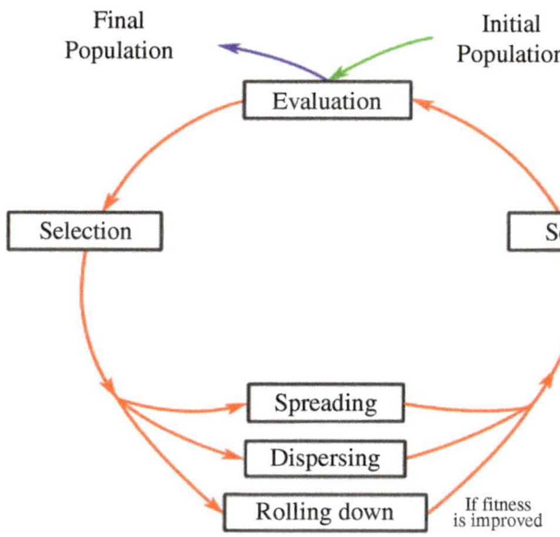

FIG. 3.14 Methodology of invasive weed optimization.

F_i is the fitness of weed i

F_{min} and F_{max} are the minimum and maximum population fitness

The seeds of each weed are defined by the weed mutation operators, which are known as the spreading, dispersing, and rolling-down operators. By means of these algorithmic operators, dissemination of the seed and the functions of exploration and exploitation of the search space are carried out.

3.4.2.1 Spreading

The operator spreads the seed randomly for the purpose of duplication, and new weeds are randomly created. Mutations can be applied at this stage to increase the sparsity of the solution space. Massive or moderate mutations can be applied as appropriate to the optimization problem. A large number of mutations cause a lot of scatter, which can hinder the systematic search for solutions. Thus a portion of the seeds is spread while a percentage of the original is kept, as shown in Fig. 3.15A [24].

3.4.2.2 Dispersion

The aim at this stage is to disperse the seeds in a controlled way, based on weighting of the distances over which they are thrown. By means of this metric, it is possible to control the dispersion zone as the simulation progresses. Another approach applies mutation to a portion of the seeds to avoid becoming trapped in a local minimum. The mutation used will depend on the problem to be solved; in the case of coordination of DOCRs, it is generally convenient to use adaptive mutations, as shown in Fig. 3.15B.

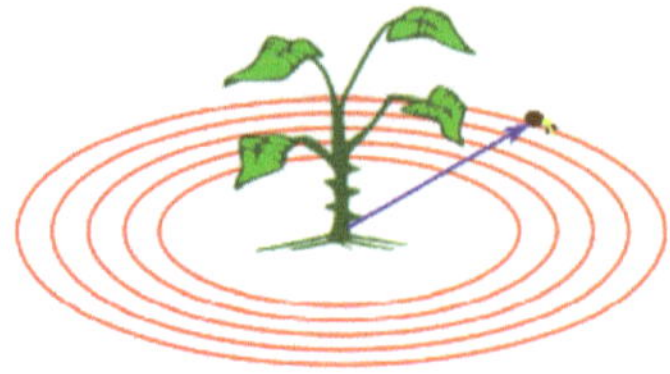

(a) Spreading operator. Seeds are spread all over the area.

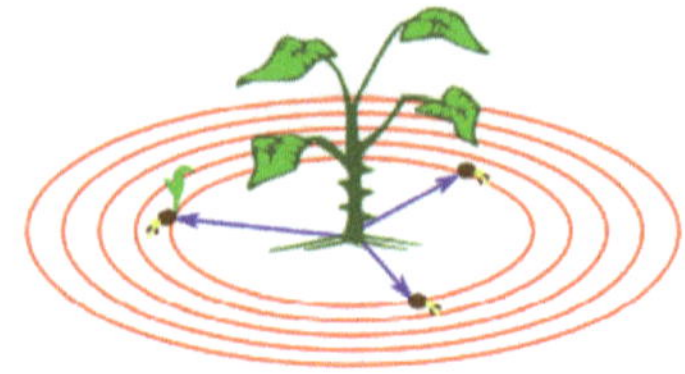

(b) Dispersing operator. Seeds are dispersed in the weed neighborhood.

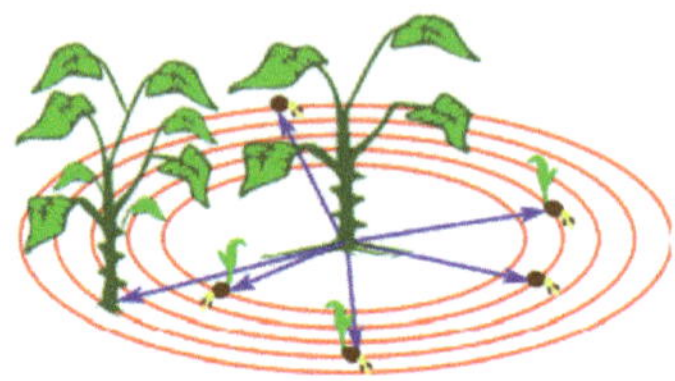

(c) Rolling-down operator. Seeds are spread and dispersed until a best solution is found.

FIG. 3.15 IWO operators.

3.4.2.3 Rolling down

In this stage, the aim is to improve the locations of the seeds within the solution space. To achieve this, the search space is evaluated and weighted, and in case of a better space, that is, that weights fitness to a greater extent, the seed is sown. The planting area will be limited to a certain space to create copies of the weed. When a more successful weed is found (that is, with better weighting), dissemination of the seeds continues, and mutations can be used to improve the search spaces. Seeds with high weights are kept, while unsuccessful seeds are discarded, as shown in Fig. 3.15C.

The behavior of invasive weeds has been applied to find solutions to very diverse problems, as it is a simple algorithm with a good search engine [26,27]. For the problem of DOCR coordination, it can be quickly implemented and has yielded good results with low requirements for computing resources. The propagation operator widens the search zone, which allows new settings to be found for the relays to minimize the operating times. When an exploration zone is found, the dispersal operator exploits it to find local solutions. If the fitness of a weed improves, it spreads in that zone with the aim of improving the solution. These operators can be used together to achieve fast exploration and exploitation of the search space.

3.4.2.4 IWO for DOCR coordination

The special stages of IWO are applied after the evaluation stage. The weeds are classified based on their fitness (F), and each weed leaves seeds depending on its ranking, as illustrated in Fig. 3.16.

FIG. 3.16 Leaving seeds based on the ranking results of weeds.

A seed is a clone of the actual weed that will be subjected to mutation operators. The total number of seeds (T_S) assigned to each weed is computed using Eq. (3.14):

$$T_{Si} = S_{rm} + \left[(F_w - F_i) \times \frac{S_{rM} - S_{rm}}{F_w - F_b} \right] \tag{3.14}$$

where S_{rm} and S_{rM} are the minimum and maximum possible quantities of weeds; F_b and F_w are the fitness values of the best and worst elements, respectively; and F_i is the fitness of the actual individual.

The implementation of all three operators for the problem of DOCR coordination is done as follows. The spreading is the first operator and it is used to create a new plant based on the current seed, in the same way as for a new chromosome. The implementation involves mutating up to 50% of the seed content for coordination proposes. The second operator disperses the seed over the neighborhoods surrounding the weed. In this work, dispersion is carried out in two stages: the first applies small perturbations to the seed by multiplying each setting by a maximum variation of $\pm 1\%$, while the second mutates up to 20% of the seed elements.

The last operator creates copies of the current seed and then combines the first two mechanisms to disperse and spread the seed. This process is repeated until a better solution is found or the seed copies are exhausted. The percentage of mutation, the magnitude of perturbation, and the settings used for mutation are randomly selected. Examples of the three operators can be seen in Fig. 3.17, and pseudocode for this process is presented in Fig. 3.18.

3.4.3 Differential evolution algorithm

The differential evolution (*DE*) algorithm is a form of evolutionary computation, which was introduced in the 1960s by I. Rechenberg [28]. This

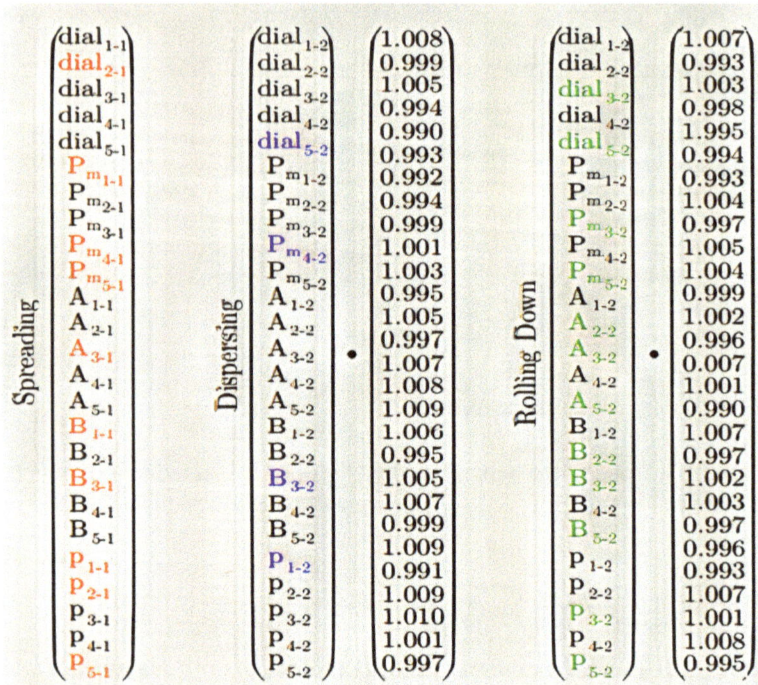

FIG. 3.17 Weed operators for DOCR coordination.

BEGIN *(Initial Population)*

Generate random initial weed matrix of feasible solutions (3.1).
Evaluate fitness of initial weed matrix per weed/individual (*OF*).
Rank all weed according to their fitness (roulette wheel, rank, and elitism).
WHILE termination criterion not satisfied
FOR 1:NP *(Total Seed Operation)*
Execute total seed (offspring) calculation scheme for each weed of the current weed matrix according to their fitness.
END
FOR 1:NP *(Mutation Operation)*
Randomly select any of the 3 operators (spreading, dispersing, and rolling down)
Each operator has certain probability of being selected.
Generate new weeds.
END
FOR 1:NP *(Selection Operation)*
Compare fitness among current weeds and mutated weeds.
Select *n* best fitness weeds.
END
END when termination criterion is satisfied

END

FIG. 3.18 Pseudocode for the IWO algorithm.

evolutionary algorithm is based on the natural selection of genes. The idea of *DE* was first reported by Storn and Price [29].

The *DE* algorithm has the same computational steps as standard evolutionary algorithms (EAs), except that it perturbs the actual members of the generation population with scaling differences, randomly selected and different population. Hence, no separate probability distribution needs to be used to generate the offspring. For this reason, the algorithm involves fewer mathematical operations, has a lower computational demand, and yields a faster execution time compared with other algorithms. The steps in the *DE* algorithm are presented in Fig. 3.19.

In this approach, the individual test solutions are called parameter vectors or genomes. The population of the algorithm is made up of genomes, which are evaluated and reproduced to form a new population. Selection is applied to diversify the solutions in order to guarantee that the population improves, and these are evaluated using an OF. In the case where the results of the function improve, the population is kept and reproduces, whereas if they worsen the solution it is discarded.

The process is terminated based on some criterion. For DOCR coordination, a limiting number of generations is used, since there is no reference for determining the best solution. The process uses three main control parameters, *F*, *Cr*, and *NP*. Mutation (*F*) allows for a better search in the solution space, which favors the exploration of the algorithm. Crossing (*Cr*) is defined for each type of problem to ensure the diversity of the solutions. *NP* is the selection operator, which is used to define the genomes in the next generation.

3.4.3.1 Mutation

In evolutionary computation, mutation involves a disturbance by a random element. The vectors or genomes are defined depending on their function, the parent vector of the generation, the mutant vector is known as the donor vector, and the test vector is the descendant formed by the combination of the donor vector and the target vector. In DE, all target vectors are mutated at each iteration.

In the *DE* population, three different vectors are randomly selected from the population as follows:

$$\vec{X}_{r_1,G}, \vec{X}_{r_2,G}, \vec{X}_{r_3,G} \tag{3.15}$$

for the *i*th target vector $X_{i,G}$. For each mutant vector, the indices r_1, r_2, and r_3 are generated as integer values in the range $[1, NP]$. The difference between two of

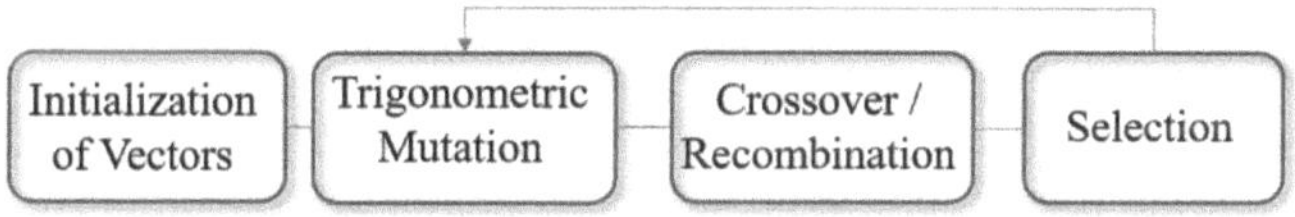

FIG. 3.19 Main stages of the DE algorithm.

the three vectors is multiplied by a scalar F in the range [0.4, 1]. The result is assigned to the third vector to obtain the donor vector $V_{i,G}$, as shown in Eq. (3.16):

$$\vec{V}_{i,G+1} = \vec{X}_{r_1} + F\left(\vec{X}_{r_2} - \vec{X}_{r_3}\right) \tag{3.16}$$

3.4.3.2 Crossover

After the donor vector is created through mutation, crossover or recombination is performed. The purpose of this stage is to improve the diversity of the population. The components of the donor vector are exchanged with the target vector $X_{i,G}$ to generate the test vector:

$$\vec{U}_{i,G} = \left[u_{1,i,G}, u_{2,i,G}, u_{3,i,G}, \ldots, u_{D,i,G}\right] \tag{3.17}$$

Two kinds of crossover methods can be used with the *DE* family of algorithms: *exponential* and *binomial* methods.

In an exponential crossover scheme, the integers nn and L are selected at random from the range [1, D]. Exchange of components with a donor vector nn is applied to initialize the target vector, where L represents the components that the donor vector contributes to the target vector. The exponential crossover scheme can be expressed as in Eq. (3.18):

$$u_{j,i,G} = \begin{cases} v_{j,i,G} & \text{if } \left(j \geq (nn)_D \text{ and } L \geq L_c\right) \\ x_{j,i,G} & \text{for all other} \quad j \in [1, D] \end{cases} \tag{3.18}$$

where $(nn)_D$ is the starting point of the crossover. L_c is used as a counter for L and can be initially expressed as $L_c = 0$. Then, we have $L_c = L_c + 1$ for each evaluation of the jth component, for $L \leq D$.

3.4.3.3 Selection

This operation is applied to determine the next generation of the test vector. The selection operation is as follows:

$$\vec{X}_{i,G+1} = \begin{cases} \vec{U}_{i,G} & \text{if } f\left(\vec{U}_{i,G}\right) \leq f\left(\vec{X}_{i,G}\right) \\ \vec{X}_{i,G} & \text{if } f\left(\vec{U}_{i,G}\right) > f\left(\vec{X}_{i,G}\right) \end{cases} \tag{3.19}$$

where $f(X)$ is the fitness of the target vector and $f(U)$ is the fitness of the trial vector. This process defines the next generation, where the aptitude of the vector is evaluated, since it can be replaced or passed to the next generation. In this way the population is not affected, since it may improve or stay the same, but will not get worse. The steps in the *DE* process when applied to DOCR coordination are shown in Fig. 3.20.

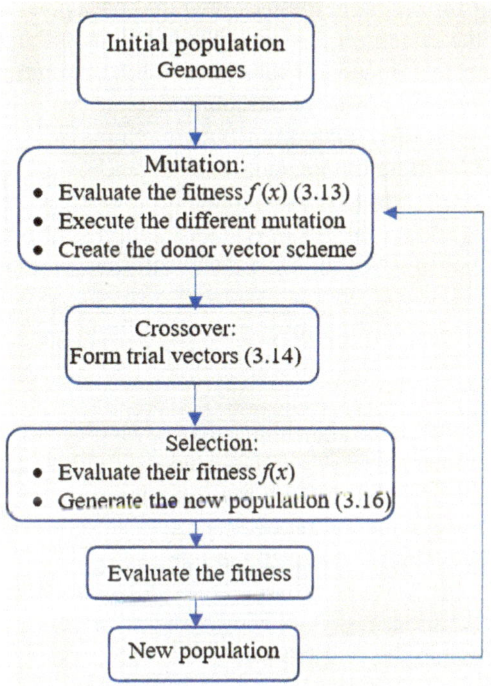

FIG. 3.20 Steps in the coordination process using DE.

3.4.4 Gray wolf algorithm

Gray wolf optimization (GWO) was proposed by Mirjalili and Lewis [30] and is based on the leadership and hunting behavior of gray wolves. The swarm intelligence of the algorithm renders the natural imitation, the simplicity of the search using a leadership and hunting mechanism, and a natural adaptation of the behavior of gray wolves. The algorithm starts with a population that represents an initial random distribution of wolves and aims to locate possible prey within the search space. In this process, the environment is scanned to determine whether a weak prey is nearby by evaluating the OF. The three best solutions are obtained by using three operators to update the position of the algorithm.

The social hierarchy of wolves is modeled by the alpha wolf (α), the beta wolf (β), and the delta wolf (δ), each of which offers a possible solution. These wolves will guide the optimization process in the search for the solution. Each wolf has attributes that are treated in the same way as chromosomes in GA, meaning that each wolf has an individual aptitude within a problem domain. The fitness value determines the closeness of each wolf to the prey, and the location for attacking the prey is actualized in each iteration of the algorithm. The process repeats three main steps till the end criterion is achieved: hunting, which involves searching for prey; encircling, in which the wolves stand around the prey; and finally the attack on the prey. When the stopping criterion is met, the final position of the alpha wolf represents the best solution.

The traits of each wolf represent the solution to the problem of coordination of DOCRs and are generated randomly within the limits established by the relay configuration. The resulting fitness value determines how far away each wolf is from its prey. Fig. 3.21 shows the evaluation and classification process for each wolf. The algorithm applies the three processes (hunting, encircling, and attacking) until it reaches a solution. The best solution is represented by the alpha wolf.

3.4.4.1 Hunting

Like other swarm intelligence algorithms, the search process starts with the creation of a random population. A pack of randomly generated wolves is positioned within the search area for the prey. As the gray wolves behave with intelligent collaboration, the OF represents the position of each wolf, and the search is led by the alpha (α), beta (β), and delta (δ) wolves with the purpose of searching the entire solution space. These wolves represent the three best solutions in each iteration. Fig. 3.22 shows a pack in which each wolf is randomly generated within the search area.

3.4.4.2 Encircling of prey

When the wolves detect a weak prey, they surround it to imitate a chase until the solution is improved. The following mathematical expressions represent this behavior:

$$\vec{D} = |\vec{C} \cdot \overrightarrow{X_p}(t) - \vec{X}(t)| \tag{3.20}$$

$$\vec{X}(t+1) = \overrightarrow{X_p}(t) - \vec{A} \cdot \vec{D} \tag{3.21}$$

where $\vec{A}$ and $\vec{C}$ are the vectors of exploration, X_p is the prey vector, X is the wolf position vector, and t is the time iteration. The distances for the three best solutions α, β, and δ are as follows:

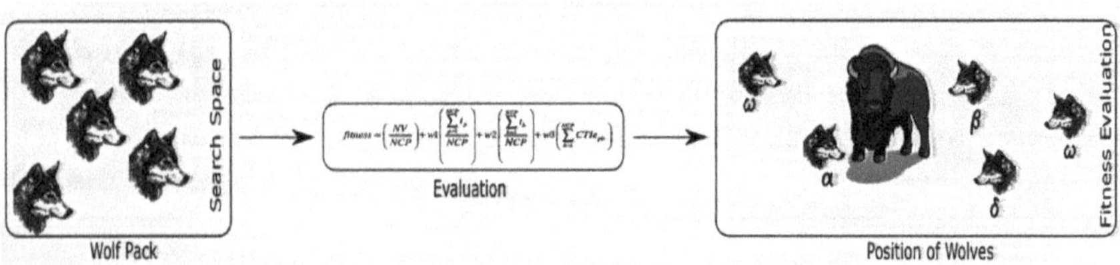

FIG. 3.21 Ranking of wolves in the search of a solution.

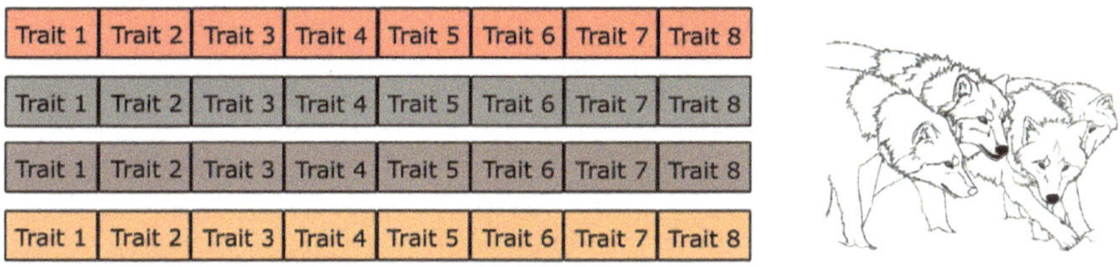

FIG. 3.22 Wolf pack representation for GWO.

$$\begin{aligned} \overrightarrow{D_\alpha} &= |\overrightarrow{C_1} \cdot \overrightarrow{X_\alpha} - \vec{X}| \\ \overrightarrow{D_\beta} &= |\overrightarrow{C_2} \cdot \overrightarrow{X_\beta} - \vec{X}| \\ \overrightarrow{D_\delta} &= |\overrightarrow{C_3} \cdot \overrightarrow{X_\delta} - \vec{X}| \end{aligned} \tag{3.22}$$

where X_α, X_β, and X_δ are the position vectors. The positions of the search agents are updated based on that of the best search agent as follows:

$$\begin{aligned} \overrightarrow{X_1}(t) &= \overrightarrow{X_\alpha} - \overrightarrow{A_1} \cdot \overrightarrow{D_\alpha} \\ \overrightarrow{X_2}(t) &= \overrightarrow{X_\beta} - \overrightarrow{A_2} \cdot \overrightarrow{D_\beta} \\ \overrightarrow{X_3}(t) &= \overrightarrow{X_\delta} - \overrightarrow{A_3} \cdot \overrightarrow{D_\delta} \\ \vec{X}(t+1) &= \frac{\overrightarrow{X_1}(t) + \overrightarrow{X_2}(t) + \overrightarrow{X_3}(t)}{3} \end{aligned} \tag{3.23}$$

where C_1, C_2, C_3 and A_1, A_2, A_3 are random vectors.

3.4.5 DOCRs coordination through GWO

Each wolf in the population represents a possible coordination solution. The set of limits on the parameter settings is randomly generated for the relay configuration. The first iterations perform an extensive exploration of the solution area. Then, the wolf agent is evaluated based on the OF to determine whether the solution is improved, where the fitness value represents the distance from the area in which the wolf searches for the prey. To ensure adequate performance of the algorithm, the OF must be continuously reduced. The final steps of the GWO algorithm represent the attack on the prey, or in other words obtaining the solution.

3.4.5.1 Enhanced hunting

The classification of the wolves during the hunting process was improved. In the evaluation process, the solutions were only saved in the first iteration. Subsequently, only the best solution α was stored, and the positions β and δ were updated. The GWO algorithm searched for the prey in the first half of the iterations using the $\vec{a}$ and $\vec{A}$ operators. In the first few iterations, the positions of the wolves diverged to find a better solution. The values of $\bar{a}$ were decreased linearly from two to zero using Eq. (3.24).

$$\vec{a} = 2\left(1 - \frac{t}{MaxIt}\right) \tag{3.24}$$

The linear behavior of $\bar{a}$ was replaced by an exponential function to allow the GWO algorithm to search for the prey faster and to center of attention when the prey is identified. The exponential decay of ā is given by:

$$\vec{a} = 2^{*}\, exp\left(-\frac{t^{*}2^{2}}{MaxIt}\right) \tag{3.25}$$

Fig. 3.23 shows the performance as $\bar{a}$ is decreased. The results of the proposed exponential function indicate that the exploration time is reduced by almost a quarter of the total number of iterations. An exponential function is used because the trend in GWO is to obtain good results in the initial steps, meaning that the search carried out by the algorithm can be enhanced.

The integration of the algorithm increases the execution time, but also increases the possibility of improving the results. A specific gene mutation (SPG) similar to that of the GA is used in the GWO. Through violation detection of CPs, a lack of relay coordination is detected, thus violation of wolves that represent the primary and backup relays are mutated, and the rest of the relays are not affected. This mutation process is specific to pairs of relays that have violations.

The SGM introduces diversity in terms of exploring the search space during execution of the algorithm. Fig. 3.24 shows the SGM for a specific violated CP (R2–R4).

3.4.5.2 *Block diagram*

The GWO algorithm was altered to include the SGM, in order to enhance relay coordination. A diagram of this implementation of GWO is shown in Fig. 3.25. The updates to the best wolf positions (X_α, X_β, X_δ, and X_ω) represent the search for a solution to the problem of DOCR coordination. The SGM is very important to ensure diversity in the solution and to look for new and better solutions.

3.4.6 DOCRs coordination via MOGWO

True Pareto-optimal solutions to multiobjective problems can approximate solutions effectively. A multiobjective GWO (MOGWO) population was created in a similar way to GWO, the only exception being that each agent was created in the form of a structured matrix. Positions (i.e., relay settings) and

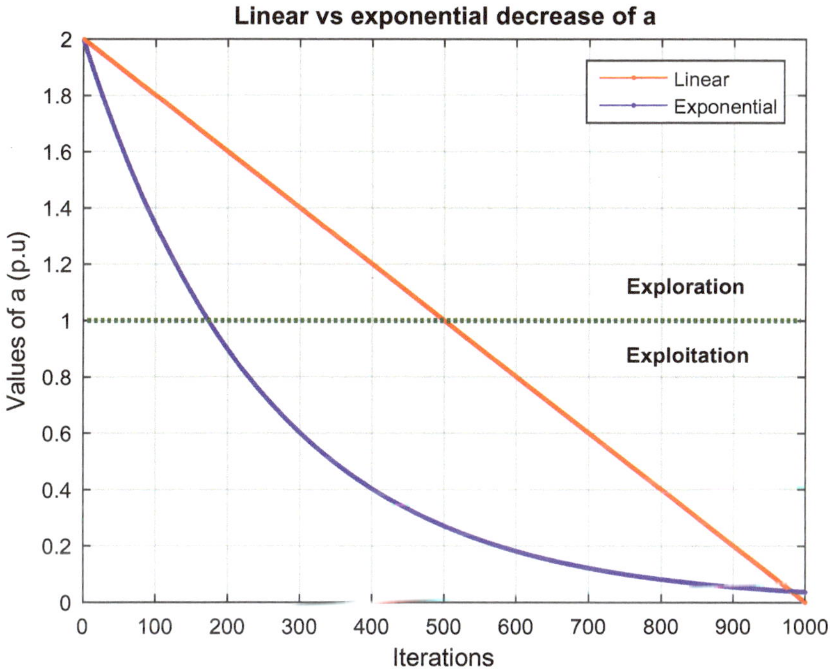

FIG. 3.23 Comparison of linear and proposed exponential behavior.

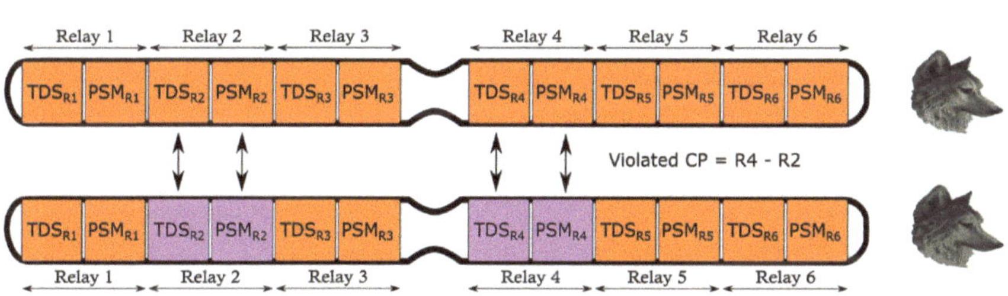

FIG. 3.24 Specific gene mutation for violated relays.

fitness values were included for each problem tested. The OF was represented by a matrix for evaluation purposes, as shown in Eq. (3.26):

$$Minimize : F(\vec{x}) = f_1(\vec{x}),\ f_2(\vec{x}),\ f_3(\vec{x}) \tag{3.26}$$

The OFs are given in Eqs. (3.9)–(3.11). Although MOGWO is based on GWO, it applies a different approach to solve the problem of DOCR coordination, as it adds three operators and the SGM. These operators are described as follows.

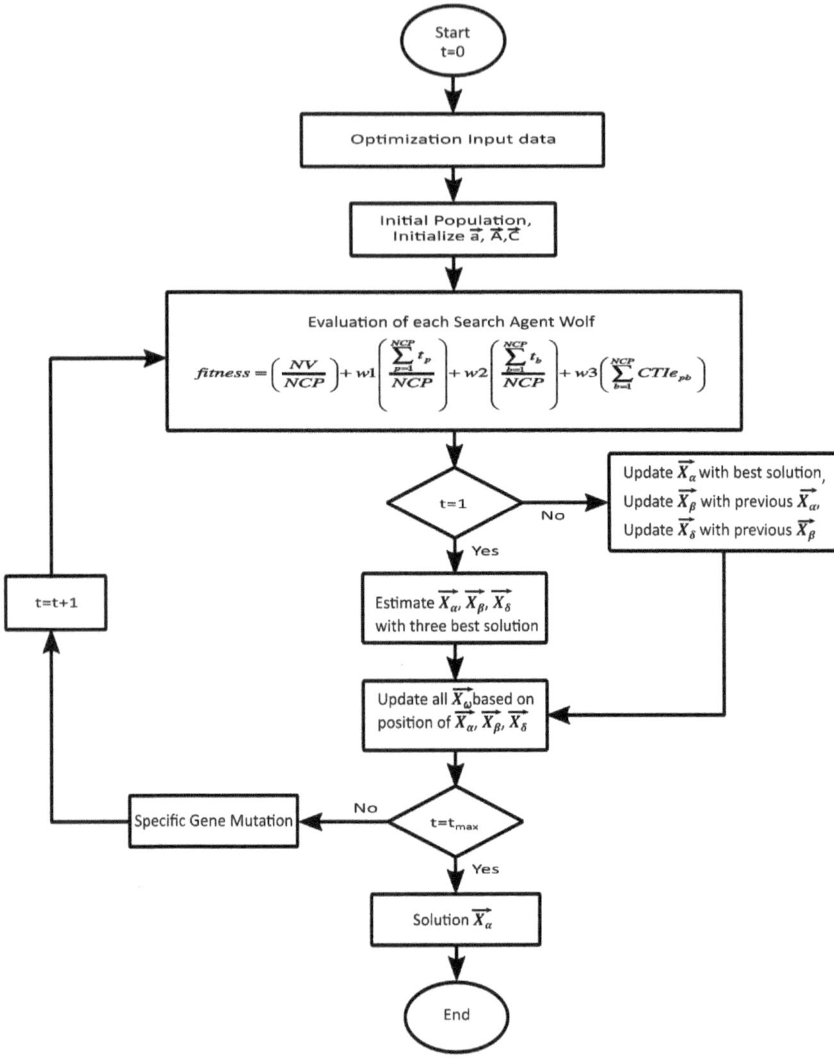

FIG. 3.25 Flowchart for the modified GWO algorithm.

3.4.6.1 Archive integration

This operator stores the nondominated Pareto solutions at each iteration. After each iteration, the elements of the file are evaluated in terms of the solutions obtained, meaning that the file is permanently active during optimization. When a solution dominates more than one element in the file, all the dominated residents are deleted because the nondominated solutions are better solutions. The best solution is defined by the set of nondominated solutions following an extraction method.

3.4.6.2 Grid integration mechanism

In order to improve the solutions, a grid method is implemented, which consists of segmenting the space if the file is complete using an inflation parameter. With this, the segmentation space is reordered, in addition, the number of grids for the destination space is established. With this, the solution is assigned to the least requested segment to increase diversity.

3.4.6.3 Leader selection integration mechanism

In GWO, the three best solutions are used to guide the omega wolves toward more feasible solution spaces in order to improve the solution. In MOGWO, a leader selection process is carried out to assist in the selection process.

3.5 Test systems and results

The test systems were an IEEE 14-bus system with 40 relays and a 30-bus system with 82 relays. The total numbers of coordination pairs were 50 and 120, respectively. Although the nonsensitive relays were coordinated, they were not included in the coordination analysis, since their functional limitations are not related to the coordination process. In this fault analysis, a maximum of three-phase close-in faults and minimum line-end phase-to-phase faults were considered. Fault currents were obtained with the remote bus-line open. The generator impedance for the positive sequence in both systems was 0.1 and 0.3, respectively, while the negative sequence impedances were 15% of these values.

PSM was in the range of continuous values of [1.4:1.6], while for *TDS* this was [0.5:2.0], and the *CTI* used was 0.3 s. The soft constraint on the OF apply a strong penalty when the *CTI* when the error was less than 0.3 s. The OF used for the coordination algorithms was (Eq. 3.8). Based on a Pareto front analysis, the weighting factors were $\omega_1 = 1$, $\omega_2 = 0.75$, and $\omega_3 = 0.25$.

3.6 Results and discussion

The results of three optimization algorithms (GA, GWO, and MOGWO) were compared. In the optimization process, a total of 500 chromosomes was used for the GA, while 60 individuals were used for GWO and MOGWO for the 14-bus network and 100 individuals for the 30-bus network, at maximum load in both cases. A total of 20 repetitions was carried out to obtain the best results and to prove the robustness of the algorithms. Each repetition of the algorithm consisted of 1000 iterations of the optimization algorithms. Three test cases were analyzed.

The best times were computed for *Tp*, *Tb*, and *CTI* and an average of 20 repetitions for At_p, At_b, and *ACTI* was considered. The number of violations (%NV), average *TDS* and *PSM* settings, and the standard deviation for the 20 repetitions are presented later. Finally, the overall performance of each algorithm is discussed. Three algorithms (GA, GWO, and MOGWO) were evaluated. Two case studies are presented: the first makes a comparison between the three algorithms mentioned before, while the second case evaluates the proposed improvement to the GWO and MOGWO algorithms.

3.6.1 Test case I

In this test case, the performance of the GA algorithm was analyzed in regard to the improvement due to the SGM, as well as the performance of the GWO algorithm with the single-objective weighted OF and the version of MOGWO with the multiobjective formulation. The tests were performed on both the 14-bus and 30-bus systems, and the results are given in Tables 3.1 and 3.2, respectively. From the graphs in Fig. 3.26, it can observed that the GWO was superior in terms of obtaining shorter operation times of the relays, although it yielded a higher standard deviation. For the 30-bus system, the trend in the results was similar, although the standard deviation was lower than the three methods (Fig. 3.27). This is because the algorithm has the ability to solve optimization problems subject to a large number of constraints. For both test systems, GWO was superior to MOGWO. The shaded table data represents the best result.

3.6.2 Test case II

In this case, the effect of the proposed improvement due to the inclusion of the $\bar{a}$ function with exponential decrement (es) was compared to the linear version (ls) for both GWO and MOGWO. The GA algorithm was not studied. Both the 14-bus and 30-bus test systems were investigated. From the results obtained, it could be observed that the proposed version of the function $\bar{a}$ with exponential decrement improved the results from GWO (see Table 3.3 and Fig. 3.28). For MOGWO, this was not the case for the 14-bus network, although the performance of the 30-bus was better with the proposed version (Table 3.4 and

TABLE 3.1 Results for Test I on IEEE 14-bus network system.

	GA			GWO			MOGWO		
	tp(s)	tb(s)	CTI(s)	tp(s)	tb(s)	CTI(s)	tp(s)	tb(s)	CTI(s)
Best	0.7858	1.0901	0.4138	0.564	0.8649	0.3998	0.6471	1.196	0.7291
Average	0.8197	1.1776	0.4541	0.6994	1.0138	0.4396	0.7165	1.2411	0.7416
std	0.0575	0.0665	0.0201	0.0951	0.1303	0.0333	0.1347	0.2695	0.1628
NV		3			3			4	
TDS		1.292			1.24			1.305	
PSM		1.445			1.444			1.448	
Overall		10.6222			9.9242			12.5914	

TABLE 3.2 Results for Test I on IEEE 30-bus network system.

	GA			GWO			MOGWO		
	tp(s)	tb(s)	CTI(s)	tp(s)	tb(s)	CTI(s)	tp(s)	tb(s)	CTI(s)
Best	0.7956	1.2041	0.5648	0.7336	1.0974	0.571	0.8587	1.3507	0.7805
Average	0.8298	1.227	0.583	0.7571	1.0799	0.5849	0.7292	1.1929	0.7919
std	0.0426	0.0568	0.0326	0.0404	0.0462	0.0276	0.136	0.222	0.1445
NV	14			14			15		
TDS	1.432			1.388			1.425		
PSM	1.466			1.468			1.48		
Overall	22.2343			21.7941			24.1114		

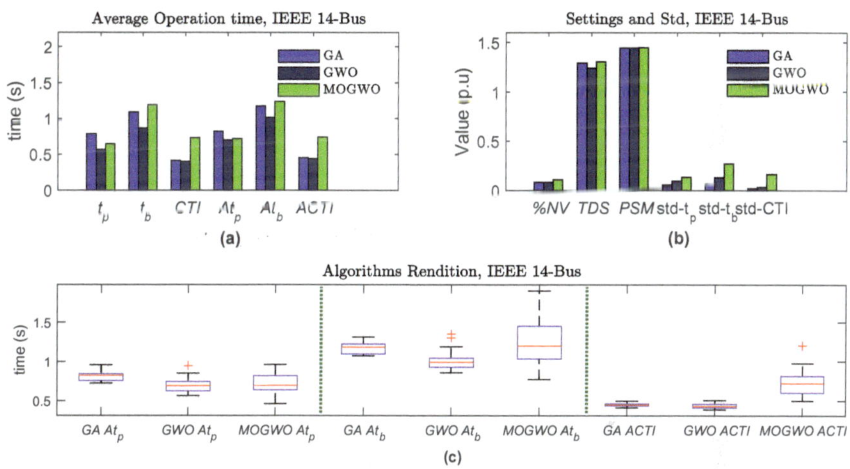

FIG. 3.26 Comparison of results from GA, GWO, and MOGWO for IEEE 14-bus network system.

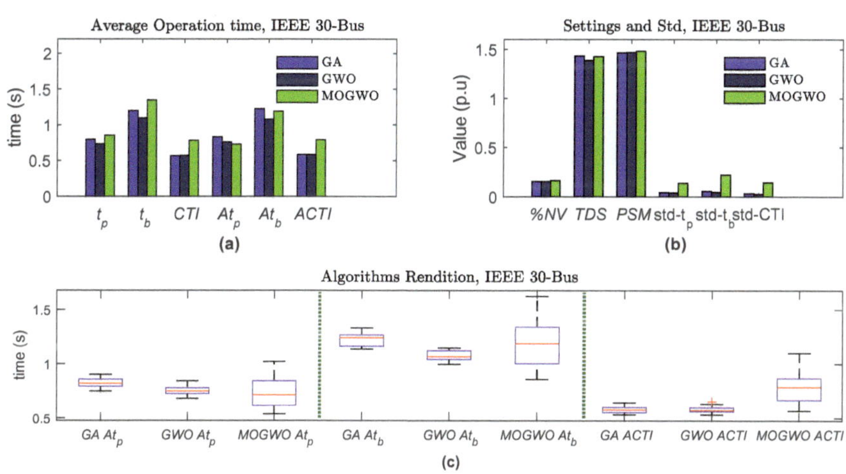

FIG. 3.27 Comparison of results from GA, GWO, and MOGWO for IEEE 30-bus network system.

TABLE 3.3 Results of Test II on the IEEE 14-bus network system.

	GWOls			GWOes			MOGWOl			MOGWOe		
	tp(s)	tb(s)	CTI(s)	tp(s)	tb(s)	CTI(s)	tp(s)	tb(s)	CTI(s)	tp(s)	tb(s)	CTI(s)
Best	0.564	0.8649	0.3998	0.5694	0.857	0.3775	0.6458	0.9983	0.5179	0.6676	1.1315	0.6023
Average	0.6994	1.0138	0.4396	0.6358	0.9228	0.4074	0.677	1.1533	0.6889	0.7102	1.2324	0.7465
std	0.0951	0.1303	0.0333	0.0833	0.0897	0.0243	0.0807	0.1509	0.0947	0.0923	0.1785	0.1303
NV		3			3			4			4	
TDS		1.24			1.209			1.251			1.332	
PSM		1.444			1.448			1.504			1.463	
Overall		9.9242			9.6242			11.7625			12.2866	

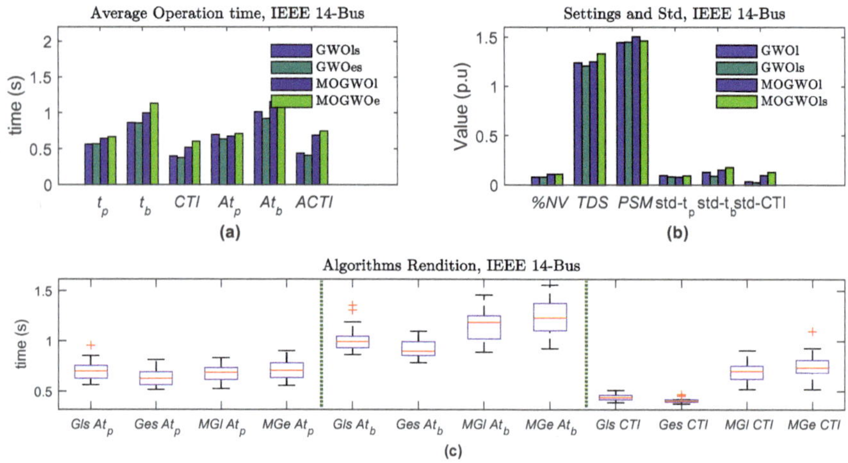

FIG. 3.28 Comparison of the results of Test II from GWO and MOGWO for IEEE 14-bus network system.

TABLE 3.4 Results of Test II on the IEEE 30-bus network system.

	GWOls			GWOes			MOGWOl			MOGWOe		
	tp(s)	tb(s)	CTI(s)	tp(s)	tb(s)	CTI(s)	tp(s)	tb(s)	CTI(s)	tp(s)	tb(s)	CTI(s)
Best	0.7336	1.0974	0.571	0.6923	1.0292	0.5338	0.7306	1.2002	0.7388	0.7135	1.1764	0.7081
Average	0.7571	1.0799	0.5849	0.7386	1.0367	0.5372	0.6886	1.1207	0.7281	0.6999	1.1326	0.7439
std	0.0404	0.0462	0.0276	0.0352	0.0342	0.0179	0.0865	0.1534	0.0929	0.0742	0.1292	0.0881
NV		14			14			15			15	
TDS		1.388			1.351			1.364			1.337	
PSM		1.468			1.484			1.471			1.454	
Overall		21.7941			21.4901			23.3748			23.2569	

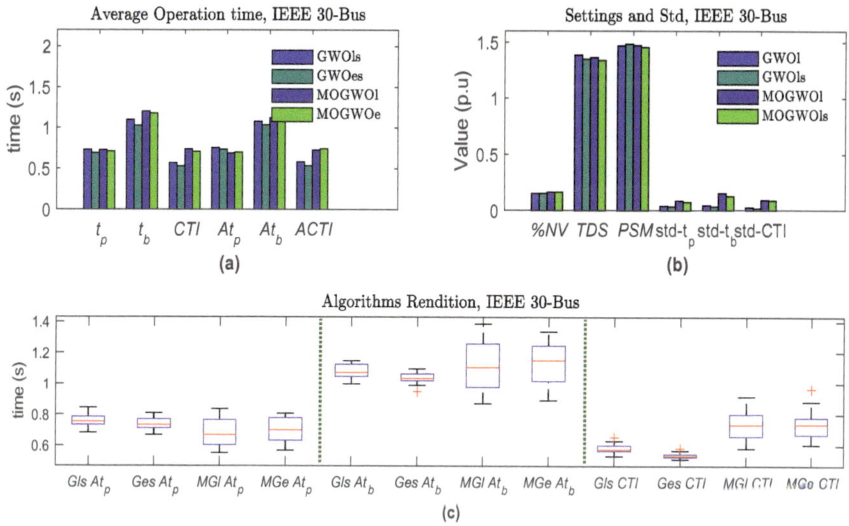

FIG. 3.29 Comparison of the results of Test II for GWO and MOGWO for IEEE 30-bus network system.

Fig. 3.29). The trend was consistent with that observed in the previous section, in that the single-objective version outperformed the multiobjective version.

3.7 Summary

Various optimization algorithms were analyzed in terms of solving a highly constrained optimization problem. This gave rise to the main contribution of this research, which is an alternative solution to the problem using GWO, a recently discovered optimization algorithm. This approach provides great results compared to others of its kind, such as GA. Two formulations were used, a multiobjective version and a weighted objective version, and were applied to the 14-bus and 30-bus IEEE test systems. The inclusion of the SGM operator and a change in the exploration process of the original GWO, in which the original linear decrease was replaced with an exponential decrease, provided better results. The range of application of this enhanced GWO algorithm is increased as it can find a good solution using fewer individuals and has an improved convergence mechanism.

References

[1] A. Urdaneta, R. Nadira, L. Perez Jimenez, Optimal coordination of directional overcurrent relays in interconnected power systems, IEEE Trans. Power Deliv. 3 (3) (1988) 903–911. [Online]. Available: http://ieeexplore.ieee.org/document/193867.

[2] A.S. Noghabi, H.R. Mashhadi, J. Sadeh, Optimal coordination of directional overcurrent relays considering different network topologies using interval linear programming, IEEE Trans.

Power Deliv. 25 (3) (2010) 1348–1354. [Online]. Available: http://ieeexplore.ieee.org/document/5491369/.

[3] V.A. Papaspiliotopoulos, G.N. Korres, N.G. Maratos, A novel quadratically constrained quadratic programming method for optimal coordination of directional overcurrent relays, IEEE Trans. Power Deliv. 32 (1) (2017) 3–10. [Online]. Available: http://ieeexplore.ieee.org/document/7154507/.

[4] C. So, Application of genetic algorithm for overcurrent relay coordination, in: 6th International Conference on Developments in Power Systems Protection, vol. 1997, IEE, 1997, pp. 66–69. [Online]. Available: http://digital-library.theiet.org/content/conferences/10.1049/cp{_}19970030.

[5] H. Zeineldin, E. El-Saadany, M. Salama, Optimal coordination of overcurrent relays using a modified particle swarm optimization, Electr. Power Syst. Res. 76 (11) (2006) 988–995. [Online]. Available: http://www.sciencedirect.com/science/article/pii/S0378779605002701.

[6] M. Barzegari, S. Bathaee, M. Alizadeh, Optimal coordination of directional overcurrent relays using harmony search algorithm, in: 2010 9th International Conference on Environment and Electrical Engineering, IEEE, 2010, pp. 321–324. [Online]. Available: http://ieeexplore.ieee.org/document/5489935/.

[7] M. Singh, B.K. Panigrahi, R. Mukherjee, Optimum coordination of overcurrent relays using CMA-ES algorithm, in: 2012 IEEE International Conference on Power Electronics, Drives and Energy Systems (PEDES), IEEE, 2012, pp. 1–6. [Online]. Available: http://ieeexplore.ieee.org/document/6484275/.

[8] T. Amraee, Coordination of directional overcurrent relays using seeker algorithm, IEEE Trans. Power Deliv. 27 (3) (2012) 1415–1422. [Online]. Available: http://ieeexplore.ieee.org/document/6180204/.

[9] G. Darji, M. Patel, V. Rajput, K. Pandya, A tuned cuckoo search algorithm for optimal coordination of Directional Overcurrent Relays, in: 2015 International Conference on Power and Advanced Control Engineering (ICPACE), IEEE, 2015, pp. 162–167. [Online]. Available: http://ieeexplore.ieee.org/document/7274936/.

[10] C.A. Castillo Salazar, M.Y. Shih, A. Conde Enríquez, Adaptive directional overcurrent relay coordination using ant colony optimisation, IET Gener. Transm. Distrib. 9 (14) (2015) 2040–2049. [Online]. Available: http://digital-library.theiet.org/content/journals/10.1049/iet-gtd.2015.0394.

[11] F.A. Albasri, A.R. Alroomi, J.H. Talaq, Optimal coordination of directional overcurrent relays using biogeography-based optimization algorithms, IEEE Trans. Power Deliv. 30 (4) (2015) 1810–1820. [Online]. Available: http://ieeexplore.ieee.org/document/7047892/.

[12] V. Papaspiliotopoulos, T. Kurashvili, G. Korres, Optimal coordination of directional overcurrent relays in distribution systems with distributed generation based on a hybrid PSO-LP algorithm, in: MedPower 2014. Institution of Engineering and Technology, 2014, p. 64. [Online]. Available: http://digital-library.theiet.org/content/conferences/10.1049/cp.2014.1697.

[13] M. Jevtic, J. Radosavljević, Hybrid GSA-SQP algorithm for optimal coordination of directional overcurrent relays, IET Gener. Transm. Distrib. 10 (8) (2016) 1928–1937. [Online]. Available: http://digital-library.theiet.org/content/journals/10.1049/iet-gtd.2015.1223.

[14] D. Solati Alkaran, M.R. Vatani, M.J. Sanjari, G.B. Gharehpetian, M.S. Naderi, Optimal overcurrent relay coordination in interconnected networks by using fuzzy-based GA method, IEEE Trans. Smart Grid (2016) 1. [Online]. Available: http://ieeexplore.ieee.org/document/7738589/.

[15] J. Sadeh, Y. Damchi, H. Rajabi Mashhadi, Optimal coordination of distance and overcurrent relays considering a non-standard tripping characteristic for distance relays, IET Gener.

Transm. Distrib. 10 (6) (2016) 1448–1457. [Online]. Available: http://digital-library.theiet.org/content/journals/10.1049/iet-gtd.2015.1087.

[16] M.H. Costa, R.R. Saldanha, M.G. Ravetti, E.G. Carrano, Robust coordination of directional overcurrent relays using a matheuristic algorithm, IET Gener. Transm. Distrib. 11 (2) (2017) 464–474. [Online]. Available: http://digital-library.theiet.org/content/journals/10.1049/iet-gtd.2016.1010.

[17] Z. Moravej, F. Adelnia, F. Abbasi, Optimal coordination of directional overcurrent relays using NSGA-II, Electr. Power Syst. Res. 119 (2015) 228–236. [Online]. Available: http://www.sciencedirect.com/science/article/pii/S0378779614003319.

[18] R. Benabid, A new formulation and solving of protective relays setting and coordination problem using multi-objective optimization and fuzzy logic, in: 2016 8th International Conference on Modelling, Identification and Control (ICMIC), IEEE, 2016, pp. 138–143. [Online]. Available: http://ieeexplore.ieee.org/document/7804288/.

[19] J. Shah, N. Khristi, V. Rajput, K. Pandya, A comparative study based on objective functions for optimum coordination of overcurrent relays, in: Presented at 7th International Conference on Power Systems (ICPS), 2017, https://doi.org/10.1109/ICPES.2017.8387260.

[20] IEEE Std C37.112-1996, IEEE Standard Inverse-Time Characteristic Equations for Overcurrent Relays, September 1996.

[21] IEC Standard 255-4, Single Input Energizing Measuring Relays with Dependent Specified Time, IEC Publication 255-4, first ed., 1976.

[22] M.Y. Shih, A. Conde, Novel coordination of DOCRs using trigonometric differential evolution algorithm, IEEE Lat. Am. Trans. 1548-0992, 13 (5) (2015) 1605–1611, https://doi.org/10.1109/TLA.2015.7112021. Impact factor 0.436 http://www.ewh.ieee.org/reg/9/etrans/ieee/issues/vol13/vol13issue5May2015/13TLA5_48SHIH.pdf.

[23] M. Affenzeller, S. Winkler, S. Wagner, A. Beham, Genetic Algorithms and Genetic Programming: Modern Concepts and Practical Applications, first ed., Chapman & Hall/CRC, London, 2009.

[24] H.G. Baker, G.L. Stebbins, The Genetics of Colonizing Species: Proceedings of the First International Union of Biological Sciences Symposia on General Biology, Academic Press, 1965.

[25] A.R. Mehrabian, C. Lucas, A novel numerical optimization algorithm inspired from weed colonization, Eco. Inform. 1 (4) (2006) 355–366.

[26] A.H. Nikoofard, H. Hajimirsadeghi, A. Rahimi-Kian, C. Lucas, Multiobjective invasive weed optimization: application to analysis of pareto improvement models in electricity markets, Appl. Soft Comput. 12 (1) (2012) 100–112.

[27] H. Josinski, D. Kostrzewa, A. Michalczuk, A. Switoński, The expanded invasive weed optimization metaheuristic for solving continuous and discrete optimization problems, Sci. World J. 2014 (1) (2006) 1–14.

[28] I. Rechenberg, Evolutionsstrategie '94, Frommann-Holzboog Verlag, Stuttgart, 1994 (in German).

[29] K.V. Price, R.M. Storn, J.A. Lampinen, The differential evolution algorithm, in: Differential Evolution: A Practical Approach to Global Optimization, Springer Science & Business Media, 2005, pp. 37–134.

[30] S. Mirjalili, S.M. Mirjalili, A. Lewis, Grey wolf optimizer, Adv. Eng. Softw. 69 (2014) 46–61.

Chapter 4

Design of nonconventional time curves for overcurrent relays

4.1 Nonconventional time curves of overcurrent relays

In active networks, the difficulty of coordinating DOCRs increases due to the intermittency of renewable sources and the operational diversity of the electrical network; however, the increase in processing capacity for solving more difficult problems has meant that the automation of solutions is becoming more realistic. Hence, the task of DOCR coordination, formulated as an online optimization problem, is becoming more feasible every day. However, at present, coordination is carried out offline to determine the adjustments that can guarantee adequate operation of the relays. Coordination is formulated as an operation time minimization problem in which equality and inequality constraints must be satisfied. In the initial investigations in this work, the *TDS* was the only parameter used by the algorithm to find solutions, and the *PSM* was later used to find more diverse solutions. The use of higher parametric variation was proposed to determine possible solutions. It is important that the shapes of the time curves obtained are preserved, so that coordination with other overcurrent protection devices is possible.

Optimization algorithms can be applied to time curves with adjustable parameters to improve the solution, and it is possible to consider discrete intervals for the *TDS*, the *PSM*, and the standardized time curve [1]. In [2,3], hybrid algorithms were combined with GAs to solve the problem of coordination in a real, highly interconnected system. The application of various standardized time curves to radial distribution systems and improvements in coordination were described in [4]. Close-in and line-end coordination allows for the determination of the sensitivity in the adjustment of relays, in conjunction with the diversity of standardized time curves using NSGA-II as proposed in [5]. In [6], an online OCR protection scheme was proposed using binary integer programming.

The application of *NCCs* in industrial electrical systems must take into consideration coordination with damage curves for transformers, cables, and induction motors. In [7], it was shown that the form of the *NCC* is suitable for

Overcurrent Relay Advances for Modern Electricity Networks. https://doi.org/10.1016/B978-0-323-91777-3.00004-0

coordination with fuses and other OCRs. In this case, flexible time curves offer advantages when a large number of damage curves and restrictions need to be accounted for in the coordination of industrial systems.

The main functions of the DOCR are as a trip for short circuits in the primary line and a backup for the adjacent line; it must operate if the primary relay is nonoperational. It is general practice to use the same type of curve for DOCRs to avoid the crossing of curves and to ensure adequate trip times for each utility. However, it is very common for I_{pickup} to be higher for the primary protection device than for the backup device in highly interconnected systems. In combination with the short-circuit values can result in crossing of relay curves, it will depend on the crossing zone if the protective function is compromised (as described in Chapter 1). Since the use of the same type of curve can avoid the problem of crossing time curves in meshed systems, several violations of coordination restrictions can be presented. The complexity of the problem increases exponentially as the size of the power system grows, meaning that finding the solution to each loss of coordination by protection engineers is very complex. The use of the proposed *NCCs* represents a possible solution to this loss of coordination. Its incorporation into optimization algorithms for DOCR coordination is very feasible, and it represents a viable solution to the violation of maximum operating times.

Traditionally, relay settings are set for maximum utility load or fault conditions. This means that for the remaining operating conditions, there are settings that are not adequate, due to a reduction in the sensitivity of the relay and an increase in operating times. This motivated a search for other solutions to improve overcurrent protection. The design of the proposed OCR makes it possible to obtain coordination by means of *NCCs*, with a shorter operating time over the entire current range. The shape of the curve is specifically designed for each coordination problem, which is very convenient for meshed systems where high operation times can be a common result of coordination.

4.2 Approach one: Proposed general model

The operating times of DOCRs are defined by the analytical expressions for the time curves. In the proposed model, the number of degrees of freedom is increased, as in addition to the variables *TDS* and *PSM*, we use the coefficients to define the type of curve. These values are defined on intervals bounded by the values specified in the standards. In Eq. (4.1), T_p is the operating time of the primary relay, while in Eq. (4.2), T_b is the operating time of the backup relay. The expression for h in Eq. (4.3) establishes the coordination criterion, which is used to ensure that the relay operation time sequence is not negative or greater than the *CTI*.

$$T_p(x) = \left[\frac{x^1}{\left(\frac{a}{b_1 \cdot x^5}\right)^{x^3} - 1} + x^2 \right] \cdot x^4 \tag{4.1}$$

$$T_b(y) = \left[\frac{y^1}{\left(\frac{a}{b_2 \cdot y^5}\right)^{y^3} - 1} + y^2 \right] \cdot y^4 \tag{4.2}$$

$$h(x, y) = \mathrm{T}_b(y) - T_p(x) - CTI \tag{4.3}$$

where x and y are variables as follows:

$$x = \left(x^1, x^2, x^3, x^4, x^5\right)^T \quad \text{and} \quad y = \left(y^1, y^2, y^3, y^4, y^5\right)^T$$

$$y, x \in \mathbb{R}$$

and a represents the interval of the fault current (I_{sc}), where $a = (a^{\min} : a^{\max})^T$.

Parameters x and y have the same interpretation, but different values, in dependence of relay coordination accomplish; however, for simplicity, we will refer only to x in the following. x^1, x^2, and x^3 correspond to the parameters of the standardized time curves [1]. Although these can have any value in theory, there is a risk that this will result in time curves with very different shapes than conventional relays, which would make it difficult to achieve coordination with other relays or with fuses and reclosers. For this reason, the variables must be bounded, and the lower and upper limits on the values are defined by the different types of *CCs*.

While x^4 represents the *TDS*, and its range should be specified with a limited maximum bound, for example, three, in order to obtain shorter operating times. x^5 represents the *PSM*, and to obtain better results from coordination, its value should be variable but limited; a suitable range is from 1.4 to 1.6, since values less than 1.4 could cause false triggering due to temporary conditions of overload, and values greater than 1.6 could affect the sensitivity of the relay.

Although parameters x^4 and x^5 do not change the shape of the time curve, larger values of their maximum bounds may increase the operation time of the DOCRs. It is therefore convenient to use limited maximum values; however, the negative effects are that the search space of the optimization algorithm may be reduced, and violations may occur. It is possible to use only four degrees of freedom, since x^4 linearly modulates x^1 and x^2, but the similarity with conventional relays and the specification of settings in the relay programming will be affected in this case. Each parameter of the time curve is limited to specific intervals, where the bounds on x^1, x^2, and x^3 are established by the maximum and minimum values of the standardized time curves. Although expansion of these intervals could avoid solution violations by expanding the search space of the optimization algorithm, this could cause a loss of the shape of the time curve. The proposed selection of parameter ranges is suitable for the type of time curves described in [8], as presented in Table 4.1.

The coordination constraint in Eq. (4.3) defines the sequence of operation between the primary and backup relays. This function of the coordination program is very important, as it ensures that coordination is obtained. The sign of the function h must be evaluated by the optimization algorithm, since larger positive values represent slower operation of the protection device and negative values indicate a coordination violation.

The objective function (OF) must establish a relationship between the different functions of the protection devices and must consider both the primary and backup protection functions, as shown in Eq. (4.4). This function aims to minimize the times at which violations are possible, although the terms are weighted in the same function; it is therefore convenient to include a term in the OF to ensure that coordination is not lost. This violation function can be included in the inequality constraints as a hard constraint. It must ensure that the coordination problem is resolved, that is, only with the reduction of the operation times the coordination is not solved, and adequate operation must be guaranteed by means of times that allow for good operation of the protection devices. Very short times can cause false trips due to transient operating conditions, whereas very long times could compromise proper fault clearance. Another important aspect is to ensure that violations are avoided, and that the protection is sensitive to faults at the remote end of the adjacent line, since a situation where areas are left without protection must be avoided. To achieve this, coordination must be done at two points, for close-in faults and line-end faults.

TABLE 4.1 Ranges for the proposed time curves.

	Boundaries	
Parameter	*Minimum*	*Maximum*
x^4(*TDS*)	0.50	5
x^5 (PSM)	1.40	2
x^1(A)	0.01	30
x^2(B)	0	0.50
x^3(p)	0.01	2

$$\text{OF} = \underset{T_{mc},f,g,h}{\text{minimize}} \frac{T_{mc}}{n} + \omega_1 \sum_{i=1}^{m} \frac{T_{p\_i}(x_i)}{m} + \omega_2 \sum_{j=1}^{n} \frac{T_{p\_i}(x_i)}{n} + \omega_3 \sum_{j=1}^{n} \frac{h_j\left(x_i, y_j\right)}{n} \tag{4.4}$$

where m is the total number of relays and n is the total number of coordination pairs of relays in the system. The factors ω_1, ω_2, and ω_3 are used to modulate each term of the OF. The problem can also be formulated to guarantee compliance with a specific term, for example. This OF is the same as in Eq. (3.8), except that the time functions T_p and T_b have five degrees of freedom. Since the time curves for the primary and backup relays may be different, it is important that coordination is verified in the close-in and line-end locations and that there is no crossing of curves. Although this verification requires an additional process, it is useful even for standardized curves, as described in Chapter 1.

The protection system must ensure coordination for a maximum level of I_{sc}, and therefore the OF result for other levels will be considered with smaller weights. The constants ξ and μ serve as weights for the different short-circuit magnitudes (Fig. 4.1). The complete fitness function is then defined as the sum of the weighted OF results for the minimum (OF_m) and maximum (OF_M) short-circuit levels, as follows:

$$OF = \xi \times OF_m + \mu \times OF_M \tag{4.5}$$

From Fig. 4.2, it can be observed that the shape of these time curves is similar to that of conventional time curves. This is favorable for relay applications, as they can be used with other conventional relays or with fuses and reclosers.

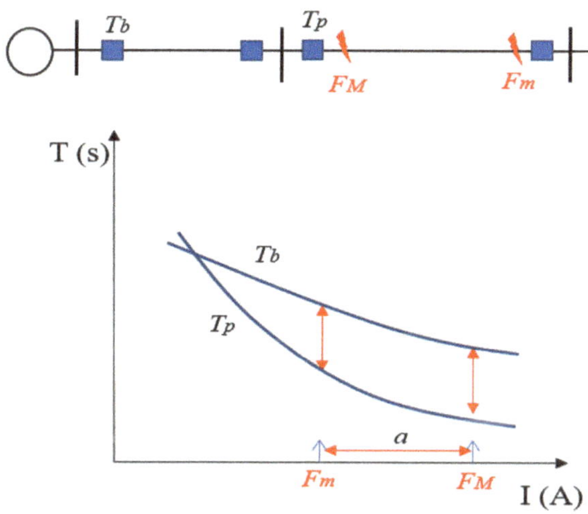

FIG. 4.1 Close-in and line-end relay coordination.

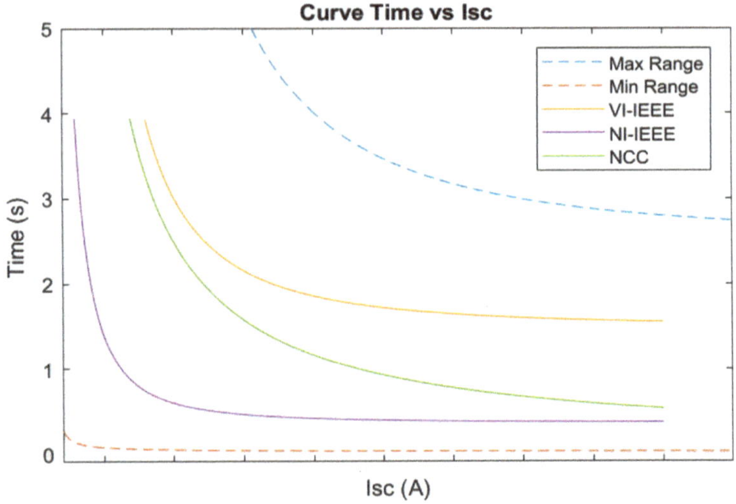

FIG. 4.2 Examples of time curves with variations in the limits on each parameter.

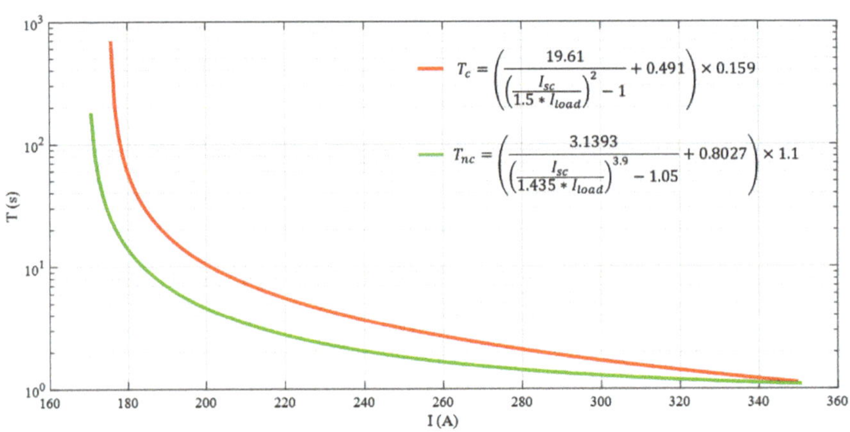

FIG. 4.3 Example of a solution with NCCs.

In addition, the qualities of the overcurrent protection principle are maintained; for example, tolerance to temporary overloads, coordination with damage curves, and expected operating times for the OCRs.

Fig. 4.3 shows a solution to the coordination problem based on the application of *NCCs*. It is notable that the application of the proposed time curves offers a solution to the loss of coordination with conventional relays and results in a decrease in operating times. This behavior can be replicated in larger electrical networks using optimization programs, where the formulation of the *NCCs* and the restrictions on each variable are defined according to Table 4.1.

The optimization of relay coordination using *NCCs* improves coordination and is based on the use of metaheuristic algorithms as described in Chapter 3.

4.3 OCR coordination by IWO

IWO was developed to solve the problem of relay coordination by incorporating the proposed time of functions of operation. With the addition of more degrees of freedom, the dimensionality of the coordination problem increases, and the optimization method must therefore be able to solve high-dimensional problems. IWO is a nature-mimicking metaheuristic algorithm that emulates the behavior of weeds [9]. The algorithm is robust and yields adequate results for the problem of coordination (Chapter 3), and the simplicity of this approach allows for the incorporation of the proposed time functions in a transparent way.

The population is configured with chromosomes (Cx) that represent the parameter settings for the *NCCs*. With five *ASs* and a total considered chromosomes (T_{Cx}) for the total number of OCRs (T_r), the structure of the matrix that represents the population is shown in Fig. 4.4. The chromosomes are arranged as columns and the ASs as rows.

The IWO operators for $T_r = 5$ are shown in Fig. 4.5. Mutations of the seed elements are defined for the 50% spreading operator and for the 20% dispersing operator. The rolling-down operator performs the combination and selects the best mutation.

By adjusting the magnitudes of each parameter, the parameters of the GA and IWO algorithms are determined as shown in Fig. 4.6. The parameters are defined in three stages. The first stage represents the adjustment criteria

$$P_{Cx} = \begin{bmatrix} x_1^4 & x_2^4 & \cdots & x_{T_{Cx}}^4 \\ \vdots & \vdots & \ddots & \vdots \\ x_{T_r\_1}^4 & x_{T_r\_2}^4 & \cdots & x_{T_r\_T_{Cx}}^4 \\ x_1^5 & x_2^5 & \cdots & x_{T_{Cx}}^5 \\ \vdots & \vdots & \ddots & \vdots \\ x_{T_r\_1}^5 & x_{T_r\_2}^5 & \cdots & x_{T_r\_T_{Cx}}^5 \\ x_1^1 & x_2^1 & \cdots & x_{T_{Cx}}^1 \\ \vdots & \vdots & \ddots & \vdots \\ x_{T_r\_1}^1 & x_{T_r\_2}^1 & \cdots & x_{T_r\_T_{Cx}}^1 \\ x_1^2 & x_2^2 & \cdots & x_{T_{Cx}}^2 \\ \vdots & \vdots & \ddots & \vdots \\ x_{T_r\_1}^2 & x_{T_r\_2}^2 & \cdots & x_{T_r\_T_{Cx}}^2 \\ x_1^3 & x_2^3 & \cdots & x_{T_{Cx}}^3 \\ \vdots & \vdots & \ddots & \vdots \\ x_{T_r\_1}^3 & x_{T_r\_2}^3 & \cdots & x_{T_r\_T_{Cx}}^3 \end{bmatrix}$$

FIG. 4.4 Population matrix for five ASs.

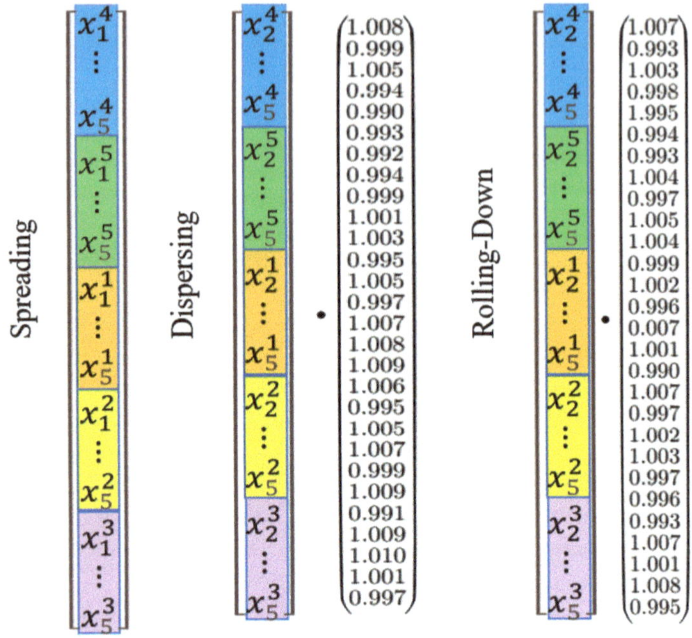

FIG. 4.5 Spreading, dispersing, and rolling-down operators in IWO.

Parameter	GA	IWO
T_i	2000	700
P_s	200	100
P(C)	0.85	-
P(M)	0.10	-
P(E)	0.05	-
P(S)	-	0.25
P(D)	-	0.05
P(R)	-	0.70

Parameter	Setting
ω_1	0.60
ω_2	0.30
ω_3	0.10
ξ	0.25
μ	0.25

	Boundaries	
Parameter	Minimum	Maximum
x_4	0.50	5
x_5	1.40	2
x_1	0.01	30
x_2	0	0.50
x_3	0.01	2

FIG. 4.6 Selection ranges of each adjustable setting.

for the optimization algorithms, such as the population (P_s) and number of generations (T_i). The probabilities used for the crossover $P(C)$, mutation $P(M)$, and elitism $P(E)$ operations are defined only for the GA. For IWO, the probabilities of the operators are also specified, such as those used in the spreading $P(s)$, dispersing $P(D)$, and rolling-down $P(R)$ operations. The weights of the functions are defined in the second stage, where the parameters ω_1, ω_2, and ω_3 are fixed for both methods and are used in the OF, and ξ and μ are the weights for each fault location (close-in and line-end). Finally, in the third stage, discrete intervals for the parameters of the proposed time curve are defined.

In the evaluation, coordination practices presented in existing research works were considered as cases. In each case, discrete intervals of the *TDS* and *PSM* were used, since this is general practice. Five coordination cases were evaluated. The first was the proposed method, where the five parameters of the time curve were considered as degrees of freedom and were adjustable. The OF was weighted and was formulated to solve the coordination problem at two fault points, close-in and line-end. In this way, settings of the relays that satisfied the coordination criteria could be obtained. In the second case, a weighted OF was used, and eight standardized curves were evaluated as a selection alternative to improve coordination [10]. In the third case, only the very inverse time curve was used for all relays. The coordination results were compared at two points on the line for cases one to three, and for a single close-in point (as in conventional coordination) for cases four and five. In cases four and five, we did not use weights for the line-end faults, the fourth case used the selection of eight standardized time curves as a criterion and the last considering the very inverse curve, and fifth case is taken as the base case, since this is general practice. A summary of all five scenarios is presented in Table 4.2.

The GA and IWO coordination algorithms were evaluated using three academic systems. Fig. 4.7 shows the results for 9-bus, 14-bus, and 30-bus IEEE systems, where the total numbers of coordination pairs were 12, 50, and 124, respectively. These values are multiplied when coordination is performed at more points; hence, for the cases studied here, the number of pairs and their restrictions were multiplied by two, as we considered close-in and line-end faults. The cases presented in Table 4.2 were evaluated to obtain results for

TABLE 4.2 Case studies used to evaluate the NCCs.

	Case				
Parameter	*1*	*2*	*3*	*4*	*5*
x_4			[0.50–5]		
x_5			[1.4–2]		
x_1	[0.01–30]	8 curves	19.60	8 curves	19.60
x_2	[0–0.50]		0.49		0.49
x_3	[0.01–2]		2		2
ω_1	0.25	0.25	0.25	0.00	0.00
ω_2	0.25	0.00	0.00	0.00	0.00
ω_3	0.50	0.75	0.75	1.00	1.00

System	Results		GA Case 1	GA Case 2	GA Case 3	GA Case 4	GA Case 5	IWO Case 1	IWO Case 2	IWO Case 3	IWO Case 4	IWO Case 5
9-Bus system		f	0.10	0.18	0.56	0.21	0.56	0.07	0.15	0.20	0.14	0.21
		mc%	0.00	0.00	6.67	6.67	6.67	0	0	0	0	0
	I_{sc}^{m}	t_m	0.27	0.46	0.97	0.49	0.98	0.19	0.46	0.69	0.44	0.73
		E_{CTI}	0.64	0.89	5.30	1.56	5.30	0.46	0.63	0.82	0.74	0.91
		t_{bu}	1.21	1.65	6.42	2.27	6.42	0.93	1.38	1.77	1.48	1.91
	I_{sc}^{in}	t_m	0.12	0.32	0.50	0.30	0.50	0.08	0.32	0.42	0.29	0.42
		E_{CTI}	0.18	0.30	2.06	0.42	2.06	0.14	0.24	0.44	0.27	0.51
		t_{bu}	0.61	0.91	2.82	1.01	2.82	0.52	0.85	1.14	0.85	1.23
	I_{sc}^{M}	t_m	0.09	0.26	0.38	0.23	0.38	0.05	0.27	0.35	0.24	0.35
		E_{CTI}	0.03	0.12	1.07	0.18	1.08	0.00	0.10	0.24	0.10	0.30
		t_{bu}	0.41	0.69	1.76	0.72	1.76	0.35	0.66	0.88	0.63	0.94
14-Bus system		f	0.14	0.18	3.67	0.25	3.72	0.10	0.13	0.30	0.14	0.29
		mc%	2.13	2.13	11.35	6.38	13.48	0	0	4.26	0	2.13
	I_{sc}^{m}	t_m	0.28	0.37	0.68	0.38	0.66	0.24	0.34	0.69	0.35	0.71
		E_{CTI}	0.77	0.83	3.70	2.18	2.81	0.42	0.59	1.36	0.86	1.55
		t_{bu}	1.31	1.48	4.48	2.75	3.61	0.93	1.21	2.30	1.48	2.47
	I_{sc}^{in}	t_m	0.15	0.24	0.47	0.21	0.45	0.15	0.21	0.52	0.21	0.52
		E_{CTI}	0.23	0.32	0.70	0.36	0.69	0.17	0.25	0.50	0.30	0.53
		t_{bu}	0.66	0.86	1.41	0.86	1.40	0.61	0.75	1.29	0.80	1.32
	I_{sc}^{M}	t_m	0.11	0.19	0.43	0.16	0.41	0.12	0.18	0.49	0.17	0.48
		E_{CTI}	0.10	0.17	0.35	0.15	0.31	0.07	0.12	0.26	0.14	0.29
		t_{bu}	0.50	0.67	1.03	0.62	1.01	0.49	0.60	1.02	0.61	1.04
30-Bus system		f	0.22	0.25	1.43	0.58	1.60	0.13	0.20	0.50	0.18	0.49
		mc%	5.37	5.93	20.34	8.47	15.25	0	0	5.93	0.28	3.67
	I_{sc}^{m}	t_m	0.43	0.50	0.87	0.59	0.88	0.37	0.57	0.95	0.49	1.11
		E_{CTI}	0.78	0.81	1.77	1.45	1.59	0.48	0.86	1.11	0.88	1.38
		t_{bu}	1.44	1.55	2.58	2.19	2.46	1.11	1.68	2.11	1.62	2.49
	I_{sc}^{in}	t_m	0.27	0.33	0.62	0.35	0.69	0.25	0.37	0.76	0.29	0.87
		E_{CTI}	0.33	0.41	1.00	0.60	0.61	0.26	0.48	0.48	0.46	0.54
		t_{bu}	0.87	1.01	1.72	1.22	1.40	0.78	1.13	1.34	1.03	1.53
	I_{sc}^{M}	t_m	0.22	0.27	0.57	0.27	0.65	0.21	0.30	0.72	0.22	0.82
		E_{CTI}	0.17	0.24	0.41	0.29	0.26	0.16	0.32	0.26	0.30	0.29
		t_{bu}	0.67	0.80	1.10	0.86	1.02	0.64	0.90	1.08	0.81	1.24

FIG. 4.7 DOCR coordination results obtained from GA and IWO.

the fitness (f), the percentage of loss of coordination (mc%), the best operating times for the primary (t_m) and backup relays (t_{bu}), and the errors in the CTI (E_{CTI}). These parameters were evaluated for three different values of fault current, coordination for close-in, line-end, and mid-line faults [11].

In all the cases studied, and for each of the three test systems, IWO yielded the best results. It is notable that for cases one and two, zero violations were obtained, and case one (the proposed approach) gave the best results. For example, in case one for the 14-bus system, an mc% of 2.13 was obtained with GA, meaning that one relay was not coordinated. As the electrical system becomes more interconnected and contains a greater number of relays, the equality and inequality restrictions on the coordination problem increase. It is therefore to be expected that in larger systems, there will be worse indicators. Despite this, it is possible to see a trend and to make a comparison between the performance of the optimization algorithms used for each case. Due to the inverse shape of the time curve, longer times will be obtained when analyzing the coordination for line-end faults.

The proposed system (case 1) and the IWO method yielded the best results. The value of E_{CTI} is important since it represents the coordination errors: when this value is negative, coordination is lost, and when it is zero, the *CTI* between relays has the specified value. The greater the value of the E_{CTI}, the more difference there is between the times of the coordination pair. In the cases studied, these values were greater than zero, indicating that the coordination was correct. However, it is possible to observe that the values of E_{CTI} were lower for IWO in all cases, that is, for each of the three test systems and each of the three coordination points. It can be seen that the proposed method in case 1 gave better results than case 3 for line-end faults; this is because the NCCs are more flexible and allow the coordination of the pair of relays to result in shorter times. A graph of the results is presented in Fig. 4.8.

The coordination results in Table 4.7 were calculated based on the average values from of 10 runs, since metaheuristic algorithms do not start from a specific initial condition and the results may therefore vary with each execution. The deviation in the data depends on the algorithm used; the results from IWO showed a much smaller deviation than those from the GA. Similarly, it is necessary to define a stopping criterion for the simulations that can be adapted depending on the slope of the fitness and the lack of response to mutations. In the cases studied here, a stopping criterion of 700 generations was used for IWO and 2000 generations for GA. The number of generations used for all the test

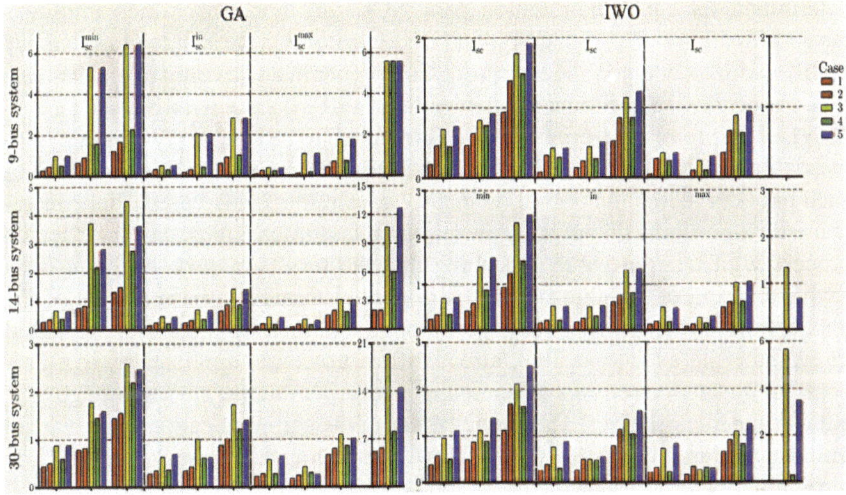

FIG. 4.8 Graphical results for relay coordination using GA and IWO.

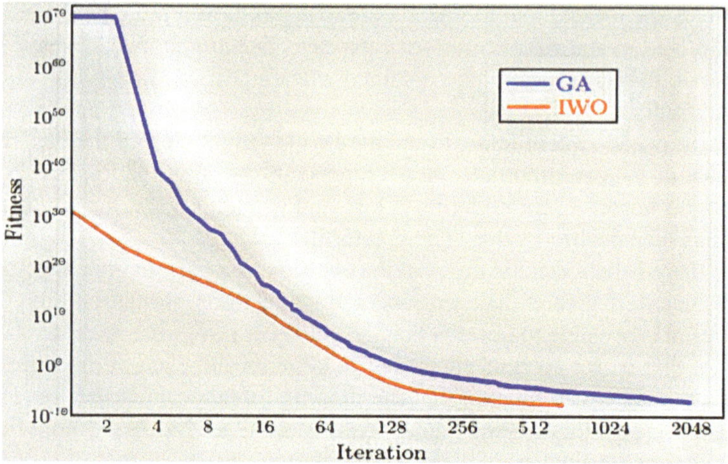

FIG. 4.9 Fitness slopes for OCR coordination using the GA and IWO algorithms.

systems was the same and was obtained based on a stopping criterion for the slope of fitness. The results were satisfactory and gave adequate operating times for the overcurrent protection function. The fitness behavior of the algorithms is shown in Fig. 4.9.

4.4 OCR coordination by GWO and DE

In meshed systems, using the same time curve for each OCR results in numerous violations of the coordination constraints. In contrast, when the time curve is determined for each coordination pair, NCCs are obtained, which improves the coordination times. The GWO and DE algorithms were adapted to solve this highly complex and nonlinear optimization problem, and a comparison was carried out of these algorithms to analyze their performance with NCCs.

The algorithmic performance was determined to be suitable for solving multivariable problems, since with the increase in the number of degrees of fit of each time curve, the number of inequality constraints is increased. Since metaheuristic algorithms do not have preset initial conditions, they start their search process with random values, meaning that the results may be different from each execution of the algorithm. It is important to make an assessment of the application of the optimization algorithms in real time; it is therefore essential to evaluate the execution time, although for protection applications it is proposed to update the settings on the scale of several minutes, having enough execution time in most cases. The latency times of the communication system are in milliseconds and have little impact. However, an important parameter affecting the evaluation criterion is the deviation between the results obtained from several runs of the algorithm.

4.4.1 Test systems

To evaluate the coordination results from the DE and GWO algorithms, we used 6-bus networks with 14 relays and 12 coordination pairs, and IEEE 30-bus networks with 78 relays and 92 coordination pairs. Each relay was identified by nodal numbering. In this study, the relays protecting the transformer were omitted, as the damage curves for the transformers were not included. A sensitivity verification was also carried out to detect relays that, due to the values of I_{pickup} and I_{sc}, failed to satisfy the fault detection criteria for the remote bus of the adjacent line. Although these relays were adjusted, these results are not shown, as their limitations arise from the overcurrent protection principle rather than from the algorithm used.

Populations of 100 individuals were considered for DE and 60 for GWO. A thousand generations were carried out for each algorithm. These values, although they may be different, that is, we can increase individuals and reduce generations and vice versa. In the studies carried out here, this combination gave the best results in terms of the relay times and a reduced execution time for the algorithm. A comparison was drawn between the relay coordination results using CCs and the proposed NCCs. The conditions of the 6-bus network are conducive to coordination, as there are fewer restrictions to comply with, whereas for the 30-bus network, more violations and longer execution times are expected (Fig. 4.10).

Case studies [12,13]:

Case 1: Six-bus IEEE system.
Case 2: Thirty-bus IEEE system.

Case 1 From the results presented in Tables 4.3 and 4.4, it can be observed that both methods achieved a solution with zero violations (NV). In general, the operation times (both the average and the best times) were lower using the *NCCs* in both test systems and both algorithms, as shown in Fig. 4.10. It can

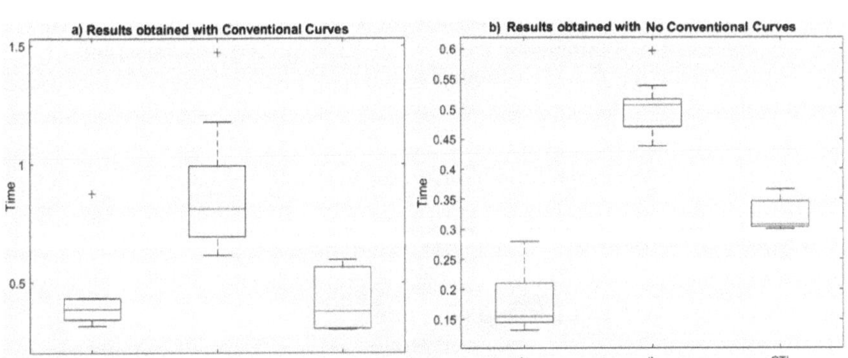

FIG. 4.10 Results for execution times using CCs and NCCs, for the six-bus IEEE system.

TABLE 4.3 Coordination results from GWO with CCs and NCCs.

	CC			NCC		
	tp	*tb*	*CTI*	*tp*	*tb*	*CTI*
Best	0.31	0.65	0.3	0.13	0.44	0.30
Average	0.46	0.89	0.42	0.19	0.50	0.32
NV	0			0		
Fit	1.972			3.615		
Time (s)	3.127			3.490		

TABLE 4.4 Coordination results from DE with CCs and NCCs.

	CC			NCC		
	tp	*tb*	*CTI*	*tp*	*tb*	*CTI*
Best	0.31	0.64	0.30	0.18	0.49	0.32
Average	0.42	0.87	0.41	0.22	0.61	0.39
NV	0			0		
Fit	1.882			4.109		
Time (s)	4.041			4.296		

be seen that the DE algorithm had slower execution than GWO. The adjustments obtained from the NCCs are presented in Table 4.5.

Case 2 The results obtained for the 30-bus IEEE system are presented in Tables 4.6 and 4.7. It can be observed that the trend in the results is similar to the previous case; this is because the use of NCCs results in better operation and execution times than CCs, as shown in Fig. 4.11. However, due to the complexity of this test system, more violations are seen, which is as expected. Although the GWO is faster than the DE, the execution times for both are less than 1 min.

A comparison of the best fitness results obtained using *NCCs* and *CCs* with the GWO algorithm is shown in Fig. 4.12. A comparison of the fitness convergence between GWO and DE is shown in Fig. 4.13.

The proposed DOCR coordination method uses *NCCs* to reduce the operating times and the number of violations. To obtain the solution, the GWO and DE

TABLE 4.5 Time curve settings for NCCs, Case 1.

	A	B	P	TDS	PSM
Relay	x_1	x_2	x_3	x_4	x_5
14	19.372	0.346	1.752	0.584	1.385
15	4.150	0.111	1.464	0.952	1.591
16	2.240	0.167	1.840	0.738	1.375
46	3.155	0.120	1.957	1.190	1.328
56	11.255	0.168	1.998	0.507	1.204
41	16.486	0.155	1.772	0.678	1.287
51	2.558	0.144	1.986	1.523	1.589
61	8.993	0.122	1.989	0.509	1.211
64	0.051	0.117	0.272	0.726	1.402
65	15.441	0.185	2.000	0.501	1.208

TABLE 4.6 Coordination results from GWO with CCs and NCCs for the 30-bus system.

	CC			NCC		
	tp	*tb*	*CTI*	*tp*	*tb*	*CTI*
Best	0.20	0.52	0.30	0.15	0.72	0.32
Average	0.79	1.10	0.43	0.66	0.82	0.34
NV	20			7		
Fit	562.10			380.67		
Time (s)	44.18			45.26		

algorithms were used, and the operation times obtained for the primary and backup protection are shown, as well as the execution times. It can be observed that both methods obtain satisfactory results, with little difference in the results. The execution times are reduced, meaning that this approach can be applied in online protection systems. The use of the proposed NCCs yields a substantial improvement in the results, with better operation times and reduced numbers of violations.

TABLE 4.7 Coordination results from DE with CCs and NCCs for the 30-bus system.

	CC			NCC		
	tp	*tb*	*CTI*	*tp*	*tb*	*CTI*
Best	0.17	0.42	0.30	0.17	0.59	0.31
Average	0.80	1.18	0.59	0.89	1.22	0.64
NV	18			8		
Fit	532.043			480.674		
Time (s)	56.861			58.260		

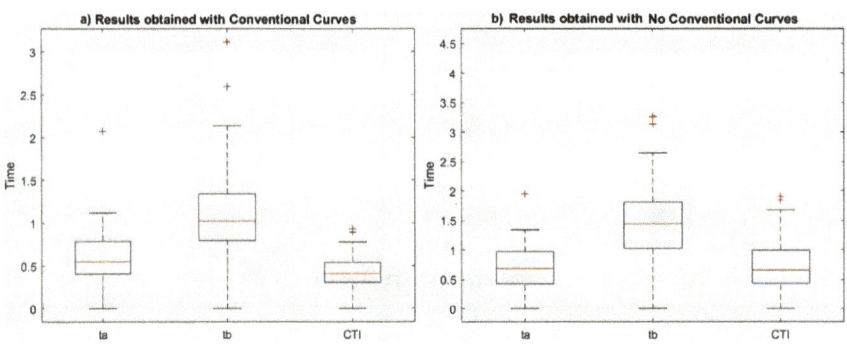

FIG. 4.11 Results for execution times using CCs and NCCs, for the 30-bus IEEE system.

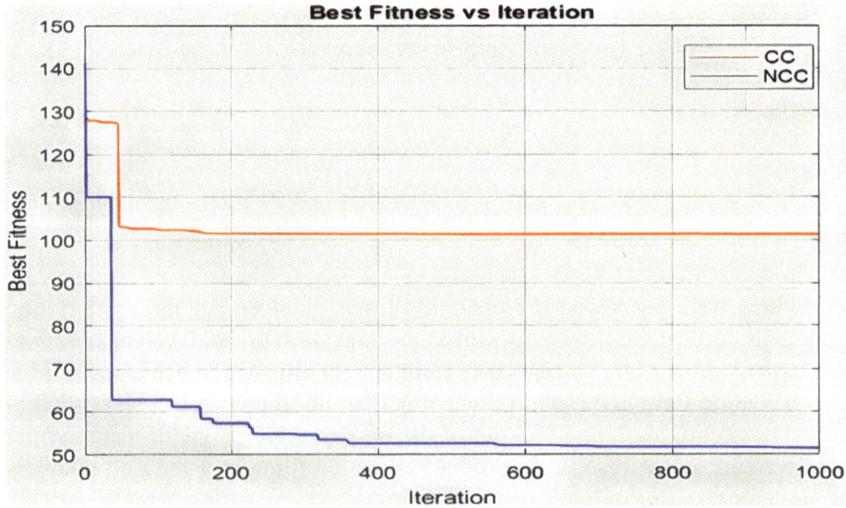

FIG. 4.12 Comparison of best fitness between CCs and NCCs for the 30-bus system.

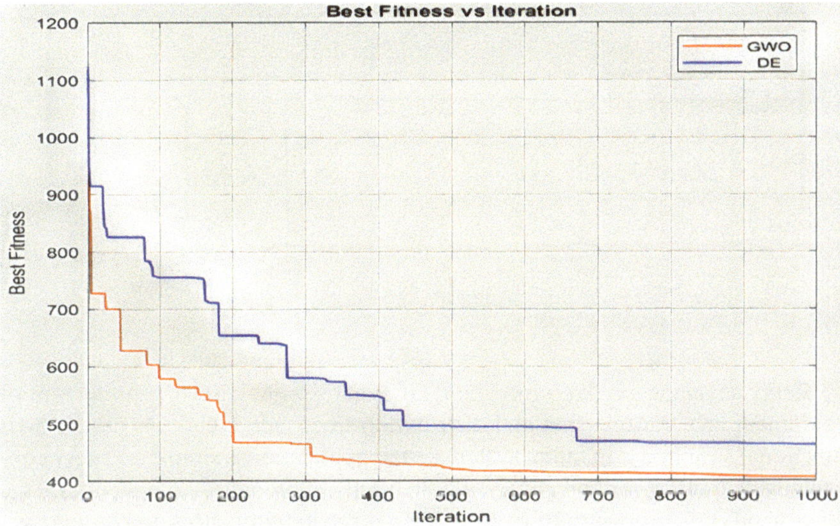

FIG. 4.13 Comparison of best fitness between GWO and DE for the 30-bus system.

4.5 Phase and ground coordination

The adjustment of DOCRs using NCCs increases the flexibility of the timing curves, allowing for better coordination of the relays. To maintain the shape of the time curve for the purpose of maintaining coordination with various protective devices, it is important that the NCC parameters are bounded to keep the parametric variation within the limits established for the standardized time curves. However, incorporating additional coefficients into the analytical expression allows for diversity that is beneficial in terms of problem solving. Traditionally, relay settings are set based on maximum grid load or fault conditions, meaning that for the remaining operating conditions, there are settings that are not suitable, and which give a reduction in the sensitivity of the relay and an increase in operating times. This motivated us to look for other solutions to improve overcurrent protection.

The design of the proposed OCR makes it possible to obtain coordination, through the use of NCCs, with lower operating times across the current range. The shape of the curve is designed for each specific coordination problem, which is very convenient given the large number of damage curves and the restrictions that apply to the coordination of industrial systems. The design presented in this chapter starts from the characteristic equations for standard inverse time OCRs, and a time curve is established depending on the coordination needs for typical cases of equipment and systems. Better coordination is obtained with the proposed design, considering the limits and damage curves of the primary equipment such as transformers, lines or cables, and motors.

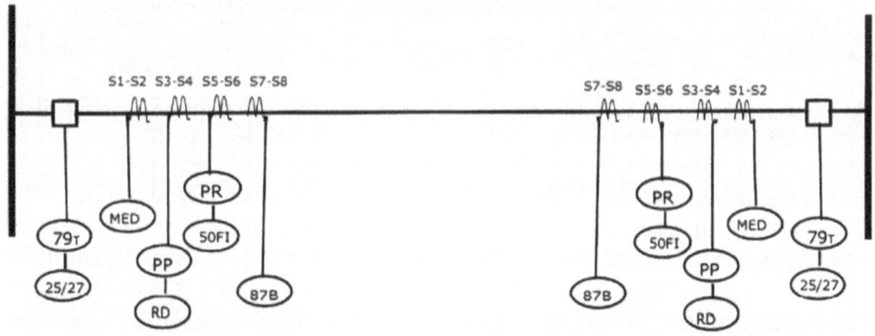

FIG. 4.14 Line protection scheme.

Relay schemes for line protection. OCRs are suitable for the protection of equipment and transmission and distribution lines (see Fig. 4.14) due to their functional simplicity. In radial systems, their application is simple and very convenient, as it offers reliable protection. In highly interconnected systems, on the other hand, it is necessary to incorporate a directionality check to obtain operational selectivity. This ensures that the tripping direction is adequate for both phase and ground protection [10].

Positive sequence relays (*PS*). Protection against faults between phases (three phase and two phase) is obtained through the use of phase relays. To carry out adjustment and coordination, the fault resistance of the electric arc can be considered without the need to include the ground impedance.

Negative sequence relays (*NS*). Phase relays often respond insensitively or with a time delay to phase-to-phase faults, and it is therefore common to use negative sequence relays to specifically respond to this type of fault. The setting and coordination criteria use the negative sequence current for phase-to-phase faults at the end of the protected line.

Zero sequence relays (*ZS*). The ground relay acts as protection for earth faults. The setting criterion for the pickup current is defined based on 10%–50% of the load current. This offers high sensitivity for the detection of ground faults and yields protection with the highest probability of operation, since more than 50% of faults involve the ground loop.

Time curves for positive sequence protection relays, PSE (67p-67b); negative sequence relays, NSE (67Qp-67Qb); and zero sequence relays, ZSE (67Np-67Nb) are shown in Fig. 4.15. Time curves for the primary and backup protection functions are presented in order to illustrate the interaction between these relays. Coordination can be achieved by incorporating the constraints and settings of each principle into the coordination algorithms.

4.5.1 Academic test system

To evaluate the implementation of the sequence relay coordination algorithms, the IEEE 11-bus electrical network in [13] was used. The total number of relays

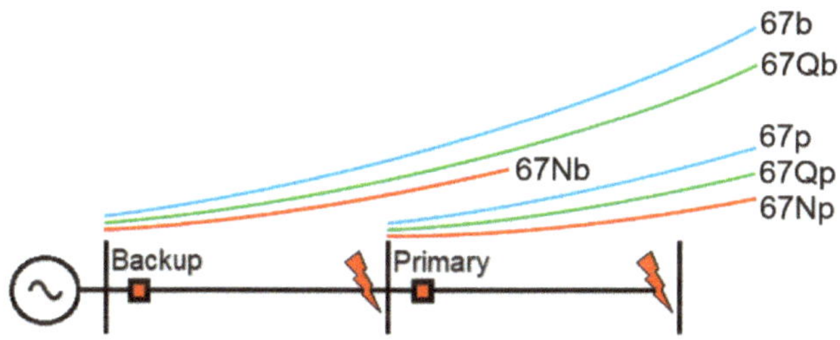

FIG. 4.15 Integration and coordination of positive, negative, and zero sequence elements.

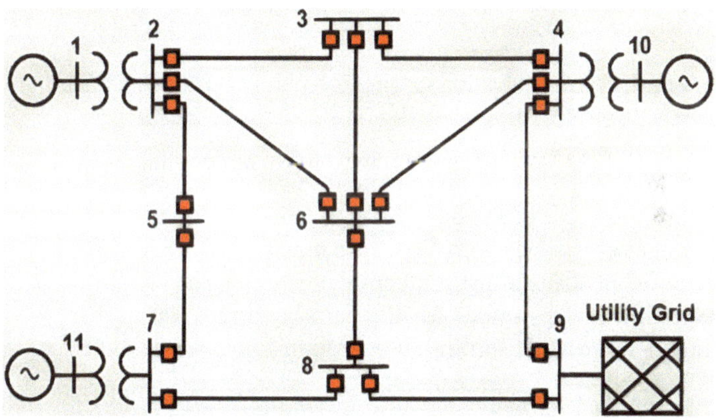

FIG. 4.16 Distribution test system.

was 22, forming 42 coordination pairs. A DG supply was added at nodes 1, 10, and 11, and the connection with the utility was at node 9 (Fig. 4.16). The objective was to evaluate the coordination between the sequence relays, and a fault analysis was carried out at each node, with the remote end of the line open, to obtain the currents for each sequence.

4.5.2 Results

Based on the short-circuit results and the relay configuration, the coordination pairs were formed for each relay sequence. The algorithm used was the DE, with a population of 60 individuals and 1000 generations, where each chromosome corresponded to the adjustment of the *NCCs* and *CCs* for each of the 22 relays separately. The results in terms of the operation times for each pair of relays and a comparison between the *CCs* and the *NCCs* are shown. Ten runs were weighted, and the resulting average times were used to plot a graph of the results.

Case 1 *Positive sequence relays using CCs and NCCs for three-phase fault.*

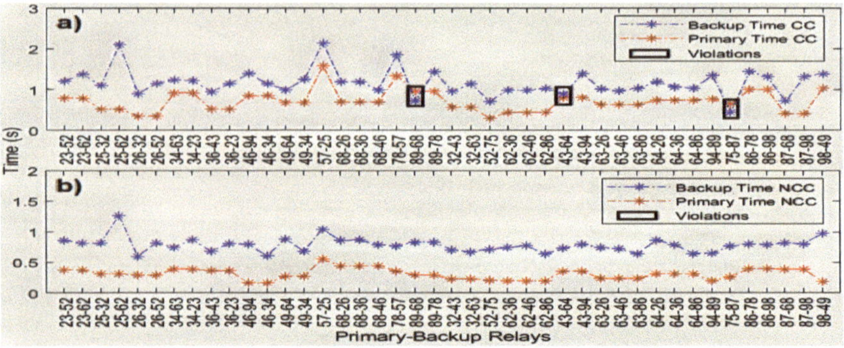

FIG. 4.17 Relay operation times for each coordination pair using (A) *CCs* and (B) *NCCs*.

The operating times of each coordination pair (primary and backup relays) at each fault location are shown in Fig. 4.17. A three-phase positive-sequence short-circuit current was used. Fig. 4.17A shows the operation times using *CCs*, while Fig. 4.17B shows the operation times using NCCs. It can be seen that there are three coordination violations with *CCs*, but no violations using *NCCs*. The operating times with NCCs are much lower than with *CCs*, resulting in great benefits in terms of the protection of the electrical network. In addition, the operation times are similar for the pairs, giving a flat profile, where with *CCs* there is a large variation in the operation times for each pair of relays.

Case 2 *Negative and positive relays coordination using NCCs for phase-to-phase faults.*

The coordination results for negative and positive sequence relays using *NCCs* are shown in Fig. 4.18. It can be observed that there are no violations, and coordination is obtained by evaluating phase-to-phase faults. The benefit of including negative sequence relays lies in a reduction in the operating time, since I_{pickup} is lower than for positive sequence relays.

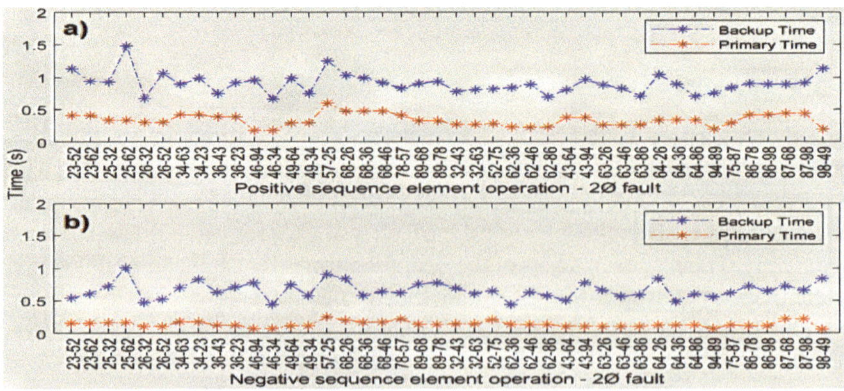

FIG. 4.18 Relay operation times for PSE and NSE for phase-to-phase faults.

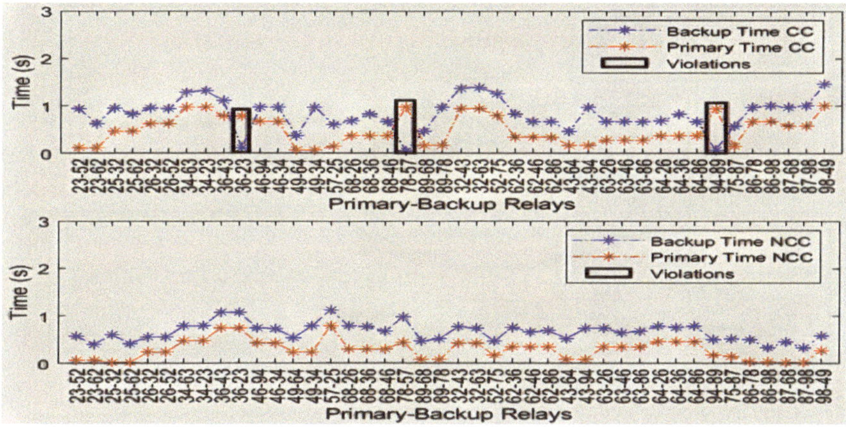

FIG. 4.19 Relay operation times for *ZSE* for one-phase ground faults using CCs and NCCs.

Case 3 *Ground protection relays for one-phase faults using CCs and NCCs.*

Fig 4.19 shows the graphical results for ground relay coordination for one-phase ground faults using *CCs* and *NCCs*. It can be observed that various violations occur using *CCs*, while with *NCCs* there are no violations. In addition, the operating times obtained with *NCCs* are much lower.

The design presented in this chapter starts from the characteristic equations for standard inverse time OCRs [8], and the time curve is established based on the coordination needs for typical cases of equipment and systems. Better coordination is obtained with the proposed model, considering the limits and damage curves for primary equipment such as transformers, lines or cables, and motors.

4.6 Approach two: Dynamic *J(I)* function

In the second approach, we sought to modify the response of the relay by incorporating a nonlinear function that was dependent on the current, with the intention of introducing an additional dynamic to the operation of the relay. Using the shape of the time curve of the relay as a base and the nonlinear function as a modulating factor of the relay allows us to increase or decrease the relay operation time. It is possible to obtain different forms of the time curve that have different applications, mainly in terms of the protection of equipment. This is in contrast to the first approach, in which the selection of the NCC was established by increasing the number of variables in the analytical expression over bounded intervals of the five parameters that define the type of time curve.

In the scheme proposed in this section, a parameter is assumed to vary with time and the value of the current. The functionality of the OCR must be maintained, since it is important for the protection of the electrical network in terms

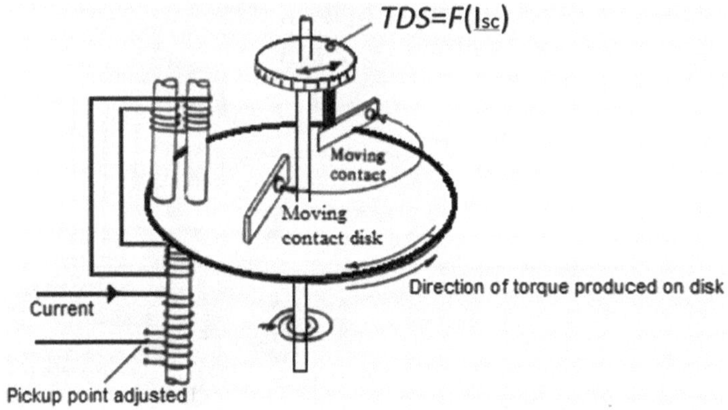

FIG. 4.20 Diagram of an analogous electromechanical device based on the proposed relay.

of the interaction between the relays and adequate coordination with fuses and reclosers. Hence, the variation in the coefficient should not have a different functional dynamic from the protection devices in the electrical network.

The proposed model is based on the interaction between the functions of the digital relay; as an analogy, the induction disk of an electromechanical relay is the only element that is displaced by the interaction of the currents and fluxes induced in the disk. In this section, we propose to modify the function A to allow it to vary depending on the input current, which makes it possible to increase the operational dynamic of the analytical expression for the relay time curve (Fig. 4.20).

4.6.1 Proposed model

Digital relays should reproduce the dynamic operation of electromechanical relays [14]. In Chapter 1, Fig. 1.26 presents a diagram of a generalized version of the DOCR based on the model functions [15]. The function $J(I)$ represents A in Eq. (1.1) and also the value that interrelates the design elements in electromechanical relays [16]. The angular distance of travel of the disk is θ, K_d is the damping of the magnet, and the force of the retention spring is τ_s, which are constant parameters that define A in Eq. (1.1) and [8,17].

$$A = \frac{K_d \theta}{\tau_s} \tag{4.6}$$

In the proposed approach, the function A is modified to be variable and dependent on the input current. The objective is to define nonlinear functions for $A(I)$ that modify the relay response, that is, increasing or decreasing its operating time in function on the protective application. For the electromechanical equivalent, it is necessary to define a dynamic behavior for the *TDS* such that θ in Eq. (4.12) depends on the current $\theta(I)$. Then, Eq. (4.6) can be expressed as follows:

$$A(I) = \frac{K_d \theta(I)}{\tau_s} \tag{4.7}$$

The dynamic expression for the proposed model uses the integrator output function G_k and the nonlinear function of the lever $A(I)$, which is dependent on the current, to represent the dynamics of the proposed digital relay. The time OCR is modified since the operation changes and the dynamic performance of $A(I)$ depends on the input current I, as shown in Fig. 4.20. By using the digital version of the inverse time OCR and comparing the output of the integrator G_k and the function of the lever $A(I_k)$, and by developing discrete forms of the analytical functions, we obtain the dynamic function in Eq. (4.8) for the proposed model:

$$\Delta t \sum_{k=1}^{k_{op}} \frac{A(I_k)}{T(I_k)} = A(I) \tag{4.8}$$

where:

$$T(I_k) = \frac{A(I_k)}{H(I_k)} + B$$

The function $H(I_k)$ is the denominator of the model defined in [18,19].

$$H(I_k) = M^P - 1 \tag{4.9}$$

In the proposed model, the function $H(I_k)$ is not modified, since it defines the desired qualities of the relay time curve. Thus the time curves obtained with the proposed model maintain the functional performance of conventional relays but contain functions that allow them greater flexibility to adapt to different coordination conditions, such as those in industrial systems.

The function $A(I_k)$ is part of the integration that establishes the behavior of the dynamic relay (Fig. 4.21). In contrast, the function $A(I)$ does not form part of the integration process, and the operation condition is set to the current value of the function. Hence, the time curve is not completely modified, and certain quantities specified by the time curve of the relay are maintained, for

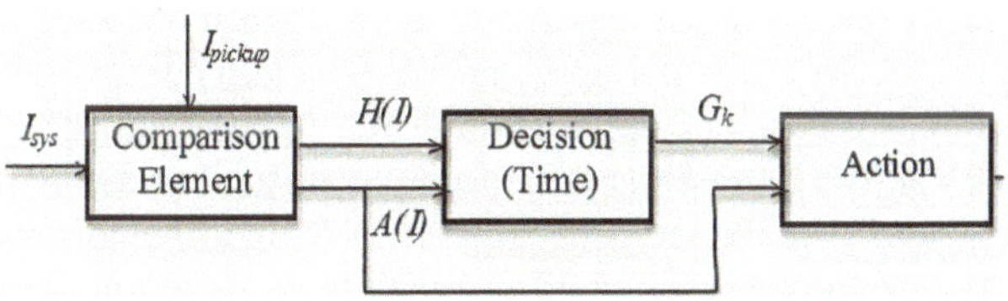

FIG. 4.21 Digital version of the inverse time OCR with dial time variable function.

example, the asymptotic behavior of I_{pickup}, the similarity in the shapes of the time curves with those of other protection devices like fuses and reclosers, and the coordination with damage curves for equipment like transformers and power cables. Consequently, the proposed model maintains the functional performance of conventional relays with the addition of time functions that modify the shapes of the curves.

The function $A(I_k)$ for the OCR can be designed based on the specific coordination problem. It is not the objective of this work to determine which expression is the most appropriate for each application, but the process may be simplified through the use of cross-functional coordination programs, either by expanding the catalog of available curves or by incorporating curve fitting routines [14] into the coordination programs. The analytical expression for the best fit to the time curve can be obtained in this way, and the time curve can then be defined in a catalog of curves for coordination and a model obtained using a computational approach. It is important that the mathematical model of the relay is simple to apply; for example, polynomial expressions offer a good alternative for relays and fuses, but they are expressions with high-degree terms, making their implementation difficult [20].

This process results in time curves that are no longer established by existing standards, and which rely on general functions that can be adapted according to the needs of overcurrent protection. In the following, three different types of expression for $A(I)$ are evaluated.

(a) Exponential type: $A(I) = Ae^{-I/c}$
(b) Logarithm type: $A(I) = A\ \ln(I/c)$
(c) Polynomial type: $A(I) = A_1 + A_2e^{-I} + A_3Ie^{-I}$

Where c is the parameter that modifies the TDS, depending on the change in the input current.

The standard form of the time curve is very inverse and is taken as a reference for comparison with the three types of curve shown before. A graphical illustration of the standard, exponential, logarithmic, and polynomial curves is shown in Fig. 4.22. The exponential form can reduce the times markedly in the middle part of the curve and slightly in the region near the pickup current, but the time is not decreased for the maximum current as it is limited by the definite time B, and this is also the case for the other curves.

The logarithmic time curve model reduces the operation times in the middle part of the curve, and these times are further reduced for currents close to the pickup current. If the logarithm parameter is changed in an attempt to further decrease the times, the curve collapses near the pickup current, and this poses a serious problem in terms of curve fitting. The polynomial form gives good time reductions across almost the entire current range, but the time cannot be reduced at the maximum current because it is defined by the coefficient B. These functions are shown in Fig. 4.22 to illustrate the different shapes obtained

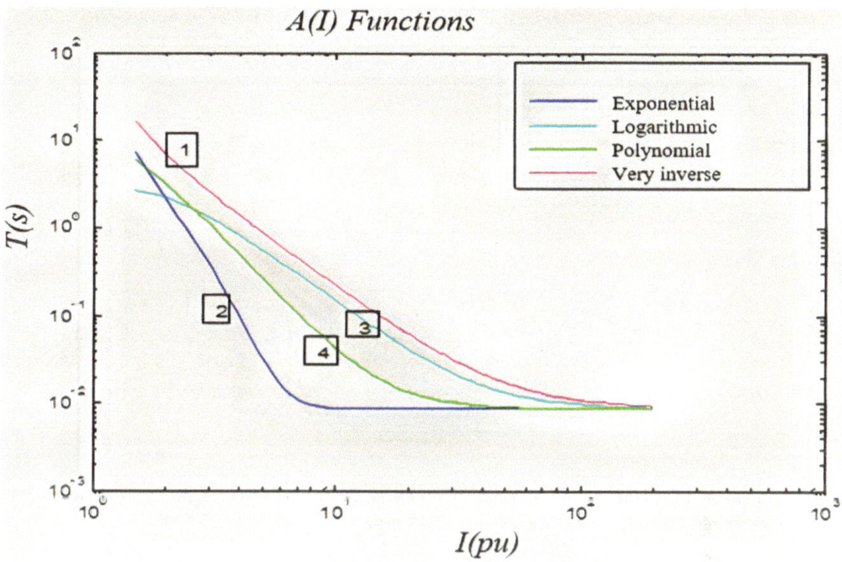

FIG. 4.22 Different forms of the numerator *A*(*I*) of the time curve.

by including simple expressions. For a specific coordination application, a catalog of functions may be available that allow the best model to be obtained for the specific coordination problem.

To obtain a new, nonstandardized time curve for the OCR, it is necessary to estimate the parameters *A*, *c*, *N*, and *B*. The proposed function for the new characteristic equation is nonlinear, and a nonlinear regression method is applied to estimate its parameters, based on the input data and the formulation given in [16]. The Gauss-Newton method is used since the fitted time curves show bounded errors within the recommendations [19].

Example 1 *Exponential time curve fitting.*

Fitting to a very inverse time curve is carried out to validate the proposed methodology and to verify that the proposed adaptive function gives the same results for both constant and variable fault currents. Fig. 4.23 shows the results of the parameter estimation and curve fitting, which can be used to obtain the coefficients of the proposed analytical expression. The expression for the nonconventional exponential characteristic is as follows:

$$T(I) = \left[\frac{A \cdot e^{-(I/c)}}{I^N - 1} + B\right] \cdot TDS \tag{4.10}$$

The values of the parameters for the new exponential type of characteristic function were determined using nonlinear regression as shown in the graph in Fig. 4.24.

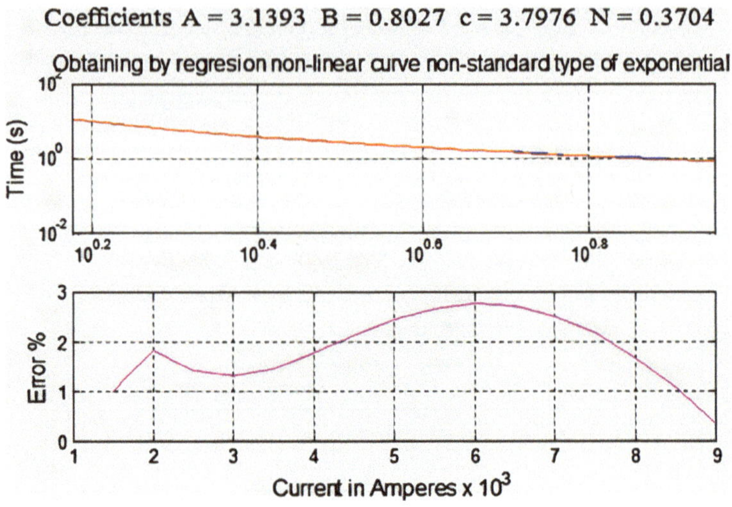

FIG. 4.23 Parameter estimation for exponential curve fitting.

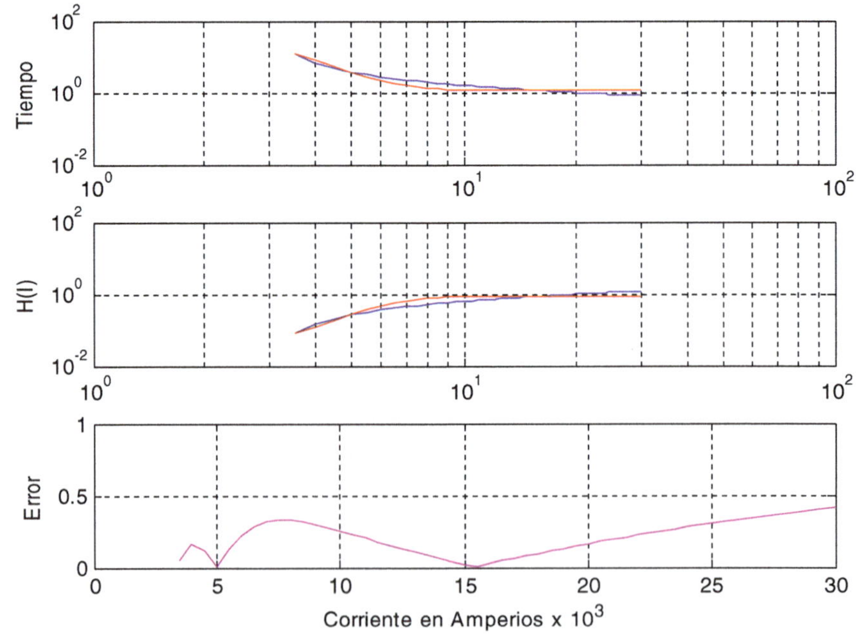

FIG. 4.24 Parameter estimation for polynomial curve fitting.

The parameters of the function *T(I)* are as follows:

$$T(I) = \left[\frac{3.1393 \cdot e^{-(I/3.7976)}}{I^{0.3704} - 1} + 0.8027\right] \cdot TDS$$

Example 2 *Polynomial time curve fitting evaluation.*

The coefficients of the time curve expressed in polynomial form provide another alternative for fitting the time curve. The expression for the nonconventional polynomial characteristic function is as follows:

$$T(I) = a_0 + a_1 e^{-I} + a_2 \cdot Ie^{-I}$$

The polynomial function used in the nonlinear regression program corresponds to the graph in Fig. 4.24. The coefficients of this polynomial are $a_0 = 1.16$, $a_1 = 281.6783$, and $a_2 = 19.992$. The expression for the time curve in its polynomial form is as follows:

$$T(I) = 1.16 + 281.6783e^{-I} + 19.992 \cdot Ie^{-I}$$

Fig. 4.24 shows the results of evaluating the curve fits for the polynomial model using *H(I)*. We can observe that this expression complies with the percentage error recommended by the existing standards and the recommendations given in [21,22].

The evaluation results for the dynamic performance of the model under constant and dynamic fault currents are shown in Figs. 4.25 and 4.26. The proposed relay model is found to comply with the standardized operation criteria. The proposed model is defined by means of the function *A(I)*, which results in a curve that is very similar to the very inverse curve of the standard in [8] (Fig. 4.25). As they have the same time curve, the integration process for both models results in the same operating times for constant and variable currents. Fig. 4.26 shows the results for a conventional relay, while Fig. 4.27 shows

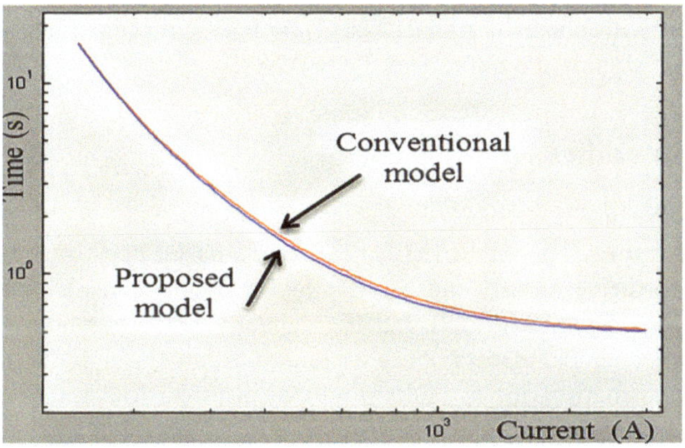

FIG. 4.25 Behavior of conventional and proposed OCR models.

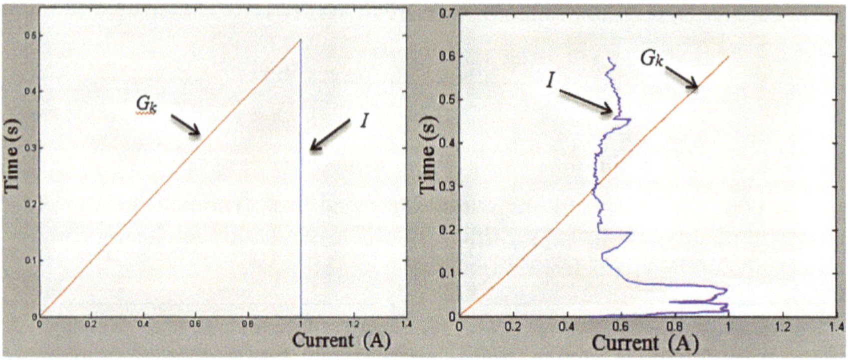

FIG. 4.26 Dynamic behavior of the conventional model.

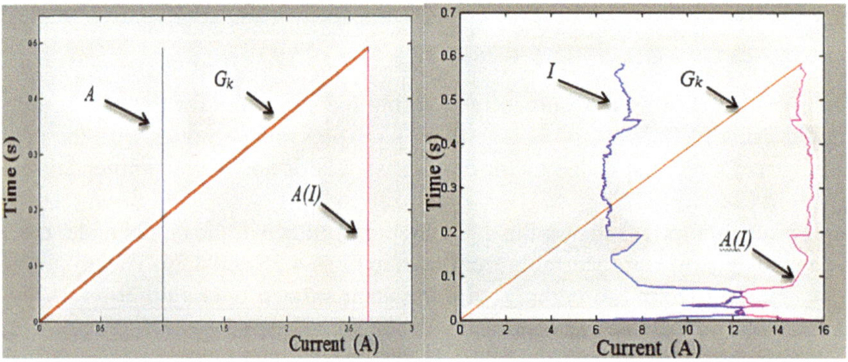

FIG. 4.27 Dynamic behavior of the proposed model.

the results for the proposed relay. It can be confirmed that the application of the proposed model will have the same functional characteristics as conventional relays, as shown in this base case. For both relays, the operation time of the integrator output with constant signals was $T_{op}=0.49$ s, and for integrator output with dynamic signals the value was $T_{op}=0.58$ s.

4.6.2 Case studies

Case I *Transformer protection.*

Damage curve. The ANSI/IEEE [23] standards set performance limits for power transformers. Initially, these operating limits only considered the thermal effects of transformer overload, but the capacity limit was later changed to include the mechanical effects of higher fault currents through the transformer. Energy transformation due to faults produces physical forces that cause compression, insulation wear, and friction-induced displacement in the coil. These effects are cumulative and should be considered over the life of the transformer.

TABLE 4.8 Categories of power transformers.

Category	Single-phase (kVA)	Three-phase (kVA)
I	5–500	15–500
II	501–1667	501–5000
III	1668–10,000	5001–30,000
IV	Above 10,000	Above 30,000

Table 4.8 presents four categories [24] of liquid-immersed power transformers, based on the transformer nameplate data.

The fault damage curve across the transformer is limited to 2 s. To recognize the cumulative nature of this damage, the standard complements the limit curve through failure capacity to reflect mechanical damage. The calculation of the transformer fault capacity limit curve is determined by I^2t and is based on the actual impedance of the transformer. For Category II transformers, the mechanical effect is considered for fault currents greater than 70% of the maximum short-circuit current, while for Category III and IV transformers, damage from currents greater than 50% of the maximum short-circuit current is considered. I^2t is calculated at the maximum short-circuit current for a time of 2 s.

After determining the limit curve according to the fault capacity through a transformer, the time-overcurrent characteristic of the relay is selected for coordination. In distribution transformer applications where few feeders are connected to a low-voltage bus, the feeder relay acts as the first line of defense. The IEEE C37.91 standard [25] recommends adjusting the inverse time-overcurrent characteristic of the feeder relay to achieve coordination with the capacitance limit curve of the fault across the transformer.

4.6.2.1 Transformer protection with standard curves

Many substations with Category III transformers are found in electrical power systems, as shown in Fig. 4.28 (reproduced from the IEEE Transformer Protection Standard [25]). The complement curve of the fault limit across the transformer corresponds to an impedance of 10% and a maximum three-phase fault current of 5540 A with reference to the secondary side. The curves of the phase protection devices on the primary side correspond to OCRs with a very inverse characteristic, and their settings are $I_{pickup}=8$ and $TDS=3$ for current transformers (CTs) with a 20:1 ratio. The phase relay on the secondary side is very inverse, and its settings are $I_{pickup}=6$ and $TDS=3$, with a 60:1 TC ratio.

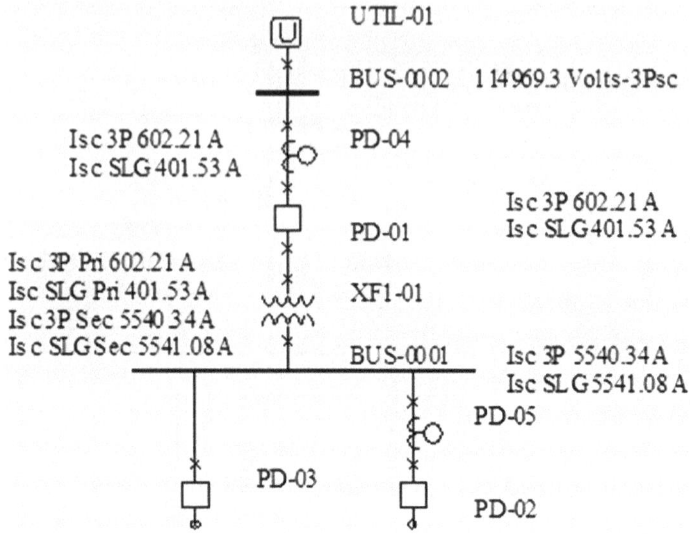

FIG. 4.28 Substation with Category III transformer.

Coordination with standard characteristic time curves is illustrated in Fig. 4.29. The *CTI* between the primary and backup protections is determined, resulting in a relatively large time value for low-value faults and overload protection.

4.6.2.2 Determination of the parameters for the proposed nonconventional curves

Defining a nonconventional curve relay makes it possible to improve the protection of a XF1-01 Category III transformer. The original PD-04 relay curve is replaced by the new PD-04ne time curve. The estimation of the parameters is performed to determine the parameters of the exponential type of function with $A(I)$ by applying nonlinear regression (Fig. 4.30).

The values of the coefficients of the new exponential type of characteristic function of the PD-04ne relay are determined using a nonlinear regression program corresponding to the graph in Fig. 4.30, as follows:

$$T(I) = \frac{3.08 * e^{-I/25.332}}{I^{1.2} - 1} + 0.27 \tag{4.11}$$

Results. Coordination of a Category III transformer damage curve with conventional and nonconventional characteristic relays is illustrated in Fig. 4.31. An improvement in the protection with a reduced time, and a decrease in mechanical

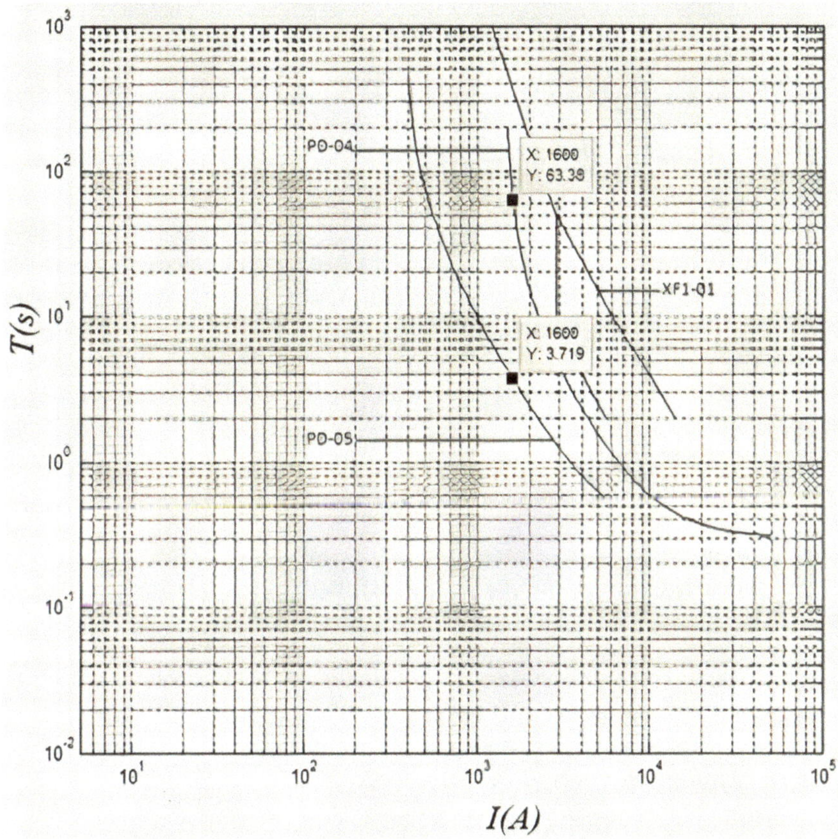

FIG. 4.29 Coordination of a Category III power transformer with primary and secondary side overcurrent protection.

stress and the allowable thermal effect in the transformer are observed with the nonconventional characteristic PD-04ne of the OCR when the fault is present.

Improvement in the operating time of the relay and reduction of mechanical and thermal stresses in the primary equipment.

Table 4.9 presents the operating times for the nonconventional OCR and the effort that is reduced in the power transformer. The time values t_R are greater before overload currents and at low fault currents, while there is a reduction in t_R at high values of the fault current. A method is presented later to quantify the reduction in the values left by a factor $\Delta I^2 t$ that corresponds to the mechanical and thermal stress in the primary equipment, under conditions of high magnitude of the fault current and long duration of the fault. In the values of the factor $\Delta I^2 - t$, there is a greater reduction of effort in high fault current and in low fault current; in the mean fault values is where there is less reduction of the factor $\Delta I^2 - t$.

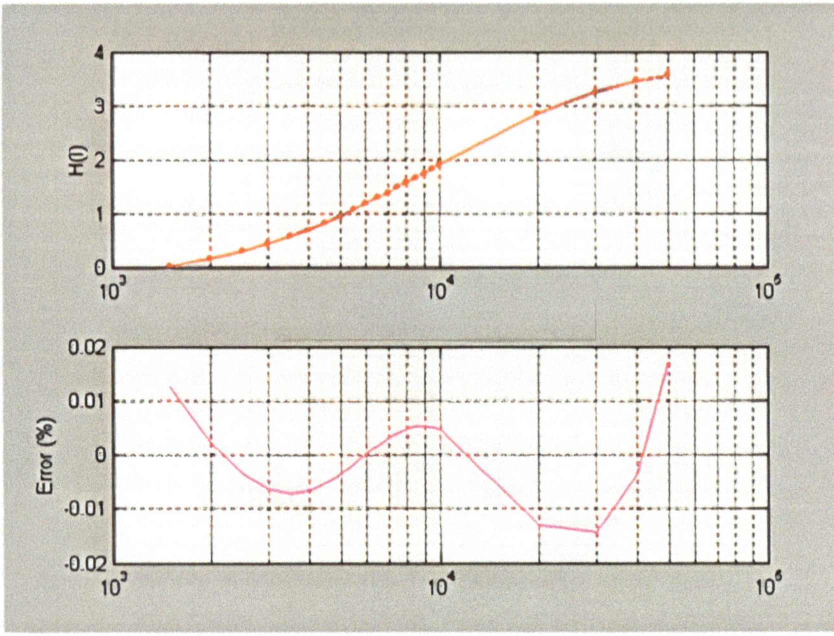

FIG. 4.30 Setting curve for a nonconventional OCR for protection of a Category III power transformer.

Case II *Transformer-cable-load main feeder.*

Industrial power systems contain arrangements of primary elements (such as a transformer-cable-motor arrangement) that use overcurrent protection devices. When coordinating the protections of this "supply-consumption" arrangement, the shape of the curve must be modified to reduce the *CTI*, reduce the protection devices, and improve the coordination with the damage curve of the protected elements, as discussed in [25]. Fig. 4.32 shows an industrial power consumption system with this problem and its characteristic protection curves. In this case, there are large time margins for the coordination of the conventional curves and a need to use two protection devices to achieve a fit with the motor operating curve.

4.6.2.3 Protection coordination with standard curves

Next, a study of short-circuit and protection coordination is presented using protection devices with CCs. Fig. 4.33 shows the coordination graph for an industrial electrical power network. There are two options for improving the protection setting: the first is to replace the PD-05 relay (MTR relay stall time) [26] with a flexible curve relay, and the second is to adapt the protection to the XF2-01 damage curve of the Category III transformer by modifying the curves of the PD-02 and PD-03 relays for the new NCC. The aim is to reduce tripping

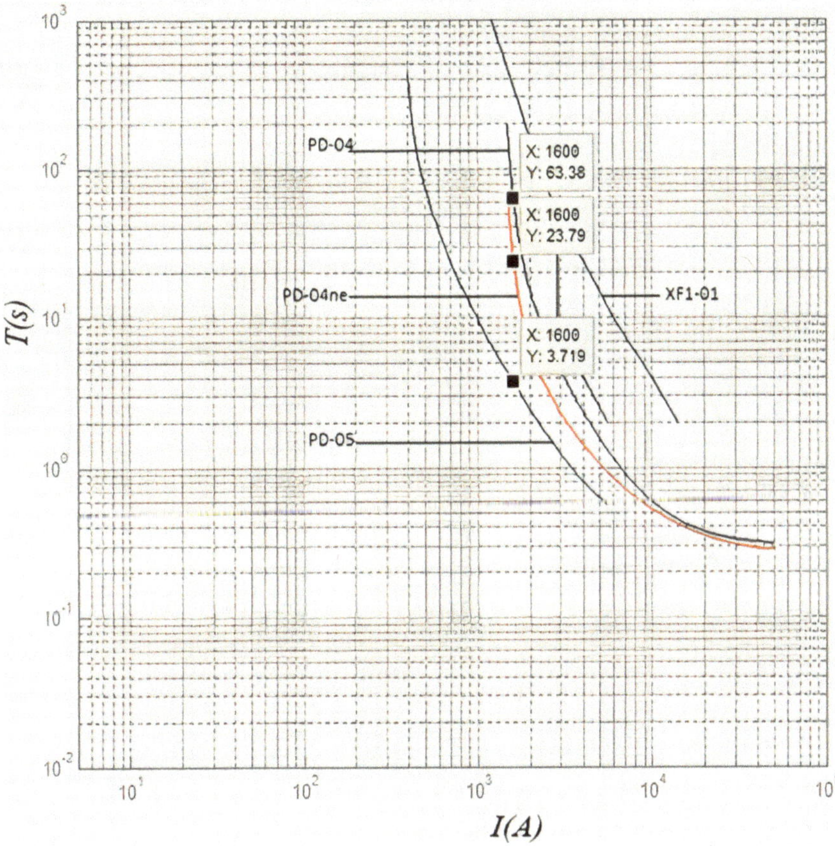

FIG. 4.31 Time comparison of the overcurrent protection characteristics of a Category III transformer.

times and to reduce the mechanical stress and thermal damage to the power equipment of the electrical system, including the motor, transformer, and cable.

4.6.2.4 Determination of the new NCCs

Fig. 4.34 shows the curve fitting for transformer protection, including the proposed time curve, the curve of the nonconventional function defined by nonlinear regression with its coefficients, and the percentage error between the two curves, which is below 3%. The coefficients of the new exponential characteristic of the time curve of the PD-5nc relay are as follows:

Coefficients: $A=3.1393$, $B=0.8027$, $c=3.7976$, $N=0.3704$, $\mathrm{TD}=0.8$

$$T(I)=\frac{TD*A*e^{-I/c}}{I^{N}-1}+TD*B \tag{4.12}$$

TABLE 4.9 Coordination results for standard (PD-05 and PD-04) and nonconventional (PD-04ne) relays with a Category III transformer.

Fault I (kA)	PD-05 t (s)	PD-05 vs PD-04ne CTI (s)	PD-04 t (s)	PD-04ne t (s)	t_R (s)	Reduced effort ΔI^2-t (A^2-s)
1.50	4.2220	67.258	195.968	71.480	124.488	**280.09**
2.00	2.4436	4.0198	15.5100	6.4634	9.0466	**36.19**
2.50	1.6541	1.7346	7.2587	3.3887	3.8700	**24.19**
3.00	1.2338	1.0741	4.4843	2.3079	2.1764	**19.59**
3.50	0.9832	0.7770	3.1442	1.7602	1.7600	**21.56**
4.00	0.8217	0.6092	2.3767	1.4309	0.9458	**15.13**
4.50	0.7114	0.5007	1.8904	1.2121	0.6783	**13.74**
5.00	0.6328	0.4239	1.5604	1.0567	0.5037	**12.59**
5.50	0.5748	0.3661	1.3254	0.9409	0.3845	**11.63**
6.00			1.1514	0.8517	0.2997	**10.79**
6.50			1.0189	0.7808	0.2381	**10.06**
7.00			0.9154	0.7234	0.1920	**9.41**
7.50			0.8330	0.6760	0.1570	**8.83**
8.00			0.7663	0.6362	0.1301	**8.33**
8.50			0.7114	0.6025	0.1089	**7.87**
9.00			0.6658	0.5735	0.0923	**7.48**
9.50			0.6274	0.5484	0.0790	**7.13**
10.0			0.5948	0.5264	0.0684	**6.84**
20.0			0.3725	0.3504	0.0221	**8.84**
30.0			0.3325	0.3072	0.0253	**22.77**
40.0			0.3180	0.2897	0.0283	**45.28**
50.0			0.3116	0.2814	0.0302	**75.50**

4.6.2.5 Coordination results for protection devices with NCCs

The nonlinear regression program of the NCC that improves coordination times is carried out. Fig. 4.35 shows the coordination graph, where the green curves of proposed relays PD-05nc, PD-02nc, and PD-03nc are of the exponential type obtained.

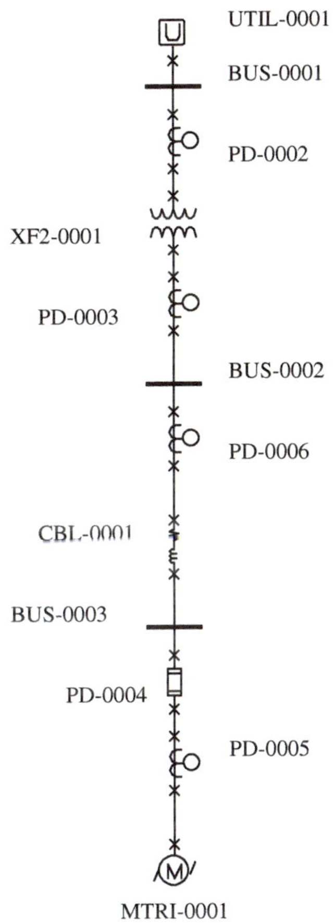

FIG. 4.32 Industrial power system, 13,800/4160 V.

In the results of these two overcurrent protection improvement opportunities with the proposed curve model, the first was able to cover protection against overload and low short-circuit values—being able to adjust close to the damage curve of the motor. The other benefit of the flexible curve is that it can achieve adequate protection in the fault current damage curve through the Category III power transformer, especially in the part corresponding to the effect of mechanical stress.

The proposed OCR model based on an *NCC* for PD-05nc motor protection is as follows:

$$J(I) = Ae^{-I/C} \tag{4.13}$$

where $A = 12$ and $C = 0.4$.

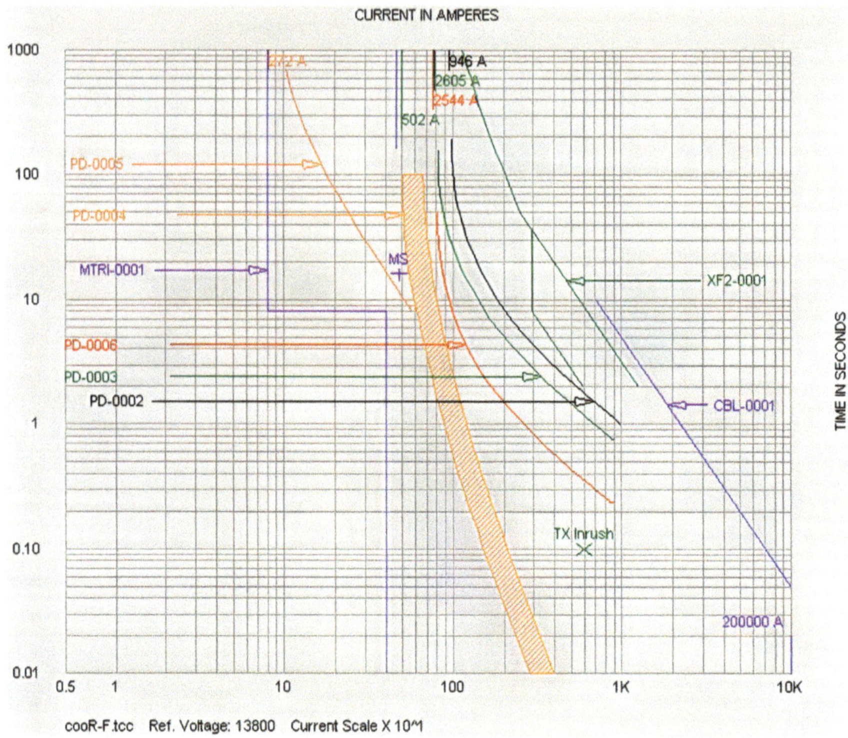

FIG. 4.33 Coordination graph for protection devices with CCs for the industrial electrical system shown in Fig. 4.32.

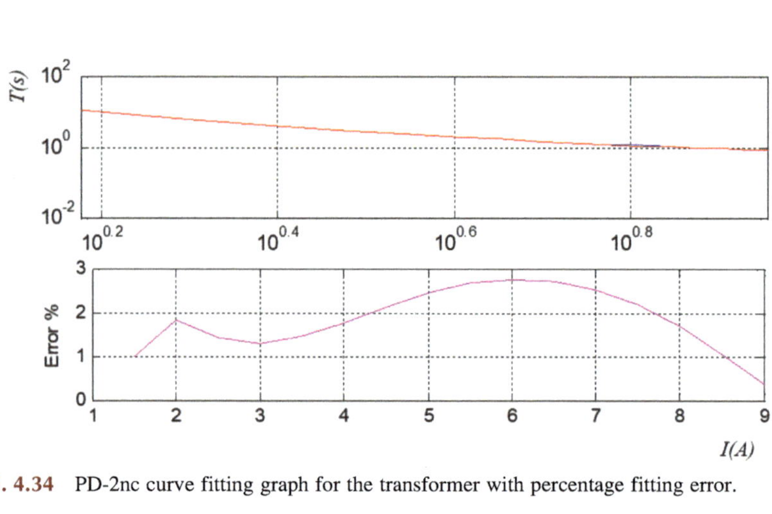

FIG. 4.34 PD-2nc curve fitting graph for the transformer with percentage fitting error.

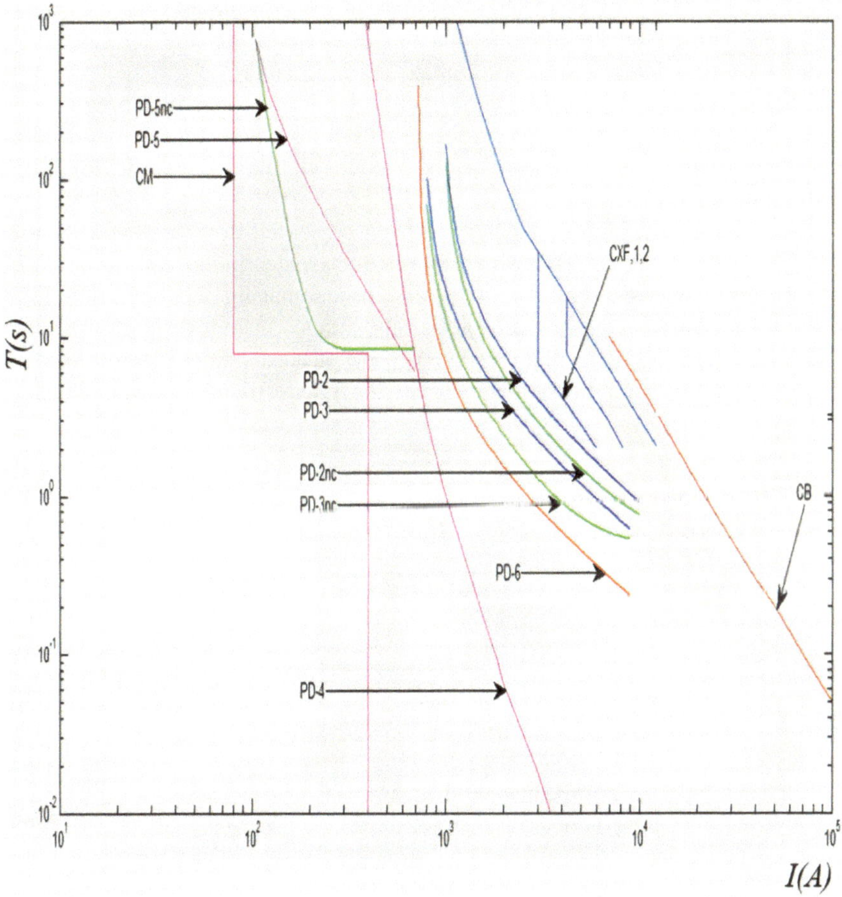

FIG. 4.35 Coordination with the proposed flexible curves for the PD-5, PD-2, and PD-3 relays.

The proposed OCR model based on NCC for protection of the PD-02nc and PD-03nc relays is as follows:

$$J(I) = Ae^{-IC} \tag{4.14}$$

where $A=2.5$ and $C=3.8$ for PD-02nc, and $A=14.5$, $C=6.5$ for PD-03nc.

4.7 Summary

The use of NCCs improves the performance and coordination of OCRs, as it reduces the relay operation time and the number of coordination violations. The design of the proposed time curves yields an increase in flexibility by increasing the number of degrees of freedom. In this way, a time curve can be designed for each specific coordination problem, thereby achieving better coordination. The number of chromosomes in the coordination algorithms is increased, but the results indicate that this does not affect the solution time, as the algorithm finds fewer coordination commitments.

The coordination of positive and negative sequence relays is improved with the use of NCCs and gives zero violations in cases where CCs present violations. Lower operating times are also obtained for all relays due to the flexibility of the time curve.

An alternative design for OCRs has been presented in which the TDS is a function of the relay current, which allows for greater functional variation in the operation of the relay, by increasing or decreasing the operation time of the relay. This design has substantial advantages in industrial applications where the time curve of the relay is subject to various restrictions in terms of coordination with fuses and the damage curves for transformers or power cables.

References

[1] M. Ezzeddine, R. Kaczmarek, A novel method for optimal coordination of directional overcurrent relays considering their available discrete settings and several operation characteristics, Electr. Power Syst. Res. 81 (7) (2011) 1475–1481.

[2] F.B. Bottura, W.M.S. Bernardes, M. Oleskovicz, E.N. Asada, S.A. Souza, M.J. Ramos, Coordination of directional overcurrent relays in meshed power systems using hybrid genetic algorithm optimization, in: Proceedings of the 12th IET International Conference on Developments in Power System Protection, 2013, pp. 1–6.

[3] F.B. Bottura, M. Oleskovicz, D.V. Coury, S. de Souza, M. Ramos, Hybrid optimization algorithm for directional overcurrent relay coordination, in: Proceedings of the IEEE/PES General Meeting, Conference & Exposition, 2014, pp. 1–5.

[4] A.A. Kida, L.A. Gallego, A high-performance hybrid algorithm to solve the optimal coordination of overcurrent relays in radial distribution networks considering several curve shapes, Electr. Power Syst. Res. 0378-7796, 140 (2016) 464–472.

[5] Z. Moravej, F. Adelnia, F. Abbasi, Optimal coordination of directional overcurrent relays using NSGA-II, Electr. Power Syst. Res. 119 (2015) 228–236.

[6] R. Corrêa, G. Cardoso, O.C.B. de Araújo, L. Mariotto, Online coordination of directional overcurrent relays using binary integer programming, Electr. Power Syst. Res. 0378-7796, 127 (2015) 118–125.

[7] O. Arreola, A. Conde, L.A. Trujillo, Overcurrent relay with unconventional curves and its application in industrial power systems, Electr. Power Syst. Res. 110 (2014) 113–121.

[8] IEEE Std C37.112-1996, IEEE Standard Inverse-Time Characteristic Equations for Overcurrent Relays, September 1996.

[9] A.R. Mehrabian, C. Lucas, A novel numerical optimization algorithm inspired from weed colonization, Eco. Inform. 1 (4) (July 2006) 355–366.

[10] IEEE Std 242-2001, [The Buff Book]: IEEE Recommended Practice for Protection and Coordination of Industrial and Commercial Power Systems, 2001.

[11] C.A. Castillo, A. Conde, M.Y. Shih, Improvement of non-standardized directional overcurrent relay coordination by invasive weed optimization, Electr. Power Syst. Res. 157 (2018) 48–58.

[12] Power Systems Test Case Archive, Univ. Washington. [Online]. Available: https://www2.ee.washington.edu/research/pstca/).

[13] H. Saadat, Power System Analysis, McGraw-Hill, 1999.
[14] G. Benmouyal, Some aspects of the digital implementation of protection time functions, IEEE Trans. Power Deliv. 5 (4) (1990) 1705–1713.
[15] M. Titarenko, I. Noskov-Dukelsky, n.d. Protective Relaying in Electric Power Systems, Second, Printing, Peace Publishers, Moscow, p. 1963.
[16] G. Benmouyal, M. Meisinger, J. Burnworth, IEEE standard inverse-time characteristic equations for overcurrent relays, IEEE Trans. Power Deliv. 14 (3) (1999) 868–872.
[17] IEC Standard 255-4, Single Input Energizing Measuring Relays with Dependent Specified Time, first ed., IEC Publication 255-4, 1976.
[18] H.A. Darwish, A.I. Taalab, H. Assal, A novel overcurrent relay with universal characteristics, in: Transmission and Distribution Conference and Exposition, IEEE/PES, vol. 1, 2001, pp. 53–58.
[19] H.A. Darwish, M.A. Rahman, A.I. Taalab, H. Shaaban, Digital model of overcurrent relay characteristics, in: Industry Applications Conference. Thirtieth IAS Annual Meeting, IAS '95, vol. 2, 1995, pp. 1187–1192.
[20] M.S. Sachdev, T.S. Sidhu, B. Chattopadhyay, B. Takuldar, G. Mc-Donald, R. Chan, Design and evaluation of an adaptive protection system for a distribution network, in: *Cigré* Paper 34–202, París, 1995.
[21] G.E. Radke, A metod for calculating time-overcurrent relay settings by digital computer, IEEE Trans. Power Syst. 82 (Special Supplement) (1963) 189–205.
[22] S.S. Begian, R.O. Bigelow, D.M. Lyle, G.E. Radke, W.F. Reinke, R.F. Silva, J.R. Truitt, H.Y. Tsien, E.L. Worthington, A computer approach to setting overcurrent relays in a network, in: Proceedings of the IEEE 1967 PICA Conference, May 1967, pp. 447–457.
[23] IEEE Standard General requirements for Liquid-Immersed Distribution, Power, and Regulating Transformers, IEEE Standard C57.12.00-2000.
[24] IEEE Guide for Liquid-Immersed Transformer Through-Fault Current Duration, IEEE Std. C57.109-1993.
[25] IEEE Guide for Protecting Power Transformers, IEEE Std C37.91™-2008.
[26] S. Khan, Industrial Power Systems, CRC Press, Taylor & Francis Group, Florida, 2007.

Chapter 5

Coordination of nonconventional DOCR based on the time interval criterion

5.1 Introduction

As part of a protection scheme, OCRs provide a protection zone that is more sensitive to the operating conditions of the power system [1] than one based on distance or differential relays. They can be used for primary and backup protection in electrical networks with practically all voltage levels, from generators to industrial networks. Relay adjustment is a difficult process owing to the functional dependence between OCRs [2]; this is the only protection principle that shows this behavior. The adjustment process consists of establishing the boundary conditions of both the load current (I_{load}) and fault currents for the pickup current (I_{pickup}) and its coordination [3]. The complexity of coordination is proportional to that of the electrical network, and for this reason, it is necessary to use optimization algorithms [4] to carry out coordination of highly interconnected networks. This process has been formulated as an optimization problem with the aim of minimizing the operation times of the relays [5]. Linear optimization methods have been proposed, such as algorithmic solutions [6], which has given rise to a new area of research. Various optimization methods based on metaheuristics and deterministic approaches have been developed to find solutions to the task of relay coordination.

Deterministic methods have been used to solve the problem of DOCR coordination. The predictive character of the algorithms depends on the identification of their inputs [7,8]. Heuristic and metaheuristic methods have been demonstrated to be capable of overcoming most of the limitations of previous methods, as they do not require initial conditions and use a population technique that provides a global search [9,10]. Hybrid algorithms that combine different techniques for similar and different categories have been used recently to obtain the benefits of both approaches [11,12].

Some potential applications have been proposed for the protection and monitoring of distribution systems using smart meters. The expectation is that these

Overcurrent Relay Advances for Modern Electricity Networks. https://doi.org/10.1016/B978-0-323-91777-3.00005-2

will provide the necessary justification for promoting future smart metering projects [13]. Flexible time curves have been formulated to improve the performance of the protection functions, in which each specific coordination problem determines the parameters of the time curve [14]. The problem of loss of coordination caused by higher levels of fault current, reduced ranges of operation of protection devices, protection blinding, and false trips has been addressed [15]. Robust and adaptive coordination approaches have been used to achieve an acceptable level of power quality and selective operation of DOCRs when the network topology changes and to find the optimal configuration of DOCRs [16]. An optimal coordination scheme has been developed using nonstandard tripping characteristics for OCRs in a power grid connected to a photovoltaic system [17].

The operation of a relay is determined by the time curve, which is weighted based on the pickup current (I_{pickup}) and the fault current (I_{sc}) to determine the time response. However, in conventional coordination, the relay time is stabilized using the following coordination criterion:

$$t_b \geq t_p + CTI \tag{5.1}$$

where t_p is the operating time of the primary relay, t_b is the operating time of the backup relay, and *CTI* is the coordination time interval. It can be observed that the operating time of one relay depends on the time of its neighbors. This makes the problem more complex in interconnected networks, where the values of I_{pickup} and I_{sc} for the primary relay may be higher than for the backup relay (Chapter 1). However, for the protection of lines, the operating times for the protection devices should be determined by fault release criteria defined by each electric company, based on indicators of the quality of energy supplied and the time of voltage sags.

An alternative to forcing relay operating times to lie within the desired ranges is to set time constraints via the coordination algorithm. In some specific applications, these time restrictions have been used to obtain results that do not violate the specific criteria for each problem. These new restrictions may cause violations due to the reduction in the coordination solution space. In other words, when the conventional formulation for relay coordination is followed, there is a greater possibility of violating the constraints on relay coordination.

In this chapter, a new formulation for relay coordination is presented, *in which Eq. (5.1) is not used*, each relay is considered to be independent, and each time curve is set to meet specific time intervals. An adjustment process is applied to find the NCC that results in compliance with the specified time intervals. In this approach, the adjustment of each relay is independent, since it does not depend on the times of the neighboring relays. The proposed relay coordination scheme is presented as an optimization problem, the new coordination method is presented based on metaheuristic optimization algorithms, and greater simplification in the coordination process is presented obtaining shorter operating times and fewer or null violations.

5.2 Novel method of OCR coordination via time intervals

A new method of DOCR coordination is proposed that is unlike conventional coordination, which is based on the coordination criterion, where the operating time of the backup relay is a function of that of the primary relay. In the proposed coordination method, each relay is set individually with predefined time intervals, without relying on the setting of the neighboring relay. For this reason, fault currents are considered only for the adjustment of each relay rather than for coordination purposes.

The operating times are bounded by upper and lower intervals to improve the operational safety. In the proposed method, the time curve is set for each time interval to guarantee that coordination is achieved between the relays at the different coordination locations (Fig. 5.1). The result is independent relay setting, as the proposed formulation only needs to satisfy the requirements for the operating times for the specified intervals. The time intervals shown for relay R_1 are as follows: first interval to obtain the operation time for close-in faults, fault F_1 are 0.1–0.2; second time interval for the bus-end fault, fault F_2 are 0.4–0.5; and the third time interval for remote line-end fault, F_4 fault are 0.7–0.8. Likewise, the setting of relay R_2 is obtained in the first interval with fault in F_3, second interval with fault F_1, and its third interval with fault in F_2.

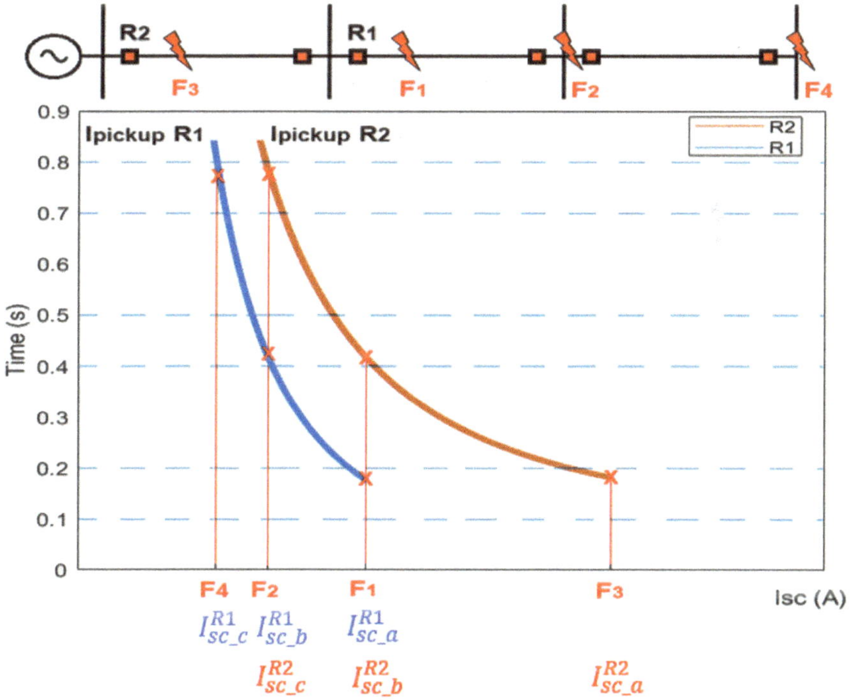

FIG. 5.1 Time intervals calculated using an NCC.

Unlike conventional coordination where only faults F_1 and F_2 are required for coordination between R_1 and R_2, in the proposed method an additional fault location is required for each relay: faults F_3, F_1, and F_2 are required for relay R_1, and F_1, F_2, and F_4 are required for relay R_2. These values are obtained using a traditional calculation of fault currents because only a known fault value is included; this criterion does not depart from the calculation, so no further information is required.

Each relay has different coordination currents to achieve compliance in each interval. For the different conditions of each relay, the algorithm calculates an NCC that guarantees compliance with the times of each interval. This simplifies the search carried out by the coordination algorithm, as only the parameters of the time curve for each relay must be found to comply with the times established for each interval, thus avoiding functional dependence between relays. The proposed approach is therefore less algorithmically demanding, since it is not necessary to find a solution that combines the operating times of all functionally dependent relays and their coordination restrictions. An approach based on a minimization optimization problem searches for operating times that are close to the lower limit of each time interval, and hence a time curve for each relay that satisfies the times required for the fault currents is obtained. Although the time required by the primary relay to react to a fault at the start of the line is not determined in conventional coordination, this is included in the proposed coordination method to guarantee coordination between interval one of the primary relays and interval two of the backup relays.

In the conventional method of coordination, if any relay is readjusted, it is necessary to check the coordination in the remaining relays, due to the functional dependency between relays. In contrast, in the proposed coordination method, only the relays affected by the load current or changes in the fault current will have to be reset. This represents a great advantage in online coordination systems and when updating the coordination.

5.2.1 Proposed optimization problem

5.2.1.1 Objective function

The coordination method minimizes the operation time of each relay to achieve compliance with the three intervals, as expressed in Eq. (5.2). The proposed method results in an objective function without weighting, which avoids the need for Pareto or multiobjective formulations.

$$OF = min\left\{\sum_{\mu=1}^{N}\left(t(I_{sc\,a})_{\mu} + t(I_{sc\,b})_{\mu} + t(I_{sc\,c})_{\mu}\right) + NV\right\} \tag{5.2}$$

where the function $t(I)$ is the time curve of the DOCRs, $t(I_{sc\ a})$ is the evaluation time for a close-in fault, $t(I_{sc\ b})$ is the time for a bus-end fault, and $t(I_{sc\ c})$ is the time for a fault in the remote line-end, for each of the N relays. NV is the number of violations recorded in the calculation of each of the times to improve the algorithm search, thereby penalizing violations.

5.2.1.2 *Time function*

The relay time is defined using *NCCs*, taking the time curve equation of the IEEE standard [18] as a reference, as shown in Eq. (5.3):

$$t(I) = \left[\frac{\zeta^1}{\left(\frac{I_{sc}}{I_{load} \cdot \zeta^5} \right)^{\zeta^3} - 1} + \zeta^2 \right] \cdot \zeta^4 \tag{5.3}$$

$$\zeta = \left(\zeta^1, \zeta^2, \zeta^3, \zeta^4, \zeta^5\right)^T, \quad \zeta \in \mathbb{R}$$

where I_{sc} is the short-circuit current seen by each relay within each interval, I_{load} is determined from an $n-1$ contingency analysis, and ζ is the vector of coefficients that allows NCCs to be obtained. The time function in Eq. (5.3) considers each coefficient to be optimized to find a nonconventional formulation of the OCR. The resulting design is individual for each relay, which increases the possibility of complying with the time interval restrictions of the relays and provides a greater flexibility for the time curves.

5.2.1.3 *Restrictions*

The OF is achieved if the restrictions of the relay parameters and their limitations are satisfied. The time curve is evaluated for each value of the fault current to comply with the predefined time intervals, as shown in Fig. 5.1. Thus the same time curve must satisfy the constraints on the three regions within a minimum time, as expressed in Eqs. (5.4)–(5.6):

$$t\left(I_{sc\ a}\right)^{min} \leq t\left(I_{sc\ a}\right) \leq t\left(I_{sc\ a}\right)^{max} \tag{5.4}$$

$$t\left(I_{sc\ b}\right)^{min} \leq t\left(I_{sc\ b}\right) \leq t\left(I_{sc\ b}\right)^{max} \tag{5.5}$$

$$t\left(I_{sc\ c}\right)^{min} \leq t\left(I_{sc\ c}\right) \leq t\left(I_{sc\ c}\right)^{max} \tag{5.6}$$

The time intervals are defined following the common criterion for the operation of relays set out by electric companies, with a *CTI* of 0.2 between the specified time ranges. These intervals are specified according to the operative conditions of the electrical network, for example, to control voltage sags or to meet utility criteria. The use of different time ranges is illustrated in the results and discussions section. The solution must ensure that the shape of the time curve is not lost, since coordination between different overcurrent devices is required, such as other OCRs, fuses, or reclosers. In view of this, the parameters ζ of the NCC must be bounded as in Eq. (5.3) (see Chapter 4).

It is necessary to highlight the difference between the proposed DOCR coordination method and the dynamic behavior of a relay. The former is a setting process that aims to determine the parameters that allow for a coordinated sequence of operation between relays. This may be achieved manually by protection engineers or using deterministic, metaheuristic, or hybrid optimization methods. The latter is a dynamic process where the time curve is mathematically integrated to obtain the relay operation time in a disturbance event. Thus the operating time of the OCR is determined by the fault current dynamics: different fault current dynamics will result in different operating times.

Hence, the operation time of an OCR is obtained only when the integration process is completed. It is widely known that the operating times obtained from the time curve are determined based on the assumptions of maximum and constant fault current. However, in a dynamic scenario, it is often the case that the operating times are longer, since fault currents are smaller and variable. This is normal, typical behavior due to the overcurrent principle [18]. In the proposed coordination method, it is expected that the operation times of the relays in dynamic conditions will be outside these intervals, since they are not valid for a dynamic state but only for coordination purposes.

The proposed method of DOCR coordination by means of time intervals is implemented using metaheuristic optimization methods, and the formulation is based on the expressions given before.

5.3 Metaheuristic algorithms for the interval method

Metaheuristic algorithms use a population-based approach to solve optimization problems. The population is formed in a similar way as in conventional methods, except that in the proposed method NCCs are used to achieve the flexibility of the time curve necessary to adjust to the time intervals. The population size can remain constant or can be modified during the iterations to improve the convergence rate of the algorithm. The length of the chromosome or search agent depends on the number of relays in the system and the number of settings.

The stages of the proposed method are illustrated in Fig. 5.2. The data required for coordination are obtained from an analysis of three-phase and two-phase faults at each node; in meshed networks, the contribution of the short-circuit current is considered with the line-end open. A power flow analysis is performed to obtain the maximum load current at each relay. Based on this information, the sensitivity of the backup relays is determined for faults at the remote node of the longest adjacent line.

The matrix of individuals required for the optimization algorithm has a population size of (NP, $D*N$), where NP represents the population size, D is the number of time curve coefficients, and N is the number of relays. The equality constraint is the time curve expression for OCRs in Eq. (5.3), which is determined using the bounded variables. The relay time is evaluated at three fault locations to comply with the three time intervals: close-in relay (first interval), bus-end (second interval), and the remote line-end relay on the longest adjacent line (third interval). The inequality constraints on the three time intervals for

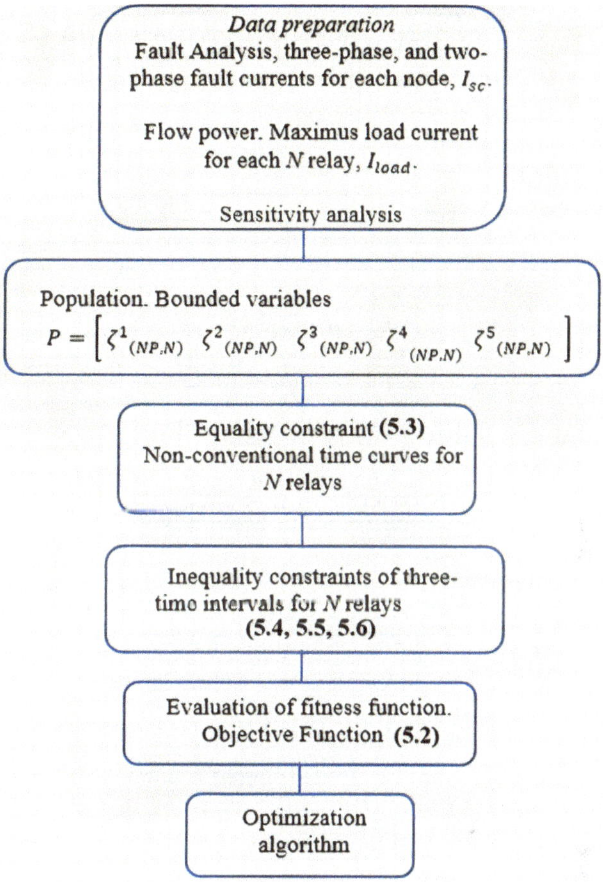

FIG. 5.2 Data preparation for relay coordination.

each of N relays are determined by Eqs. (5.4)–(5.6). Minimization of the problem is achieved by the OF in Eq. (5.2). The lower limits of each interval are obtained to reduce the operating time of each relay. The outputs of the optimization algorithm are the relay coefficients ζ.

5.3.1 DE for proposed coordination method using time intervals

5.3.1.1 Initial population

A new random population is created in which all the genes of all the parameter vectors are initialized somewhere within the desired numerical range for the corresponding relay fit.

$$P = \left(\zeta^1, \zeta^2, \zeta^3, \zeta^4, \zeta^5\right)_{\min} + rand(1, D) * \left[\left(\zeta^1, \zeta^2, \zeta^3, \zeta^4, \zeta^5\right)_{\max} - \left(\zeta^1, \zeta^2, \zeta^3, \zeta^4, \zeta^5\right)_{\min}\right)\right] \tag{5.7}$$

5.3.1.2 Mutation

$$\vec{X}_{i,G+1} = \vec{X}_{r1,G} + F\left(\vec{X}_{r2,G} - \vec{X}_{r3,G}\right) \tag{5.8}$$

and

$$\vec{X}_{r1,G} = Best\,P\left(\zeta^1, \zeta^2, \zeta^3, \zeta^4, \zeta^5\right)$$
$$\vec{X}_{r2,G} = P_{old}\left[randperm(NP)\right]$$
$$\vec{X}_{r3,G} = P_{old}\left[randperm(NP)\right]$$

where $r_1, r_2, r_3 \in [1, 2\ldots, NP]$. The mutation factor F is a constant in the range $[0, 2]$. $\vec{X}_{i,G+1}$ is called the donor vector of generation $G+1$.

5.3.1.3 Crossover

$$U_{j,i,G} = \begin{cases} V_{j,i,G} & \text{if } \left(rand_{i,j} \le CR\right) \text{ or } \left(j = j_{rand}\right) \\ X_{j,i,G} & \text{if } \left(rand_{i,j} > CR\right) \text{ and } \left(j \ne j_{rand}\right) \end{cases} \tag{5.9}$$

$$V_{j,i,G} = P_{old}(rand(NP,\ D) < CR) + V_{j,i,G}(rand(NP,\ D) < CR)$$

where *rand i, j* [0,1] is a uniformly distributed random number. A randomly chosen index *jrand* $\in$ [1, 2, …, D] is used to ensure that the test vector $U_{i,G}$ obtains at least one component of the donor vector $V_{i,G}$.

5.3.1.4 Selection

$$\vec{X}_{i,G+1} = \begin{cases} \vec{U}_{i,G} & \text{if } f\left(\vec{U}_{i,G}\right) \le f\left(\vec{X}_{i,G}\right) \\ \vec{X}_{i,G} & \text{if } f\left(\vec{U}_{i,G}\right) > f\left(\vec{X}_{i,G}\right) \end{cases} \tag{5.10}$$

$$P(NP, D) = U_{i,G}(NP, D)\ \ Re\,place\ old\ vectorwith\ new\ one$$

where $f(X)$ is the fitness of the target vector and $f(U)$ is the fitness of the trial vector. If a lower or equal value for the fitness is obtained from the new trial vector, then the target vector will be replaced in the next generation; otherwise, the target vector is kept in the population.

There is a substantial difference between the conventional coordination method and the proposed approach, where the number and complexity of the constraints impact the results and computation times. In the case of the number of restrictions, while the conventional method uses the coordination pairs as shown in Eq. (5.1) to guarantee coordination, in addition to the number of restrictions there is a dependency between them, the algorithm must perform a more intense search looking for the fulfillment of them. In contrast, in the proposed method, the restrictions on each time curve are applied independently, where each degree of freedom of the curve represents a restriction; however, since the coordination criterion is not applied, the search process of the algorithm focuses on meeting the three time constraints for each relay. This results in solutions with shorter execution times than in conventional coordination methods.

5.3.2 Relay coordination verification based on time intervals

To verify the proposed coordination method, a compact system is used to enable a better visualization of results. The test system used to evaluate the proposed approach is the six-bus IEEE system in Fig. 5.3, with selected voltages of 138/20 kV, with 10 relays and 16 coordination pairs. The three-phase currents are calculated for close-in and line-end faults, and minimum phase-to-phase faults are calculated for the remote line-end. All faults are computed based on the open-end assumption. The fault current values are updated before any topological change. In addition, the nodal voltage is measured and used to calculate the fault currents.

The coordination pairs are presented in Table 5.1. The directionality of the relays causes two independent coordination pairs to be formed; this directionality is required in networks with bidirectional power, which is why coordination with nondirectional protection devices such as fuses and reclosers is not feasible. The fault current and I_{load} for the six-bus IEEE system are given in Table 5.2.

The results of the proposed DOCR coordination method are presented in Table 5.3, and a comparison with GWO and DE is carried out. DE algorithm

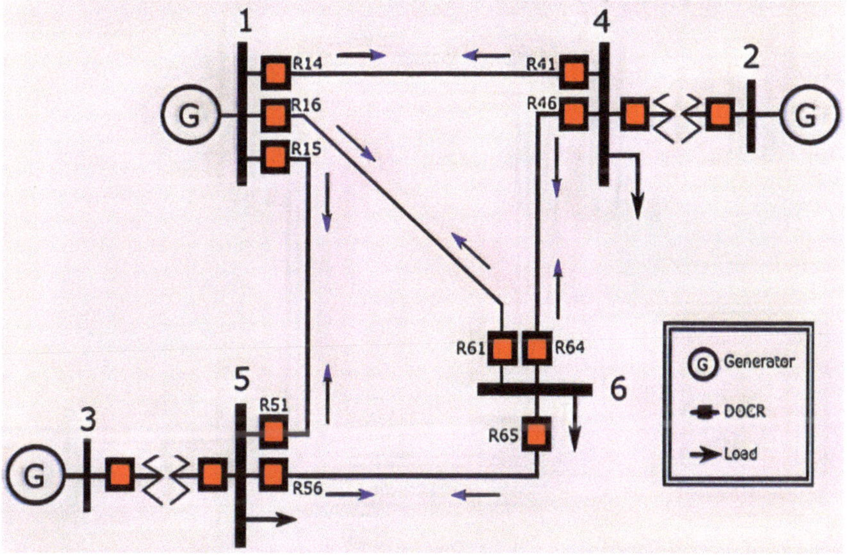

FIG. 5.3 Six-bus IEEE test system.

TABLE 5.1 Coordination pairs of relays in the six-bus IEEE test system.

Clockwise coordination		Counterclockwise coordination	
Primary	*Backup*	*Primary*	*Backup*
14	51	15	41
14	61	15	61
46	14	16	41
51	65	56	15
61	46	41	64
65	16	61	56
65	46	64	16
16	51	64	56

TABLE 5.2 Power flow and fault currents for conventional coordination of the six-bus IEEE system.

Relay		I_{sc}		I_{load}	
Primary	*Backup*	*Primary*	*Backup*	*Primary*	*Backup*
14	51	3950	1230	106.7	199.5
14	61	3950	1780	106.7	199.5
15	41	3776	910	199.5	106.7
15	61	3776	1780	199.5	199.5
16	41	4226	910	199.5	106.7
16	51	4226	1230	199.5	199.5
46	14	3428	1172	319.2	106.7
56	15	3245	1819	243.2	199.5
41	64	3469	1215	106.7	319.2
51	65	2610	1188	199.5	243.2
61	46	2716	1480	199.5	319.2
61	56	2716	1201	199.5	243.2
64	16	2162	1108	319.2	199.5
64	56	2162	1201	319.2	243.2
65	16	2586	1108	243.2	199.5
65	46	2586	1480	243.2	319.2

TABLE 5.3 Relay operating times obtained with the proposed coordination method for the six-bus IEEE system.

Relay	Primary Time (Close-in)		Backup Time			
			Bus-end		Remote line-end	
	GWO	DE	GWO	DE	GWO	DE
14	0.189	0.183	0.407	0.495	0.771	0.731
15	0.151	0.177	0.402	0.407	0.747	0.726
16	0.186	0.104	0.405	0.428	0.715	0.723
46	0.157	0.184	0.404	0.442	0.717	0.748
56	0.180	0.100	0.401	0.429	0.775	0.704
41	0.179	0.143	0.403	0.488	0.723	0.743
51	0.194	0.148	0.400	0.405	0.746	0.700
61	0.185	0.197	0.407	0.401	0.700	0.799
64	0.148	0.129	0.401	0.401	0.762	0.700
65	0.169	0.107	0.403	0.419	0.771	0.703
Num. Violations (NV)			GWO: 0		DE. 0	
Computation Time (s)			GWO: 2.23		DE: 3.15	

is shaded for enhancing the data identification. Although results show some variation, zero violations are obtained by both optimization methods. GWO is faster, but the results in terms of the operation times are slightly larger. The time profile results from the proposed coordination method are flat, meaning that the operation times between the primary and backup functions for different relays are very similar to each other (Fig. 5.4).

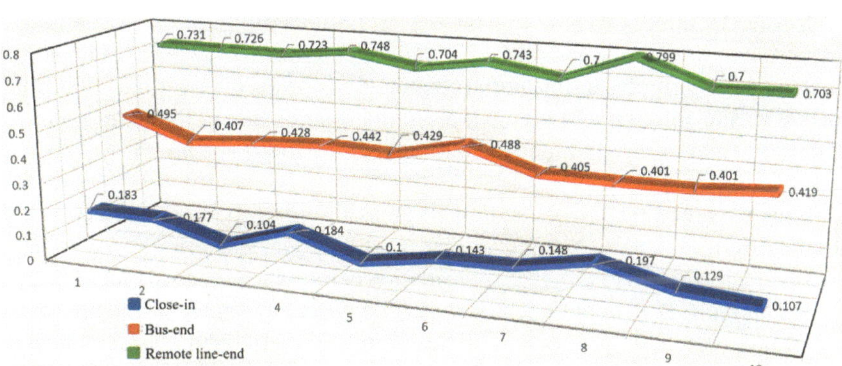

FIG. 5.4 Operation times for DOCRs obtained with the proposed coordination method.

TABLE 5.4 Power flow and fault currents used in the proposed coordination method.

Relay	Close-in	Bus-end	Remote line-end	I_{load}
14	3950	1172	867	106.7
15	3776	1819	1274	199.5
16	4226	1108	858	294.5
46	3428	1480	1080	319.2
56	3245	1201	939	243.2
41	3469	910	691	106.7
51	2610	1230	864	199.5
61	2716	1780	1240	294.5
64	2162	1215	960	319.2
65	2586	1188	928	243.2

The values of I_{load} and the fault current for each relay location for the proposed method are presented in Table 5.4. By comparing the results with those in Table 5.2, it can be observed that in general, the data are the same, but they are arranged differently because the coordination approach is different. In the proposed method based on time intervals, it is necessary to know the fault at the remote line-end. It is not necessary to modify the fault study, since this value is known since the fault calculation is performed in all the nodes of the electrical network; then only a previously calculated value is added. The results are shown for each direction of coordination to enable better visualization.

In this section, we show that in the proposed time interval coordination method, coordination of the relays can be obtained without using the coordination criterion. This offers substantial advantages in terms of both the algorithmic search for the solution and the operation times, since due to the independence of the adjustments, the obtained operation times are lower and have a flat profile, as shown in Fig. 5.4.

Figs. 5.5 and 5.6 show the results of clockwise and counterclockwise relay coordination. A table of the relays that must be coordinated is shown in each graph. Flags are used to indicate coordination compliance for each pair of relays for which the *CTI* is maintained between the operating times. In the proposed method, coordination is obtained by designing NCCs to meet the time interval constraints, without reference to the coordination criterion in Eq. (5.1). The main advantage of this is the independence in the adjustment of the relays.

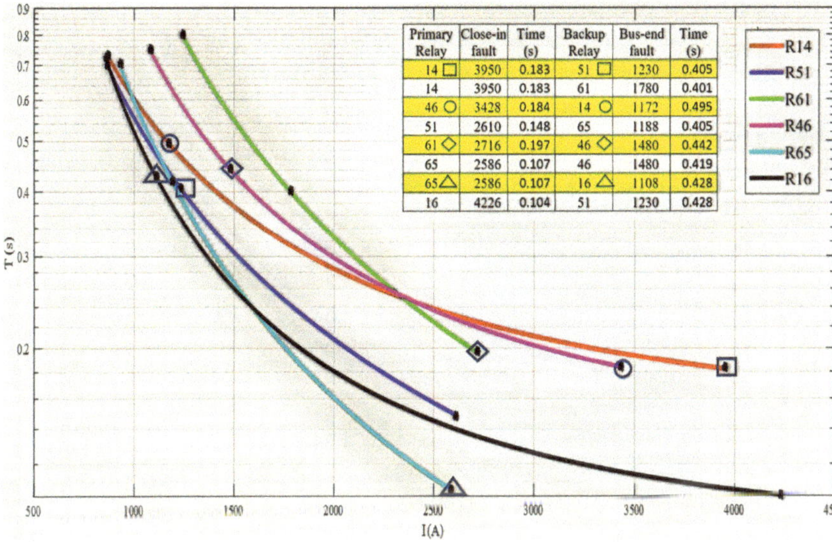

Primary Relay	Close-in fault	Time (s)	Backup Relay	Bus-end fault	Time (s)
14 □	3950	0.183	51 □	1230	0.405
14	3950	0.183	61	1780	0.401
46 ○	3428	0.184	14 ○	1172	0.495
51	2610	0.148	65	1188	0.405
61 ◇	2716	0.197	46 ◇	1480	0.442
65	2586	0.107	46	1480	0.419
65 △	2586	0.107	16 △	1108	0.428
16	4226	0.104	51	1230	0.428

FIG. 5.5 Clockwise relay coordination for the six-bus IEEE test system.

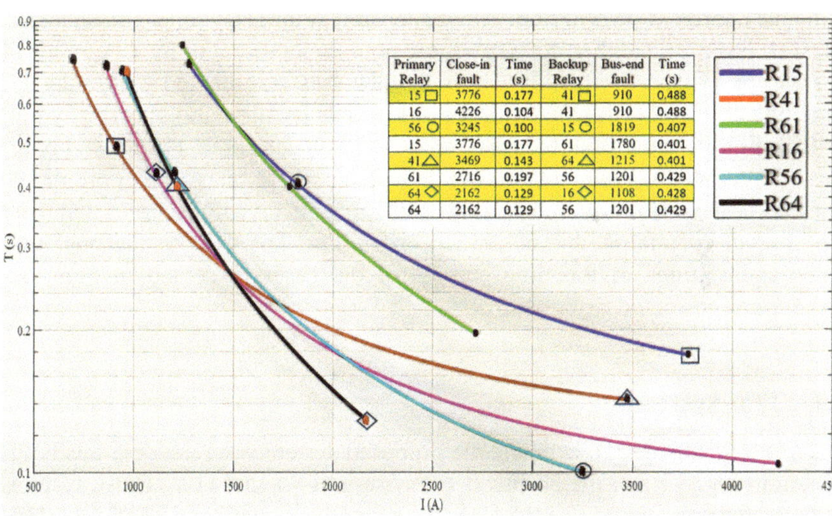

Primary Relay	Close-in fault	Time (s)	Backup Relay	Bus-end fault	Time (s)
15 □	3776	0.177	41 □	910	0.488
16	4226	0.104	41	910	0.488
56 ○	3245	0.100	15 ○	1819	0.407
15	3776	0.177	61	1780	0.401
41 △	3469	0.143	64 △	1215	0.401
61	2716	0.197	56	1201	0.429
64 ◇	2162	0.129	16 ◇	1108	0.428
64	2162	0.129	56	1201	0.429

FIG. 5.6 Counterclockwise relay coordination for the six-bus IEEE test system.

In this approach, a relay with a slow time resulting from the coordination process does not affect its neighboring relays, and the operating times can be adjusted for each interval. It is therefore expected that the operating times will be similar for each relay, regardless of the location and its adjustment.

From Figs. 5.5 and 5.6, it can be seen that coordination between the relays providing the primary and backup protection functions is achieved, without

TABLE 5.5 Settings obtained from the method of intervals for nonconventional relays for the six-bus IEEE test system.

#	Relay	Ta	Tb	Tc	*TDS* (ζ^4)	*PMS* (ζ^5)	A (ζ^1)	B (ζ^2)	p (ζ^3)
1	14	0.183	0.495	0.731	1.001	1.404	8.851	0.130	1.570
2	15	0.177	0.407	0.726	0.458	1.468	22.394	0.232	1.949
3	16	0.104	0.428	0.723	0.216	1.435	8.611	0.379	1.923
4	46	0.184	0.442	0.748	0.442	1.461	4.588	0.260	1.710
5	56	0.100	0.429	0.704	0.251	1.548	12.200	0.193	1.899
6	41	0.143	0.488	0.743	0.365	1.480	21.452	0.295	1.754
7	51	0.148	0.405	0.700	0.146	1.435	14.913	0.113	1.293
8	61	0.197	0.401	0.799	0.159	1.594	20.384	0.132	1.690
9	64	0.129	0.401	0.700	0.113	1.557	14.213	0.112	1.830
10	65	0.107	0.419	0.703	0.164	1.416	24.555	0.170	1.957

using the coordination criterion. The shaded area is the relay times for each time interval. For example, Fig. 5.5 shows that for a fault current of 3950 A, relay 14 operates as primary protection within 0.183 s. It is coordinated with relay 51, which has a current of 1230 A for the same fault location and operates within 0.405 s. For the same fault location, relay 61 has a current of 1780 A and operates within 0.401 s. We can therefore deduce that these relays are coordinated. The coefficients of the *NCC* time curves in Table 5.5 give time curves that meet the interval constraints and achieve coordination. The data on the operation times and settings of the time curves for the cases presented in Figs. 5.5 and 5.6 are presented in Table 5.5.

5.4 Test systems

The test systems used to evaluate the proposed system were the nine-bus IEEE system in Fig. 5.7 and the 30-bus IEEE system in Fig. 5.8 [19]. These systems had selected voltages of 13.8/16.5/18/230 kV with 18 relays and 132/33/11/1 kV with 68 relays, respectively.

Each relay was assigned a name derived from the bus on which it was located and the protected line. The three-phase currents and minimum phase-to-phase faults on each bus were calculated in the fault analysis, based on the open-end assumption. The values of the fault current were updated before topological changes were made. In addition, the nodal voltage was measured and used to calculate the fault currents.

The number of relays that met the sensitivity criteria was 11, and the total number used in the proposed coordination method was 55.

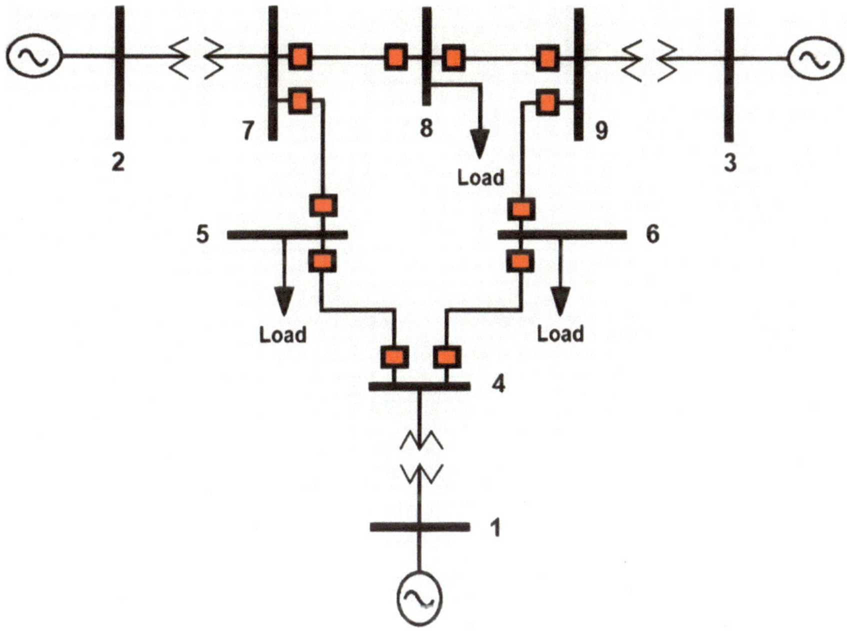

FIG. 5.7 Nine-bus IEEE test system.

5.5 Results and discussions

Using the information on the test systems described in the previous section, we defined the initial conditions for the optimization algorithm and searched for the necessary solutions. In this section, we will not discuss the characteristics and particularities of each algorithm in detail, as these are well documented elsewhere. Furthermore, the proposed coordination method is independent of the optimization method used. In the optimization process, a total of 60 and 100 individuals was considered in test networks consisting of nine and 30 buses, respectively. A total of 1000 iterations was used, and the runtime was evaluated in terms of its feasibility for online applications. Matlab software was used for the simulations.

Various formulations of the OF have been used in previous studies to solve the problem of DOCR coordination, and different metaheuristic algorithms and hybrid algorithms have been used. A comparison of these approaches is not the objective of this study; the goal is to enable a comparison with the proposed method. The OF used in conventional coordination [20], which is used for comparison with the proposed method, is expressed as follows:

$$OF = min\left\{\frac{NV}{NCP} + \omega_1\left(\frac{\sum_{p=1}^{NCP} tp}{NCP}\right) + \omega_2\left(\frac{\sum_{b=1}^{NCP} tb}{NCP}\right) + \omega_3\left(\sum_{pb=1}^{NCP} CTIe\right)\right\} \tag{5.11}$$

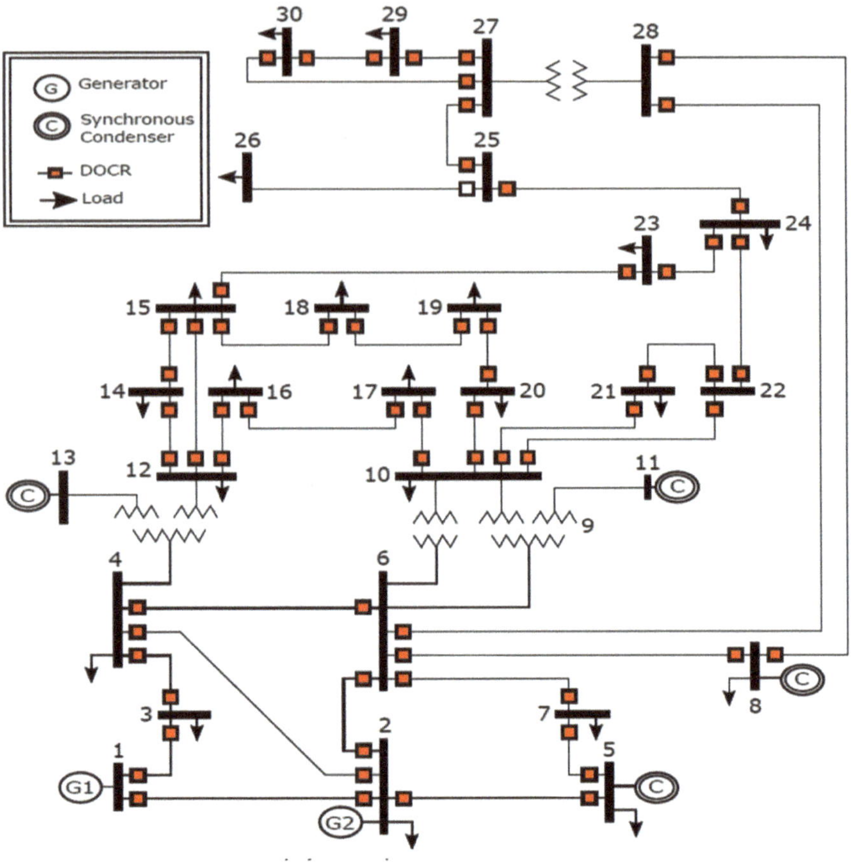

FIG. 5.8 Thirty-bus IEEE test system.

where *NCP* is the number of coordination pairs in the system; *NV* is the number of coordination pairs that violate the coordination criterion; and *CTIe* is the coordination time interval, which is interpreted as a soft-constrained *CTI* error for each coordination pair. The values $\omega_1 = 1$, $\omega_2 = 0.75$, and $\omega_3 = 0.25$ are the weighting factors determined by Pareto frontier and used to define the most important terms of the OF. The best weights were obtained through parametric variations of overcurrent coordination, mainly by the pickup and short-circuit currents.

5.5.1 Performance of the optimization algorithm with the proposed method

To illustrate the NCCs output by the optimization algorithm, the solutions obtained for two relays (R_{57} and R_{45}) of the nine-bus IEEE test system are shown in Fig. 5.9. Note that all of the relays comply with the specified ranges. Each curve

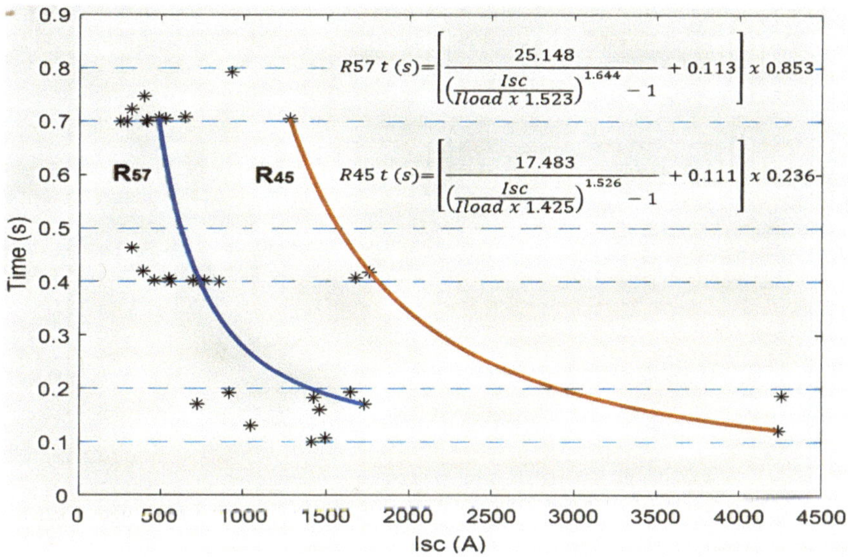

FIG. 5.9 NCCs for relays R_{57} and R_{45}.

is shown with the corresponding analytical expression to verify the operating times. The asterisks represent solutions to the operating times of other relays.

As an example, the solutions that satisfy the constraints for relays R_{45} and R_{89} are shown in Fig. 5.10. The form of the Pareto solution results in nonconvex functions; thus, the selection of the best solution is obtained only from the lowest value of the sum of the functions. Another alternative to particularize the

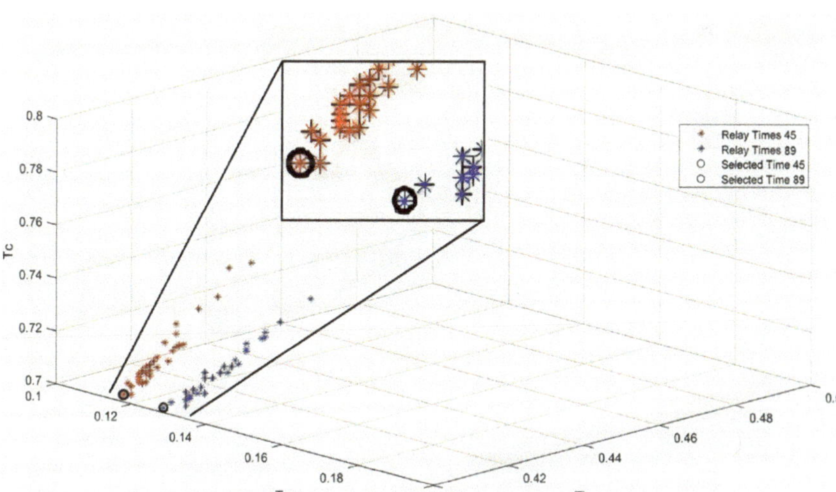

FIG. 5.10 Selection of the best solutions for relays R_{45} and R_{89}.

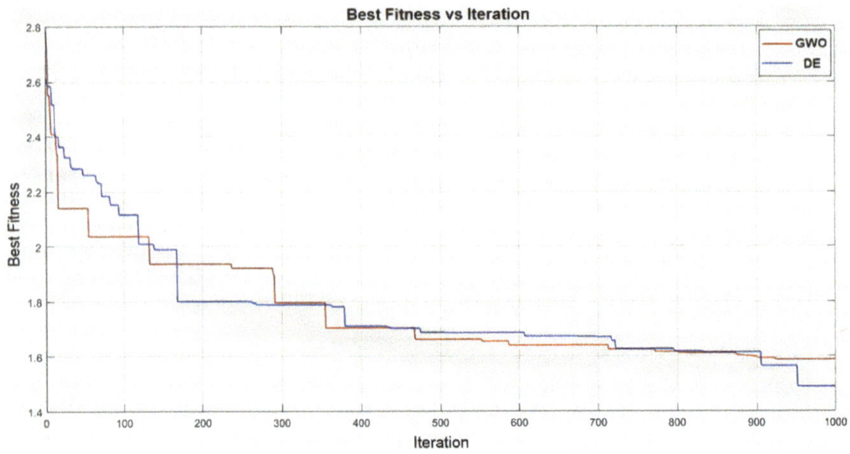

FIG. 5.11 Best fitness for the six-bus IEEE system.

results to the desired solutions is to assign weights to modulate the terms in the proposed OF. The best fitness functions obtained from GWO and DE for the nine-bus IEEE system are shown in Fig. 5.11.

5.5.2 Case 1: Comparison of conventional and interval coordination methods

Table 5.6 presents the results for the primary and backup operating times for the nine-node IEEE system obtained from the conventional coordination method.

TABLE 5.6 Results of conventional coordination with standardized curves for the nine-bus IEEE system using GWO.

Relay		tp (Dial)		tb (Dial)		CTI (Dial)	
Primary	Backup	(0.1)	(0.01)	(0.1)	(0.01)	(0.1)	(0.01)
45	64	0.270	0.055	0.953	0.577	0.683	0.522
57	45	0.319	0.519	0.624	0.118	0.305	-0.401
69	46	0.427	0.488	0.739	0.816	0.311	0.328
78	57	0.058	0.271	0.360	0.573	0.302	0.302
89	78	0.488	0.017	0.148	0.755	-0.340	0.738
54	75	0.622	0.064	0.124	0.648	-0.498	0.584
64	96	0.379	0.244	0.693	0.550	0.315	0.307
75	87	0.061	0.324	0.482	0.832	0.421	0.509
96	89	0.427	0.345	1.188	0.042	0.760	-0.303
87	98	0.306	0.518	0.606	0.821	0.300	0.303
98	69	0.396	0.520	0.696	0.820	0.300	0.300
Number of Violations (NV)						2	2
Computation Time (s)				GWO: 12.26		DE: 13.07	

There are 11 coordination pairs in the system. In the simulations, the conventional IEEE curve (very inverse) was used with two different values for the *TDS* in the ranges (0.1:0.01:3) and (0.01:0.001:3), and the behavior of the operating times was analyzed. Note that reducing the minimum value of *TDS* from 0.1 to 0.01 to minimize the operating times is not always advantageous, as violations or loss of coordination may occur and times can be obtained that are detrimental to the protection scheme. The red numbers of the table represent violations of the relay coordination. It is possible to lose the security of the network because the resulting operating times for small dials are only a few cycles.

The same test system was evaluated using the proposed time interval coordination method. The following operating intervals were established: close-in (0.1–0.2 s), bus-end (0.4–0.5 s), and remote line-end (0.7–0.8 s). The results are presented in Table 5.7. Note that both the GWO and DE algorithms generate a configuration in which all the relays comply with the specified intervals. The computation times required by the GWO and DE algorithms are also included in the table. Since the results from DE and GWO are very similar to those in

TABLE 5.7 Results of the proposed coordination method for the nine-bus IEEE system.

	Primary time (close-in)		Backup time			
			Bus-end		*Remote line-end*	
Relay	*GWO*	*DE*	*GWO*	*DE*	*GWO*	*DE*
45	0.129	0.122	0.400	0.418	0.716	0.705
46	0.173	0.186	0.405	0.407	0.798	0.793
57	0.194	0.170	0.470	0.400	0.773	0.707
64	0.122	0.170	0.401	0.420	0.701	0.700
69	0.195	0.193	0.400	0.402	0.781	0.746
75	0.129	0.101	0.473	0.463	0.701	0.700
78	0.114	0.108	0.439	0.401	0.723	0.722
87	0.192	0.191	0.403	0.400	0.701	0.702
89	0.155	0.130	0.408	0.406	0.706	0.700
96	0.184	0.183	0.405	0.401	0.702	0.705
98	0.192	0.160	0.476	0.400	0.715	0.708
Number of Violations (NV)			GWO: 0		DE: 0	
Computation Time (s)			GWO: 9.23		DE: 9.85	

TABLE 5.8 Power flows at maximum demand and fault currents for relay coordination Case 1.

Relay		I_{sc}		I_{load}	
Primary	*Backup*	*Primary*	*Backup*	*Primary*	*Backup*
45	64	4232	390	249	83
57	45	1735	1761	36	249
69	46	1651	1675	84	83
78	57	1499	737	69	36
89	78	1046	457	115	69
54	75	736	323	249	36
64	96	720	764	83	84
75	87	1413	560	36	69
96	89	1430	547	84	115
87	98	918	852	69	115
98	69	1460	694	115	84

Table 5.6, we only present the results from DE in the following. Table 5.8 presents the values of I_{sc} and I_{load} used in this analysis.

The use of NCCs in the proposed coordination method is necessary because standardized curves do not have sufficient flexibility to fit the defined time intervals. The values in Table 5.9 indicate that there are very few relays with settings that guarantee a time curve with sufficient flexibility to fit each of the preestablished intervals, which is why 17 violations appear in this system. The red numbers of the table represent that the CC relay is no accomplished with the time intervals.

Fig. 5.12 shows a comparison of the operating times of both coordination methods, for different minimum values of *TDS* in the conventional method. The *NCC* is used in the following test for the proposed coordination method because the *CC* presents a clear advantage. Note that the reduction in the minimum range with the conventional method does not guarantee shorter operating intervals, and there is a risk that more violations will occur. In Fig. 5.12B, relays with a very low backup time are marked with a box; these constitute violations as the coordination with the primary element is lost. The times obtained with the interval method are shown in the same figure. Note that the algorithm finds an *NCC* for each relay that guarantees both good operating times and coordination between the primary and backup elements. Table 5.10 presents the relay settings,

TABLE 5.9 Results of the proposed method for a nine-bus IEEE system using a standardized curve.

Relay	Primary T.	Backup T.	
	Close-in	Bus-end	Remote line-end
45	0.180	0.4171	0.700
46	0.132	*0.1736*	*0.239*
57	0.101	*0.1313*	*0.158*
64	0.102	0.4114	0.705
69	0.103	*0.1830*	*0.370*
75	0.106	*0.1225*	*0.157*
78	*0.224*	*0.1240*	*0.112*
87	0.104	*0.1145*	*0.169*
89	0.100	0.4315	0.714
96	0.103	*0.1566*	*0.239*
98	0.106	*0.1963*	*0.281*
Number of Violations (NV)			17
Computation T. (s)	GWO: 10.09		DE: 10.45

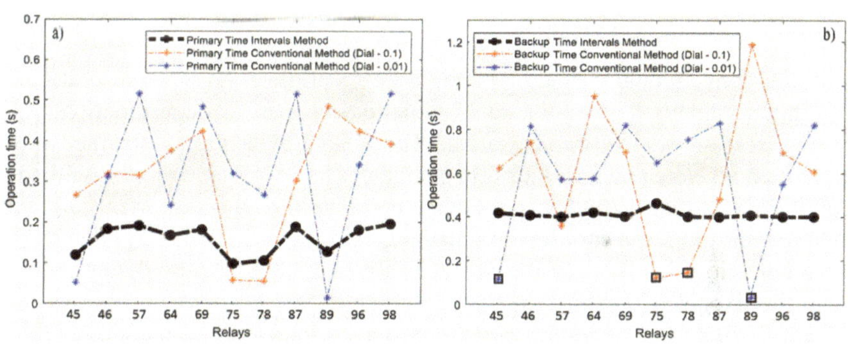

FIG. 5.12 Comparison of operation times obtained from both coordination methods: (A) Primary times; (B) backup times.

obtained using the proposed coordination method, that comply with the time ranges.

Fig. 5.13 shows the average values of the operating parameters after analyzing both methods. Note that in addition to giving zero violations, an independent adjustment of each relay was obtained with the proposed method. As described in Section 5.5.4, the main advantage of this setting is that when modifications are made to some relays, recoordination of the others is not necessary.

An evaluation of the proposed interval method was performed using the 30-bus IEEE system. In this system, there are 55 relays that comply with the sensitivity filter. Owing to the large size of the analyzed system, only the average values of the operating parameters are shown. Fig. 5.14 shows a comparison

TABLE 5.10 Settings for the time curve obtained from DE, Case 1.

Relay	A	B	p	TDS	PMS
45	17.483	0.111	1.526	0.236	1.425
46	26.776	0.264	1.298	0.359	1.404
57	14.482	0.075	1.449	1.044	1.453
64	8.777	0.183	1.661	0.265	1.424
69	7.460	0.248	0.922	0.211	1.298
75	22.720	0.145	1.890	0.537	1.418
78	10.070	0.150	1.755	0.459	1.419
87	16.654	0.111	1.991	0.626	1.412
89	18.253	0.143	1.848	0.168	1.446
96	21.698	0.266	1.657	0.263	1.576
98	194.159	1.951	2.183	0.030	1.560

with the conventional method. It should be mentioned that the conventional coordination method gives rise to 17 violations or losses of coordination.

Fig. 5.15 shows the primary and backup times for each relay obtained with each method. Note that the algorithm guarantees an *NCC* that complies with the preset time intervals, and hence there are no violations. It is important to highlight that the proposed coordination method yields times for the relays that show a flat behavior; that is, since the operation times are bounded by defined intervals, the expected operation times are obtained for all relays. This has the advantage of having operation times expected by the DOCRs; in contrast, conventional coordination results in large differences in the operation times of the relays, with large uncertainties in the fault release times. In the conventional method, the use of the coordination criterion results in functional dependency, where a long operation time for one relay can affect the operation time of a neighboring one. It can be observed from Fig. 5.15 that the operation times for the relays obtained with the conventional method are very different.

5.5.3 Case 2: Evaluation of the proposed method with different intervals

The time intervals can be made shorter to reduce the operating times, although the risk of violations is high. In the nine-bus IEEE system, to achieve shorter operating times, the intervals shown in Fig. 5.16 were used. As can be seen, the third operation interval ($I_{sc\ c}$) is extended to guarantee an *NCC* that complies

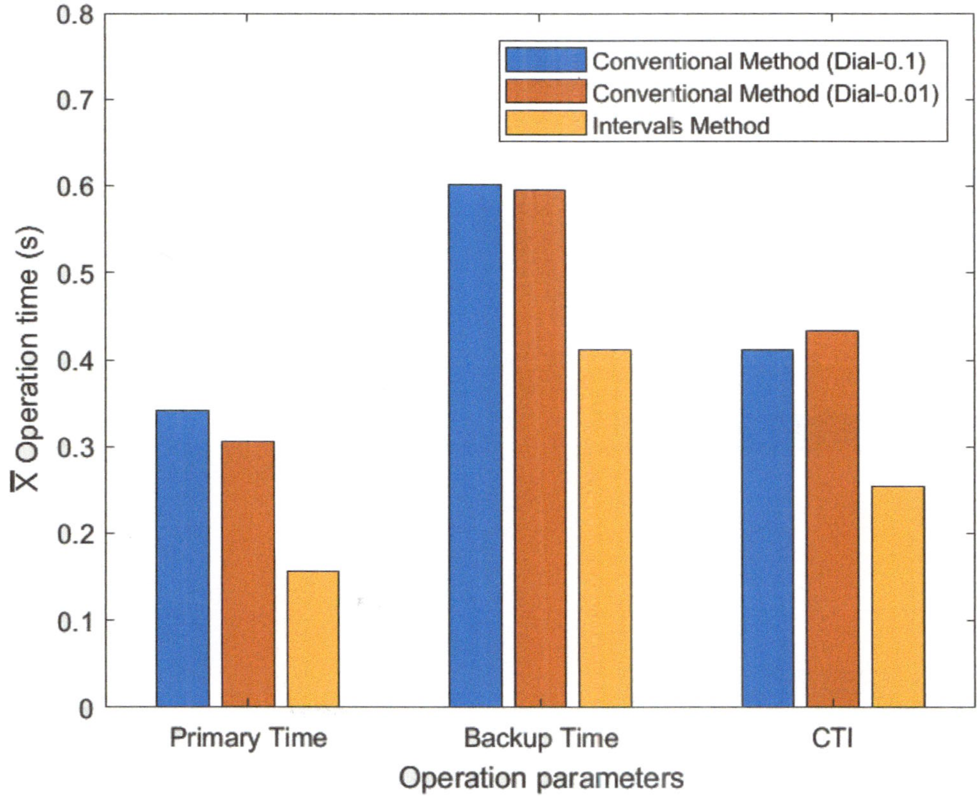

FIG. 5.13 Average values of operation parameters after analyzing both methods.

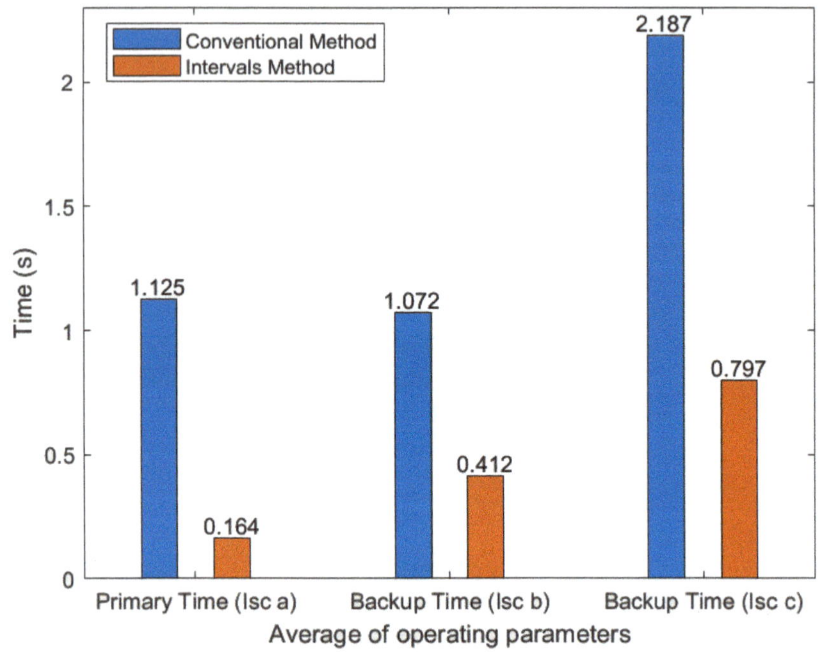

FIG. 5.14 Intervals and operating times for the 55 relays in the 30-bus IEEE system.

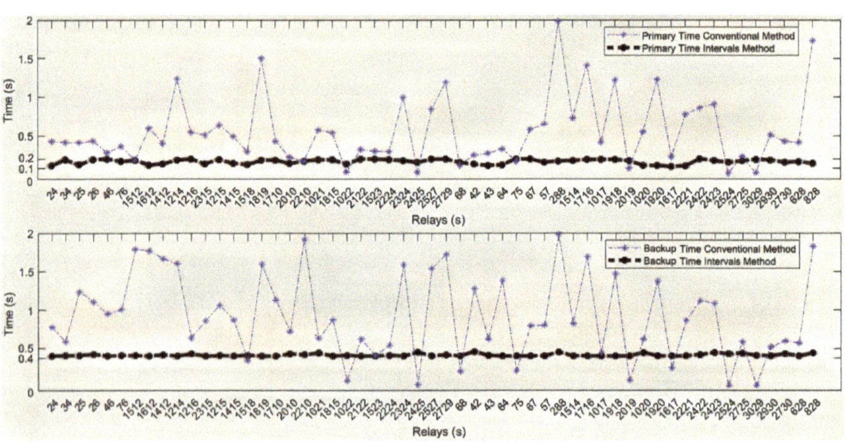

FIG. 5.15 Comparison of operation times obtained with both coordination methods for the 30-bus IEEE system, for primary and backup relays.

with the first two intervals and consequently with the corresponding coordination, maintaining reasonable times for the third interval, which guarantees the security of the network in addition to the coordination. The operation time is shown with an asterisk for simplicity, as the time curves do not allow the operation times for all relays to be shown to be within the time intervals.

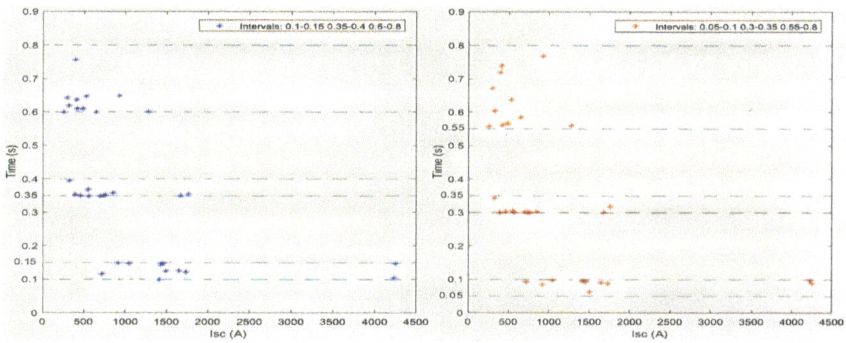

FIG. 5.16 Results for smaller time intervals.

Time intervals can also be defined in independent groups of relays or in a specific way for each relay, to avoid overlapping and meet the dependability and safety requirements.

5.5.4 Case 3: Performance of the proposed method for topological changes to the nine-bus IEEE system

In this case, the operation of the relays is assumed to occur within an online coordination system that responds to topological changes in the network. The response of the coordination method when contingencies arise in terms of operational or topological changes needs to be evaluated, and it is important to determine the robustness of the proposed interval coordination method in these scenarios. The aim here was to determine the variation in the number of relays with violations and the increase in the operation times for dynamic changes in the electrical network.

5.5.4.1 Output of the G-2 generator

In this case, the output of the G-2 generator was evaluated as a contingency, and the coordination of the relay changes according to the variations in the fault current and load current. The variation in I_{sc} seen by each of the relays under this condition is shown in Fig. 5.17. Note that there is a decrease in the value of I_{sc}, which may be higher or lower depending on the network topology.

Table 5.11 presents the results of the conventional coordination method in the event of contingencies. Note that there are two cases of violation or loss of coordination, in addition to the cases where coordination between the primary and backup elements is achieved. The red numbers of the table represent violations of the relay coordination. The resulting times are high and are unproductive for protection purposes as they can negatively influence the operation and stability of the network.

The proposed coordination method was used to compute the operation of the relays for the established time intervals under contingency conditions, as

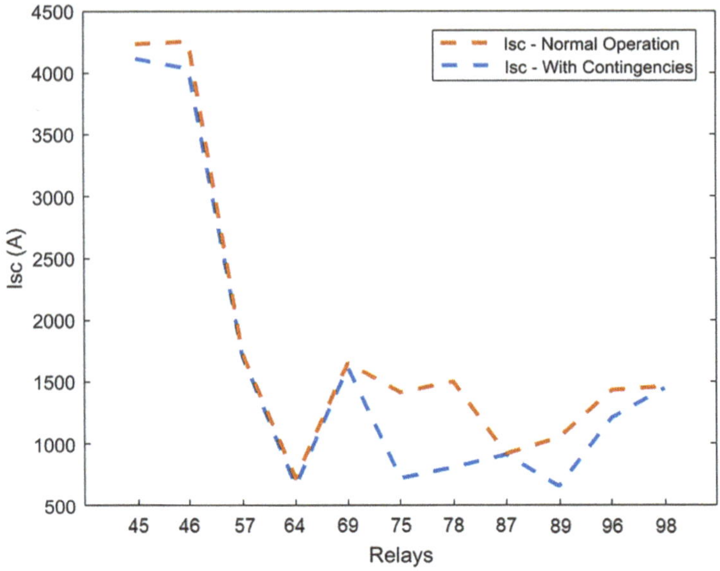

FIG. 5.17 Fault current seen by relays during normal operating conditions and contingencies (G-2 Out).

TABLE 5.11 Results of the conventional method for the nine-bus IEEE system under contingency conditions (G-2 Out).

Relay		t_p	t_b	CTI
Primary	Backup			
45	64	0.162	0.465	0.303
57	45	0.067	0.371	0.304
69	46	0.437	0.755	0.318
78	57	0.616	0.074	-0.542
89	78	0.427	0.747	0.319
54	75	2.053	0.070	-1.983
64	96	0.226	0.527	0.301
75	87	0.060	0.362	0.302
96	89	0.318	0.623	0.305
87	98	0.302	0.602	0.300
98	69	0.400	0.702	0.302
Summation (Σ)		5.068	5.154	2.754
Average ($\bar{X}$)		0.461	0.573	0.306
Number of Violations (NV)				2
Computation Time (s)				12.28

TABLE 5.12 **Results of the proposed method based on intervals for the nine-bus IEEE system under contingency conditions (G-2 Out).**

		Primary time	Backup time		
No.	Relay	*(Close-in)*	*(Bus-end)*	*(R.Line-end)*	CTI
1	45	0.165	0.400	0.739	0.235
2	46	0.194	0.402	0.794	0.208
3	57	0.193	0.404	0.701	0.211
4	64	0.183	0.423	0.706	0.240
5	69	0.180	0.402	0.762	0.222
6	75	0.139	0.426	0.712	0.287
7	78	0.180	0.405	0.796	0.225
8	87	0.179	0.410	0.772	0.231
9	89	0.183	0.454	0.722	0.271
10	96	0.169	0.403	0.780	0.234
11	98	0.194	0.424	0.705	0.230
Summation ($\sum$)		1.959	4.553	8.189	2.594
Average ($\overline{X}$)		0.178	0.414	0.744	0.236
Number of Violations (NV)					0
Computation Time (s)					9.54

presented in Table 5.12. Note that there is a decrease in the operating times for faults in the primary and backup zones, which enhances the security of the network. The total and average relay times are included in the table to allow for a comparison of the two methods in statistical terms.

The method of coordination by intervals allows the desired times for faults at the remote line-end to be obtained; this fulfills the sensitivity criterion. Table 5.11 presents that violations in coordination occur and that sensitivity is not guaranteed. It is often the case that the coordination results reported in various works do not indicate whether the relays comply with the sensitivity criterion. It is clear that compliance is not a trivial aspect, since it affects the security of the protection scheme. When the proposed coordination method is used, this compliance is clearly shown, since the time curve for each relay is designed to fulfill this criterion. The operation times obtained from both coordination methods are shown in Fig. 5.18.

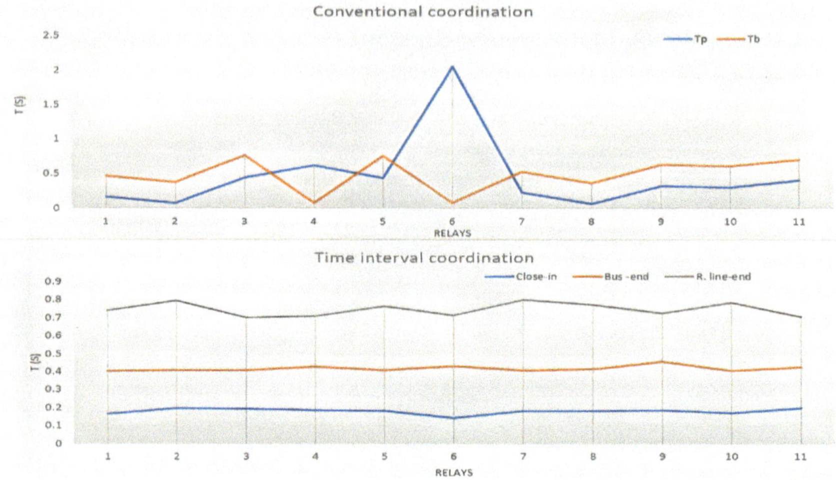

FIG. 5.18 Coordination results from the conventional and proposed methods for the output of a generator.

5.5.4.2 Output of line 6-4

In this case, the output of line 6-4 is evaluated as a contingency. When the proposed method is used, it is only necessary to readjust the relays on those lines that suffer significant variations in the I_{sc}. In this case, a variation of more than 100 A in the fault current is assumed to perform the recoordination. The relays that were disabled by the line output are marked in Table 5.13. The relays are indicated in Table 5.13 with a red cell. In contrast, 18 DOCRs must be recoordinated with the conventional method.

5.5.4.3 Time intervals coordination under recoordination

In conventional coordination, setting changes are not usually related to the complexity of the network. In the face of topological changes, coordination must be maintained, and it is therefore important to coordinate considering contingency analysis. However, it is widely known that the quality of protection is degraded when a topological event occurs in the network, since the power flows change the adjustment criteria and hence the coordination is different. This results in reduced sensitivity and possibly longer run times.

There are certain operational scenarios that require the readjustment of relays, such as expansion of the electrical network, either through the connection of microgrids or by new loads, and the potentiation of lines and transformers. In the conventional approach to coordination, it is necessary to adjust and coordinate all the relays due to the need to meet the condition in Eq. (5.1).

TABLE 5.13 DOCR recoordination with the proposed method in the nine-bus IEEE system under contingency conditions (line 6-4 out).

	Relay	Isc Normal	Isc 6-4 Out	Variation	Re-Coordinate
1	**45**	**4232**	**3838**	**394**	**X**
2	46	4252	Disabled		
3	57	1734	1719	15	
4	64	720	Disabled		
5	69	1651	Disabled		
6	75	1413	1237	176	
7	78	1499	1557	-58	
8	**87**	**918**	**608**	**310**	**X**
9	89	1046	1105	-59	
10	96	1430	1504	-74	
11	**98**	**1460**	**764**	**696**	**X**

In the proposed time interval method, it is possible to modify the settings of only the affected relays, which avoids the need to recoordinate all the relays due to the fulfillment of the time intervals defined. This offers advantages for adaptive protection applications that perform modifications online.

To exemplify the setting change of the protection relays, modifications to the 30-bus IEEE system were considered in which the line impedance was changed to 50% due to the potentiation and the %Z of the transformer 6–10 was modified with one at 75%. Given the impedance variation of each element, those relays that needed reconfiguration due to a current variation greater than 100 A were identified, and the rest did not need to be reconfigured.

From Table 5.14, it can be seen that in the event of parametric changes to the electrical network, the need for recoordination affects only a few relays. This represents a substantial difference between the two coordination methods, since in the conventional method, recoordination of all the relays is necessary.

5.6 Summary

This study proposes a coordination method that guarantees the operation of each relay within three preestablished time intervals. The use of these time intervals ensures that each relay performs independent primary and backup protection functions, with no dependence on the other relays. NCCs are applied to reduce the operating time for each relay and to provide flexibility such that the times lie within preestablished ranges that are suitable for the problem in the electrical network. It is also possible to observe that with the method of coordination

TABLE 5.14 Topological changes to the 30-bus IEEE system.

Relay	Line 1-2	Line 1-3	Line 2-4	Line 3-4	Line 2-5	Line 2-6	Line 4-6	Line 5-7	Line 6-7	TF. 6-10
24			X							
34		X		X						
26						X				
46		X	X				X			
57					X			X		
67									X	
68							X			X
21	X	X								
31		X	X							
42			X							
43	X		X	X						
52					X					
62						X				
64						X	X			
75								X	X	
76					X			X		

by intervals, the desired times for faults at the remote line-end are obtained, which fulfills the sensitivity criterion.

The proposed approach is advantageous for online protection schemes, since only the settings for relays affected by a change in load and fault currents must be modified. The results obtained for the test systems were acceptable and were within the defined ranges with zero violations. In the studied cases, the total operating times for the primary and backup relays were shorter in the proposed method than the conventional method. The operating times for the adjusted relays were similar in the proposed coordination method, resulting in smaller dispersion of the operation times of the relays than in the conventional method.

References

[1] A. Alvarez, D. Della, G. Massaa, A. Dedè, F. Ramos, A. Barbato, IEC 61850-based adaptive protection system for the MV distribution smart grid, Sustain. Energy Grids Netw. 15 (2018) 26–33, https://doi.org/10.1016/j.segan.2017.09.003.

[2] N. Torshizi, H. Najafi, A. Noghabi, J. Sadeh, An adaptive characteristic for overcurrent relays considering uncertainty in presence of distributed generation, Sustain. Energy Grids Netw. 128 (2021), https://doi.org/10.1016/j.ijepes.2020.106688.

[3] R. Hamidi, A. Ahmadian, R. Patil, A. Asadinejad, Optimal time-current graded coordination of multistage inverse-time overcurrent relays in distribution networks, Int. Trans. Electr. Energy Syst. 29 (5) (2019), https://doi.org/10.1002/2050-7038.2841.

[4] V. Rajput, F. Adelnia, K. Pandya, Optimal coordination of directional overcurrent relays using improved mathematical formulation, IET Gener. Transm. Distrib. 12 (9) (2018) 2086–2094, https://doi.org/10.1049/iet-gtd.2017.0945.

[5] A. Korashy, S. Kamel, A. Youssef, F. Jurado, Modified water cycle algorithm for optimal direction overcurrent relays coordination, Appl. Soft Comput. 74 (2019) 10–25, https://doi.org/10.1016/j.asoc.2018.10.020.

[6] A. Urdaneta, R. Nadira, L. Perez, Optimal coordination of directional overcurrent relays in interconnected power systems, IEEE Trans. Power Deliv. 3 (3) (1988) 903–911, https://doi.org/10.1109/61.193867.

[7] S. Khond, G. Dhomane, Optimum coordination of directional overcurrent relays for combined overhead/cable distribution system with linear programming technique, Prot. Control Mod. Power Syst. 4 (9) (2019), https://doi.org/10.1186/s41601-019-0124-6.

[8] V. Papaspiliotopoulos, G. Korres, N. Maratos, A novel quadratically constrained quadratic programming method for optimal coordination of directional overcurrent relays, IEEE Trans. Power Deliv. 32 (1) (2017) 3–10, https://doi.org/10.1109/tpwrd.2015.2455015.

[9] A. Wadood, C. Kim, T. Khurshiad, S. Farkoush, S. Rhee, Application of a continuous particle swarm optimization (CPSO) for the optimal coordination of overcurrent relays considering a penalty method, Energies 11 (4) (2018) 869, https://doi.org/10.3390/en11040869.

[10] T. Khurshaid, A. Wadood, C. Kim, S. Farkoush, K. Kharal, S. Rhee, A protective optimal coordination scheme for directional overcurrent relays using modified seeker algorithm, in: International Conference on Computing, Power and Communication Technologies (GUCON), 2018, https://doi.org/10.1109/GUCON.2018.8674957.

[11] A. Korashy, S. Kamel, F. Jurado, A. Youssef, Hybrid whale optimization algorithm and grey wolf optimizer algorithm for optimal coordination of direction overcurrent relays, Electr.

Power Compon. Syst. 47 (6–7) (2018) 644–658, https://doi.org/10.1080/15325008.2019.1602687.

[12] Y. Damchi, M. Dolatabadi, Hybrid VNS–LP algorithm for online optimal coordination of directional overcurrent relays, IET Gener. Transm. Distrib. 14 (23) (2020) 5447–5455, https://doi.org/10.1049/iet-gtd.2019.1811.

[13] S. Chakraborty, S. Das, T. Sidhu, A. Siva, Smart meters for enhancing protection and monitoring functions in emerging distribution systems, Int. J. Electr. Power Energy Syst. 127 (2021), https://doi.org/10.1016/j.ijepes.2020.106626.

[14] H.C. Kılıçkıran, H. Akdemir, İ. Şengör, B. Kekezoğlu, N. Paterakis, A non-standard characteristic based protection scheme for distribution networks, Energies 11 (5) (2018) 1241, https://doi.org/10.3390/en11051241.

[15] S. Zarei, S. Khankalantary, Protection of active distribution networks with conventional and inverter-based distributed generators, Int. J. Electr. Power Energy Syst. 129 (2021), https://doi.org/10.1016/j.ijepes.2020.106746.

[16] M.H. Sadeghi, A. Dastfan, Y. Damchi, Robust and adaptive coordination approaches for co-optimization of voltage dip and directional overcurrent relays coordination, Int. J. Electr. Power Energy Syst. 129 (2021), https://doi.org/10.1016/j.ijepes.2021.106850.

[17] F. Alasali, N. El-Naily, E. Zarour, S. Saad, Highly sensitive and fast microgrid protection using optimal coordination scheme and nonstandard tripping characteristics, Int. J. Electr. Power Energy Syst. 128 (2021), https://doi.org/10.1016/j.ijepes.2020.106756.

[18] IEEE Standard for Inverse-Time Characteristics Equations for Overcurrent Relays, IEEE Std C37.112-2018 (Revision of IEEE Std C37.112-1996), 2019, pp. 1–25, https://doi.org/10.1109/IEEESTD.2019.8635630.

[19] Power Systems Test Case Archive, Univ. Washington. [Online]. Available: https://www2.ee.washington.edu/research/pstca/.

[20] M. Shih, A. Conde, Z. Leonowicz, L. Martirano, An adaptive overcurrent coordination scheme to improve relay sensitivity and overcome drawbacks due to distributed generation in smart grids, IEEE Trans. Ind. Appl. 53 (6) (2017) 5217–5228, https://doi.org/10.1109/TIA.2017.2717880.

Chapter 6

Overcurrent relay protection for active microgrids

6.1 Introduction

Traditional electrical systems had a vertical structure connected power plants through transmission lines to distribution networks that directly fed loads. For several decades, technological advances in electrical systems have been renewed to adapt to changes in the increase in generation with renewable sources (RES), in demand response techniques and load management. In response, they have been developed computerized advances for monitoring, control, and protection, for the network planning and the network operation in real time.

With the RES connection, environmental sustainability has increased having operational benefits by mitigating problems of congestion in the electrical network and reducing energy losses. But they also complicate the management of the network due to intermittent generation since the load flows and fault currents are modified due to topology changes. Currently, the microgrid (MG) operation needs to balance their generation and demand using the Utility as a backup or energy reserves when their local generation is not available. This transformation of the electrical system has caused changes in the operation of distribution systems.

An MG is an electrical system that is composed of a set of loads connected to the network and the RES in medium and low voltage, complying with the limits and well-defined electrical parameters that act as a controlled system. In the case of an MG interconnected to the main grid, the frequency variations can be controlled by the Utility, if the transformer at the point of common coupling (PCC) is of sufficient capacity to allow the flow of active power for frequency control. In the case of an isolated MG, it is necessary to establish local control mechanisms for the RES and local controllable sources to guarantee limited variations in frequency and voltage. A weak connection from an MG occurs in systems located geographically far from the Utility with weak links or where the demand capacity is higher than that of the transformer in the PCC and that has dependence on local generation such as RES, battery storage, or microturbines. This MG requires fast response frequency control elements due to the low inertia of the network and local reactive power sources.

Overcurrent Relay Advances for Modern Electricity Networks. https://doi.org/10.1016/B978-0-323-91777-3.00006-4

The integration of MG in the electrical systems has required the implementation of control schemes and protection relaying. Architectures of MG protection systems can be with local or distributed architectures. The highly dynamic loading combined with minimal faults due to RES, results in more difficult for fault detection, thus lack of sensitivity can be very common in these networks. The low sensitivity causes a high operation time of relays; the tolerable operation time must be evaluated. The weak source contribution can be very common, thus the recloser operations are affected. Also, the islanding detection schemes must be activated and island salvages condition should be considered.

The sensitivity of the DOCR is defined by the fault currents (I_{sc}) and the pickup current (I_{pickup}), in the first it is not feasible to increase its magnitude substantially to improve the sensitivity, so efforts should be directed at the adaptive control of the I_{pickup}. The conventional setting criteria must be different, for the consideration of the maximum demand in conjunction with the contingency analysis results in values that it is difficult to obtain good sensitivity. Adaptive schemes aimed at reducing the I_{pickup} of the overcurrent relays (Chapter 7) can be applied to improve overcurrent relay (OCR) sensitivity. Furthermore, it is desirable that MG protection schemes avoid false tripping and that operating times are acceptable for adequate protection of the electrical network. Recognizing the high operational dynamics of MGs, it is possible to adapt the coordination method presented in Chapter 5 to improve the operation times of the relays and at the same time simplify the coordination process.

6.2 Architecture of a microgrid

These microgrids are normally connected low-voltage systems, but it is also possible to connect them in medium voltage and they normally work interconnected to the Utility. However, they can also operate in isolation when there is a disturbance in the system or when it is required to provide power to a rural area isolated from the main grid.

The requirements that a network must have to operate continuously in island mode are that it must have sufficient generation and storage capacity to supply energy to the loads when required; or where appropriate, manage generation alongside the demand. When an MG is being operated through active monitoring and efficient control, there is an optimization of the available resources within the same network to achieve an optimal solution in the different stakeholder involved. This has a great advantage over passive networks with intermittent distributed generation.

From the point of view of an electric network, an MG can be considered as a controlled entity within the power system, which can be operated as an electrical network added to the system, used to improve the resilience of the power system. An independent energy source distribution management system (*DMS*) to control the frequency and voltage using resources available in the MG. It is important to have adequate protection systems for the operating conditions affected by high RES penetration.

6.2.1 Models of renewable sources

The mathematical model of RES must consider both the energy losses and the network efficiencies. In many of the electrical network studies, the converter devices are considered like branch impedances since the transitory states are not considered and only the power contribution is evaluated. However, when there is a change in frequency the converters must remain connected according to IEEE 1547 [1] and UL 1741 standards [2]. The connection time and the converter trip must be included in the simulation of contingency. For protection, setting the dynamic evaluation of converters is not required, only its branch impedance. However, for protection evaluation purposes, the converters must be evaluated in a transitory state; the fault dynamics and the frequency components are important and must be included in the dynamic operation of relays.

A. *Power wind model*

The electrical power at the output of a power wind turbine (*PW*) is obtained by four different equations related to each other, through the efficiencies in each of the power transmission modules. These modules are referred to as wind power, mechanical power, transmission power, and electrical power.

The wind power P_w is the force that drives the rotor blades of the wind turbine, and this depends on the wind speed v (m/s), the air density ρ (kg/m^3), and the area swept by the rotor blades in m^2. This is as follows:

$$P_w = \frac{1}{2}\left(\rho A v^3\right) \tag{6.1}$$

The wind turbine blades extract the power from the wind by converting it into mechanical power as shown in Eq. (6.2).

$$P_m = C_p\left(\frac{1}{2}\rho A v^3\right) \tag{6.2}$$

where C_p is called the power coefficient and it is a portion of the power extracted from the power of the wind. The theoretical maximum value of C_p is defined by the Betz limit with a value of 59.3%, which indicates that a wind turbine can never extract more than this amount of power from a wind stream. This coefficient varies with the wind speed, the rotational speed, and the angle of inclination (pitch) of the blades.

The mechanical power is shown in Eq. (6.2), which is the power on the shaft, it is coupled to the transmission box. One way to represent the transmission power is through its efficiency η_m.

$$P_t = \eta_m P_m = \eta_m C_p\left(\frac{1}{2}\rho A v^3\right) \tag{6.3}$$

Finally, the output power of the wind turbine P_e is determined by the product of the output power of the transmission box and the efficiency of the electric generator η_g.

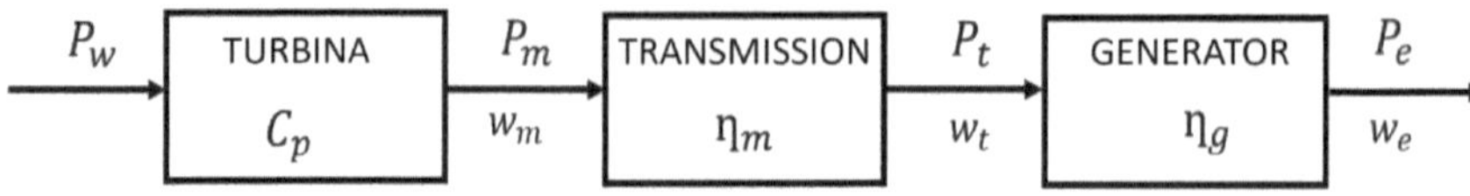

FIG. 6.1 Modules of wind power.

$$P_e = \eta_e P_t = \eta_g \eta_m C_p \left(\frac{1}{2}\rho A v^3\right) \quad (6.4)$$

In general, the block diagram that represents the powers since the force of the wind to the output electrical energy is represented as follows [3] (Fig. 6.1).

For a nominal wind speed, the nominal electrical power can be expressed as follows:

$$P_{eN} = \eta_0 \left(\frac{1}{2}\rho A V_N^3\right) \quad (6.5)$$

where η_0 determines the global efficiency of the wind turbine and this can be represented as follows:

$$\eta_0 = C_{pN} \eta_{mN} \eta_{gN} \quad (6.6)$$

So, to find the electrical power at the output of a wind turbine, the following are required: the efficiencies of each of the modules, the measurements of the wind speed vector, the area swept by the wind turbine blades, and the density of the wind. The test of the previous models for wind speeds obtained from meteorological records is shown as follows.

Test. For simulation purposes, a total efficiency of $\eta_0 = 0.3348$ was considered in the model of the wind turbine used, and it was calculated with $\eta_g = 0.90$, $\eta_m = 0.93$, and $C_p = 0.4$, giving an acceptable value for the wind system. Considering an average wind density of $\rho = 1.24\,\text{kg/m}^3$, since in winter the wind approximately has a density of $1.33\,\text{kg/m}^3$ and in summer $1.15\,\text{kg/m}^3$ [4], and a sweep area of the wind turbine blades $A = 254.46\,\text{m}^2$, considering blades of 9m. The nominal electrical power of the wind turbine is determined by Eq. (6.7).

$$P_{eN} = 0.04068\left(V_N^3\right)\ \text{kW} \quad (6.7)$$

where the efficiency values η_m, η_g, and C_p were taken from [4,5]. For simulation purposes, six different profiles of wind speeds obtained from the meteorological station were considered. These profiles were taken in different periods of a year; these curves are shown in Fig. 6.2.

To obtain a single approximate wind speed curve, a cumulative random sampling analysis of Monte Carlo method was considered, with each time interval of 5min. This approximate curve will be used to calculate the electrical power of the wind system.

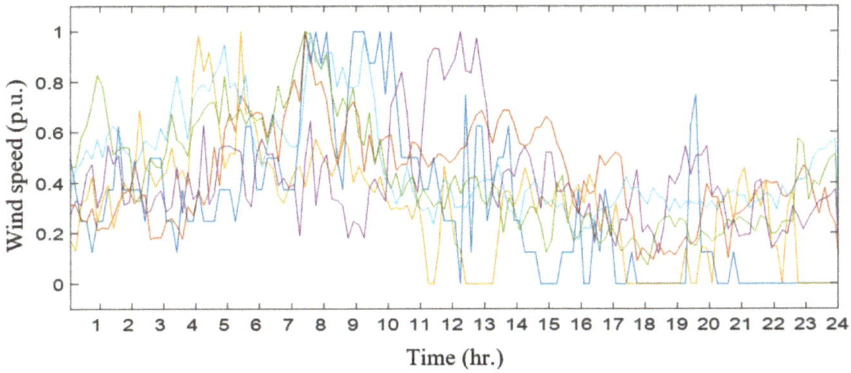

FIG. 6.2 Wind profiles.

The Monte Carlo analytical method uses statistical sampling techniques to obtain a probability approximation to the solution of a mathematical model that indicates a better performance of the system to be evaluated. Considering the six curves obtained by measurements and each time interval of 5 min, the maximum and minimum values of the curves within that interval were chosen to carry out the random approximation, and a probabilistic value was obtained that represents the approximate value of these curves. Fig. 6.3 shown the results obtained within a 24-h period, the red band indicates the interval, which are the maximum and minimum values each Δt using $n = 1000$ iterations, within each interval. The curve that passes within these intervals (black curve) is the approximation of the wind profile of the curves obtained from the meteorological station.

Once this approximate curve is obtained, it is subjected to Eqs. (6.1) and (6.7) to obtain the result of the wind power and the electrical power of the wind turbine system. This is visualized in Fig. 6.4. This graph shows the theoretical energy loss that occurs when the power of the wind is linearly transformed into electrical power (PW efficiency).

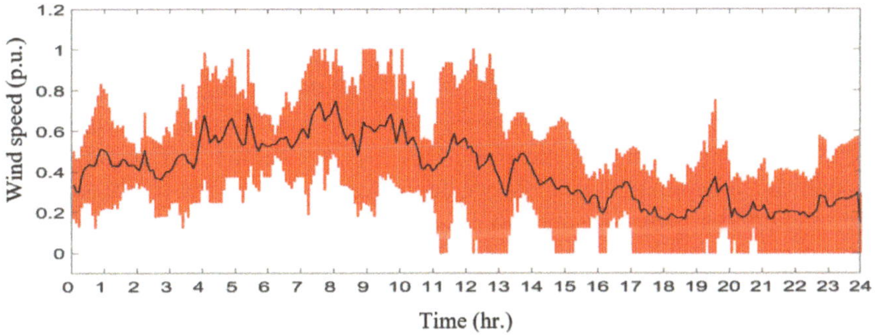

FIG. 6.3 Monte Carlo approximation of the wind curves.

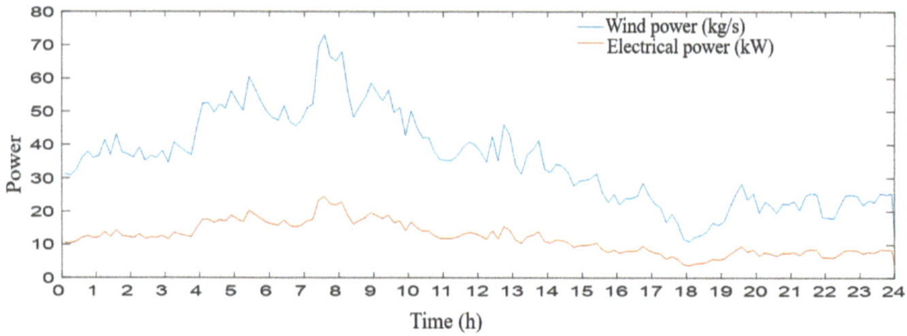

FIG. 6.4 Comparison of wind power and electrical power.

B. *Photovoltaic model*

The power produced by photovoltaic plant (PV) is dependent on solar irradiation and random environmental changes, leading to uncontrollable variations in power output. The PV connected to the distribution system is a source of electricity generation formed by a set of photovoltaic cells, providing the power in the hours coinciding with the solar irradiation. The PV operating conditions depend on the characteristic curves of solar radiation and ambient temperature [6].

The temperature coefficient of the maximum power is represented by the coefficient T_{CO} (1/°C), and the efficiency of the inverter is multiplied by the DC output, as represented in Eq. (6.8).

$$P(t)_{pv} = n_{pv} P_{rate\ \mathrm{PV}} \left(\frac{G(t)}{G_o} \right) (1 - T_{CO}(T_A - 25°)) \eta_{inv} \eta_{rel} \tag{6.8}$$

where η_{pv} are the photovoltaic module number, $P_{rate\ PV}$ is the electrical power (W), $G(t)$ is the global irradiation of PV (W/m^2), G_O is the insolation of the photovoltaic modules (W/m^2), T_A is the ambient temperature, η_{rel} is the relative efficiency of the PV modules, and η_{inv} is the efficiency of the inverter.

Test. A total power P_{ratePV} *of 310 kW* was considered, the temperature coefficient is $T_{CO}=0.0045$ and the relative efficiency of the modules is $\eta_{rel}=0.12$ [2], the ambient temperature is $T_A=32$°C and the efficiency of inverter is $\eta_{inv}=0.99$ [5]. The photovoltaic electrical power depends on the irradiance vector $G(t)$ and the module temperature. However, the relative efficiency of the module (η_{rel}) decreases by 0.5% per degree centigrade as the temperature of the module increases. For the test, six irradiance curves were taken from the same meteorological system in order to make an approximation of the measured data. Fig. 6.5 shows the irradiance curves.

Applying the Monte Carlo random approximation method with $n=1000$ iterations, the following graph of Fig. 6.6 is obtained.

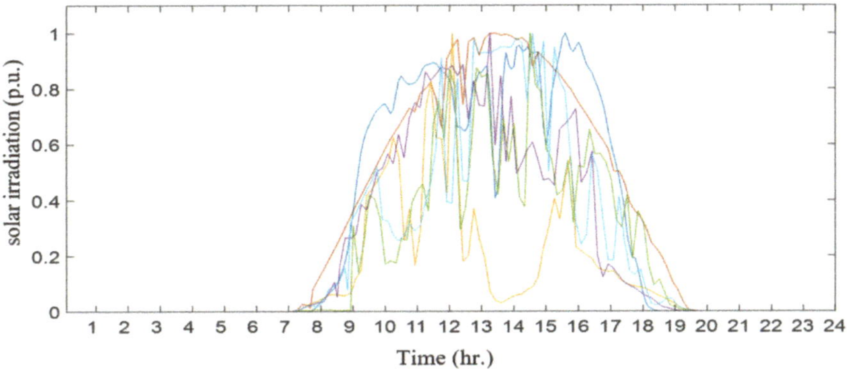

FIG. 6.5 Solar irradiation profiles.

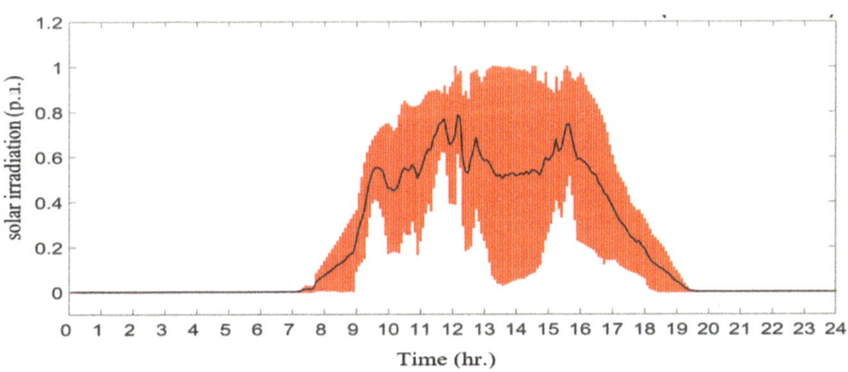

FIG. 6.6 Solar irradiation approximations.

Substituting the values of the approximate solar irradiation vector equation (6.8), the curve of electrical power supplied to the electrical network vs the irradiance of the photovoltaic system is obtained (Fig. 6.7).

The loss of power is appreciated when the energy radiated on the photovoltaic modules is transformed into electrical energy. In this model of photovoltaic cells, the converters' mathematical model was not considered, only the efficiencies taken from the indicated literature.

C. *Microturbine model*

Microturbines (MT) have been classified as distributed generation sources with less pollution emissions than centralized conventional power plants and have

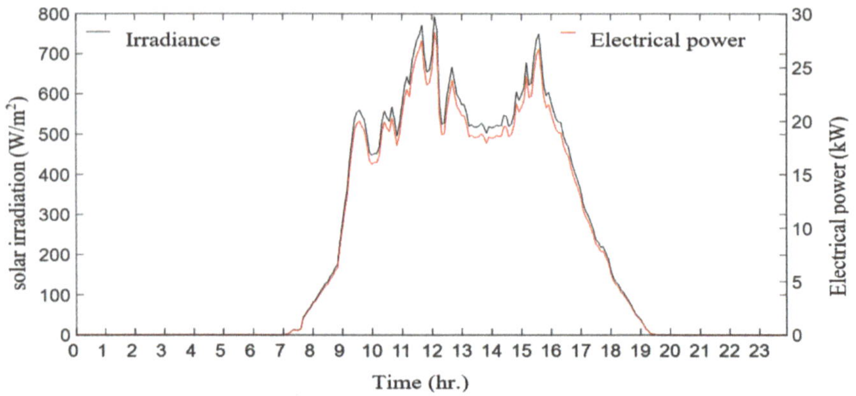

FIG. 6.7 Electrical power vs irradiation.

been participating in the operation of MGs to support renewable sources. A simplified model of an MT has the following expression [7]:

$$P(t)_{gT} = \frac{Q(t)_{MT} * \eta_e}{(1 - \eta_e - \eta_l)} \tag{6.9}$$

$$Q(t)_{MT} = \frac{K_{he}}{Q(t)_{he}} \tag{6.10}$$

where $P(t)_{gT}$ is the output power of the turbine in kW at the period Δt, $Q(t)_{MT}$ is the residual heat from the MT, η_e is the MT generation efficiency, η_l is the heat loss coefficient, $Q(t)_{he}$ is the heat provided by the microturbine, and K_{he} is the heat coefficient of the cooler.

D. *Battery technologies*

The type of storage systems most used in an MG are batteries. There are different types of battery technologies, of which the three most used in the electricity grid market will be mentioned in this section due to their longer useful life, high efficiency, and low maintenance.

Lead-acid batteries [8] are characterized by the fact that a lead is used in both electrodes as an active material. At the negative electrode, lead is oxidized by discharging into the divalent ion Pb^{2+} forming lead sulfate $PbSO_4$ in the dilute sulfuric acid electrolyte, while, at the positive electrode, the charged active material is based on four valent ions Pb^{4+} which when discharged are also reduced to Pb^{2+}.

Nickel/cadmium (NiCd) batteries belong to a complete family of secondary batteries that are based on aqueous but alkaline electrolytes, usually dilute potassium hydroxide. A common feature of these battery systems is that they use the nickel hydroxide electrode positive and when the electrode is discharged, the Ni^{3+} ions are reduced to Ni^{2+}.

Ion/lithium (Li-ion) batteries [4,8,9] are widely used since *Li* is a very light material that can be used for different applications. The positive electrode is $LiCoO_2$, and the negative electrode is sandwiched carbon, which can store *Li* up to C_6 of *Li*. This type of battery is also used for very high powers and energies.

Table 6.1 presents the types of batteries with their technical characteristics that are used on a large scale in MGs [4,10].

The *Li-ion* technology battery presents a higher degree of maturity and its wide range of applications in solving regulation problems in an MG. Since this system can operate bidirectionally with the grid, either as a load for the system or as an energy source, this helps to balance demand and conventional generation.

The structure of a complete battery energy storage system (*BESS*) is shown in Fig. 6.8. The first section is the battery pack which is made up of several lithium-ion battery cells that have a series or parallel arrangement, depending on the application required for the MG. The second section is the power conversion system; it is an electronic device that allows the conversion of energy from the output of the batteries to the power supply of the electrical network or vice versa. The third section consists of the battery management system (*BMS*), which provides protection to cells from malfunction in terms of voltage, temperature, and current to achieve optimal operation and proper functioning of the battery pack, as well as balancing cell states of charge (*SoC*) to avoid battery saturation by not allowing deep discharge.

Other *BMS* applications has been under the control of active and reactive power, where it uses the decoupling method of the direct quadrature *dq0* reference system. This last element receives the actions of a *DMS* that always supervises and monitors the entire network.

The power capacities of the BESSs that have been connected in the networks have varied from 2 to 50 kW with energy capacities from 5 kWh to MWh [11]. This shows that BESS can provide support at all different levels in an electrical power network.

6.2.1.1 Power and energy of BESS

The power capacity of a BESS is defined as the speed at which the storage system is capable of exchanging (charge/discharge) its nominal power to the distribution network or MG in a certain time.

In normal operation, the dynamics of the injected/absorbed power are given by the manufacturer of the device. However, some types of storage have the characteristic of discharging more power than its nominal value during a short period of time to solve any contingency. Also, in most types of storage, the upload rate is usually lower than the download rate. The nominal power is represented by the following equation:

TABLE 6.1 Technical characteristics of large-scale storage systems.

Battery technology	Power capacity	Discharge time	Response time	Charge/ discharge efficiency (%)	Lifetime (Years)	Life cycle (Cycles)	Applications
Li-ion	W to MW	1 min–8 h	10–20 ms	85–98	5–15	1000–10,000	Frequency regulation and intermittency compensation of wind power
Ni-Cd	<50 MW	1 min–8 h	N/A	60–70	15–20	2000–2500	Rolling reserve and stabilization of power system
Lead-acid	<50 MW	1 min–8 h	<1/4 cycle	85	3–12	500–1000	Frequency control and rolling reserve

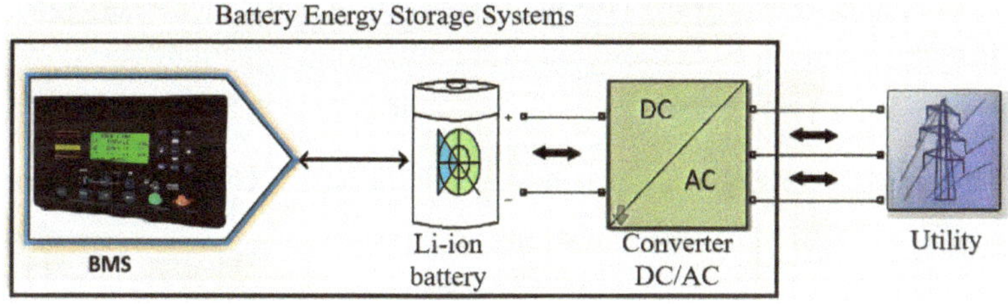

FIG. 6.8 Structure of the BESS.

$$P_{bess}^{nom} = \max_{0 \le t \le T} |P_{bess}| \tag{6.11}$$

where T is the time interval and $max|P_{bess}|$ is the maximum magnitude of the power of the *BESS* (charge/discharge) in a time t.

The nominal size of a *BESS* is defined as the maximum energy that can be stored/released during a complete cycle and is expressed in kWh or MWh. It can also be represented in *Ah,* when the voltage is considered fixed in the storage system. The expression for energy depends on the power capacity of the battery and is represented as follows:

$$E_{bess}(t) = \int_0^t P_{bess}(\tau)d\tau \tag{6.12}$$

The relationship between power and energy for a certain energy storage technology is known as the *E*/*P* ratio. This relationship is used in two cases, the first when energy storage systems are used for power quality applications, the relationship must be $E/P < 1$ because the maximum charge/discharge power is more important than the energy. And the second case when storage systems are used for energy management, the ratio must be $E/P > 1$ due to a need for a higher energy capacity.

The BESS mathematical model [12] is defined by the charge/discharge states in time intervals, dynamic restrictions such as maximum charge and minimum discharge, power capacity, and charge/discharge times are considered. The battery storage system model is represented as indicated in (6.13).

$$P_{BAT}(t) = P_{BAT}(t-1) + \left[P_{Charge}(\tau) * \eta_C - P_{Discharge}(\tau) * \frac{1}{\eta_D}\right] \tag{6.13}$$

where $\tau = (t_2 - t_1)$ is the charging/discharging time in the network in a time interval, η_C is the charging efficiency, and η_D is the discharge efficiency. These efficiencies are related to ambient temperature, depth of charge and discharge, and internal resistance. $P_{BAT}(t-1)$ is the power of the storage system in each time t, $P_{Discharge}(\tau)$ is the released power, and $P_{Charge}(\tau)$ is the absorbed power in the interval τ.

The available power of the BESS system is limited by the storage power at each instant of time t, as represented in the following equation:

$$P_{BAT\_min} \leq P_{BAT}(t) \leq P_{BAT\_max} \tag{6.14}$$

In a period, the BESS can only have a single status, charge, or discharge:

$$P_{BAT\_soc} = P_{Charge}(\tau), \ P_{BAT\_soc} = -P_{Discharge}(\tau) \tag{6.15}$$

where P_{BAT_soc} is the charge/discharge status of the batteries.

The amount of energy released or absorbed by each storage system is calculated as the difference between the energy stored in two consecutive intervals. Each interval used here is a few minutes in a 24-h horizon. Therefore, in a time interval τ, the amount of energy in each defined time interval is represented as follows:

$$P_{Charge\_SOC}(\tau) = \int_{t1}^{t2} P_{BAT}(t)dt, \ P_{Discharge\_SOC}(\tau) = -\int_{t1}^{t2} P_{BAT}(t)dt \tag{6.16}$$

The aging batteries have been considered to optimize the duration or useful life of the different technologies of the BESS. However, in many studies, aging is considered negligible in the study period. The restriction of the amount of charge/discharge power is mainly modeled for the purposes of the availability of this resource in the complete 24-h cycle; with this consideration, deep discharges are avoided to reduce premature aging of the battery.

Test. Fig. 6.9 shows the demand variation and injected power in an MG node. The demand data in these simulations were taken from a distribution substation.

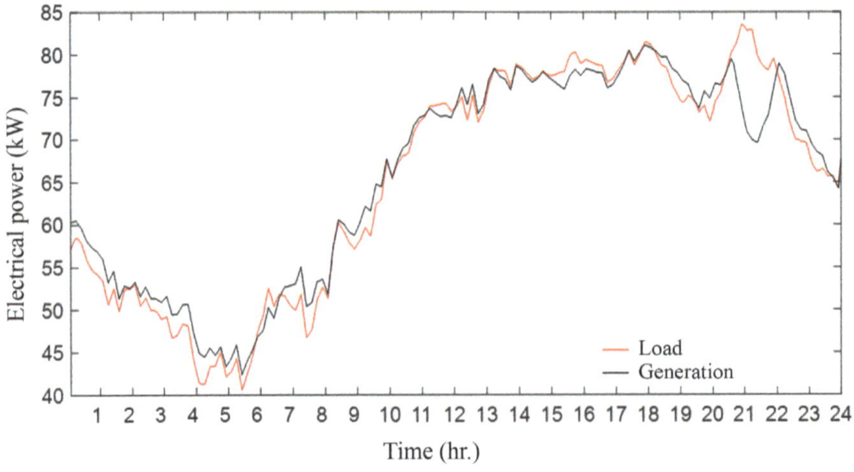

FIG. 6.9 Demand and generation in an MG node.

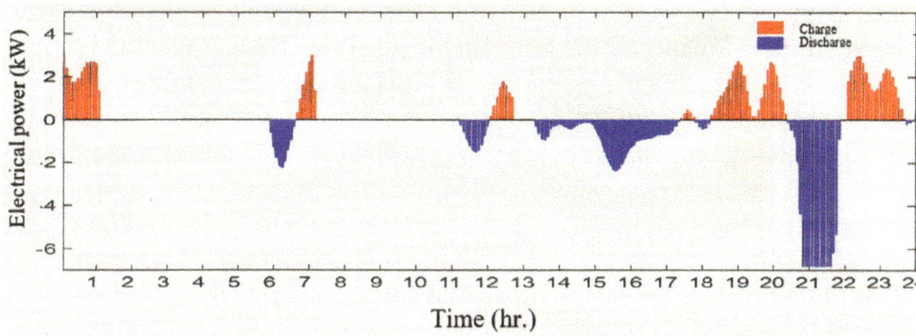

FIG. 6.10 Dynamic behavior of batteries.

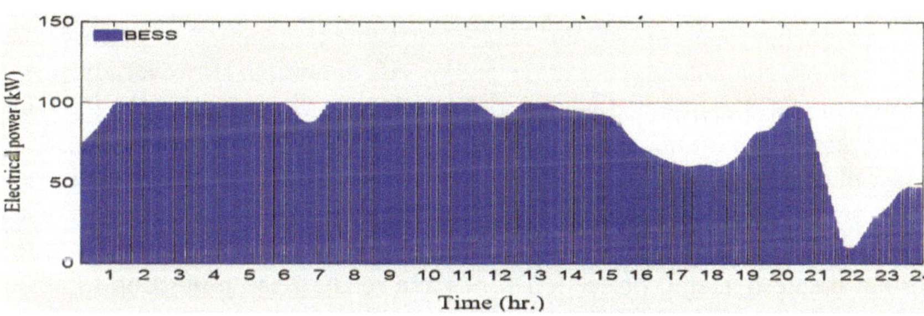

FIG. 6.11 BESS state of charge (SoC).

The dynamics of battery charges and discharges is observed in Fig. 6.10. Here it is observed that the storage system begins to balance the unbalanced power at the node. The battery operation is activated when the energy imbalance exceeds a specified amount; this action avoids continuous operation and premature wear of the batteries.

Fig. 6.11 shows the charging status of the BESS. When the curve has a positive slope, it means that the batteries are charging the energy available from renewables or from the Utility itself, and when the curve has a negative slope, the batteries are discharging active power toward the charging node. This dynamic presents charge and discharge intervals at different times reaching its maximum load limit.

At approximately 21:50 h, the BESS discharge reached its minimum in three consecutive instants of time; since the BESS is limited in this example to a minimum discharge of 10 kW, this limit did not allow it to continue discharging. The discharge time for this BESS reached 465 min (93 operations during the 24 h) and the charge time was 400 min (80 operations throughout the period).

6.3 MG protection

The operation of the electrical network must ensure that the supply of the MG is reliable, ensuring sufficient quality for continuous operation. The development of distributed control systems, either with local or distributed architectures, can improve the performance of the MG. On the other hand, the protection systems are affected by the dynamic nature of the MGs, because the traditional philosophy for the adjustment of the protections is carried out through constant settings, considering only the worst operating scenarios. In this way, the dynamic nature of the MG is not considered, often resulting in topological changes and dynamic operating conditions in low sensitivity, lack of selectivity, or high operating times.

Traditionally, low-voltage networks have been designed to operate radially; therefore the protection principle that has been widely used is nondirectional overcurrent. The adjustment of the OCRs is made considering the maximum value of the load current without monitoring the operating state; the *TDS* is adjusted to guarantee the coordination. Although the operation of the relay depends on the load current, the maximum value is assumed to perform the coordination, guaranteeing a reliable and/or safe operation. However, the operation of the OCR is affected when topology changes occur in the MG, the fault contribution with bidirectional flows and intermittent operations of the RES. Under these scenarios, it is difficult to guarantee adequate operation of conventional protection. In response to these new demands of the protection system, online protection schemes have been proposed that, through continuous monitoring, create dynamic protection zones and readjust the protection to adapt to the operation of the MG. Some schemes will be presented in Chapter 9.

The protection must recognize when the MG operates during connected or isolated condition. The main fault current contribution is from the Utility in connecting condition; fault current profiles can be flat when line impedances are small, making fault location difficult; sensitivity is not affected but operating times can be high. In disconnected condition, island detection schemes are required in case the MG cannot operate permanently disconnected. In this mode of operation, the requirement of the protected system is great for two reasons: the fault current contributions can be very low, compromising the sensitivity of the protections and the loss of frequency regulation by the Utility, causing low inertia scenarios increasing the dynamic and intermittent events of the MG.

In the design of the protections, the use of communication channels and measured input signals is contemplated. In case of having phasor measurements by PMU, the calculation of power flows can be omitted. However, it is necessary to consider that for the contingency analysis, it must be necessary to calculate the flows and fault currents. For updating protection scheme settings, the phasor data must satisfy the demand times and be updated in case of topological changes. It is necessary to consider the latency times of the communication system for real-time applications if required, but most relay settings update applications are done in steady state. For protection settings update, the latency of data can be at intervals of a few minutes for load changes or in few seconds for topological changes.

The functional limitations of the simplest protections, such as overcurrent, are mostly affected by the dynamic operation of the MG. On the other hand, protections that have more system information, such as differential protection that has two or more measurement points or with more signals such as distance, will be less affected. It is important consider the coordination of overcurrent relays with fuses or reclosers.

Overcurrent relay. In networks with feeds at both ends or in highly interconnected networks, directional functions must be incorporated to obtain selective operation. For protection in networks with a poor fault contribution, the performance and detection of DOCRs must be evaluated, fault detection must be not possible, or operation times can be inadequate for protection proposes. The standardized time curves have been an acceptable solution for the relays setting in radial systems, both medium- and low-voltage systems; the I_{pickup} should be set with the maximum value of the I_{load}; and the coordination is done in only one location fault. However, the coordination criteria may be inappropriate due to the coordination commitments of a meshed system. The crossing curves are possible, also power transfer conditions and contingency analysis must be considered. For sensitivity, the coordination must be done in two fault locations, close-in and line-end.

A possible adaptive solution to minimize the MG protection limitations is the use of relay communications channels of digital relays to enable setting groups in periods of time or when topological changes are detected in the electrical network. Another alternative is to change the traditional form of coordination where any coordination problem must be solved with the same standardized curves. Instead, it is possible by using nonconventional time curves (NCC) to design the time curve that suits the specific coordination problem. By incorporating more degrees of freedom, the time curve is more flexible, adapting to the specific coordination requirements in mesh networks. In an MG protection, the time curve may change according to the coordination commitments under specific operation conditions (Chapter 4). Also, updating the I_{pickup} increases the relay sensitivity (Chapter 7).

The DOCRs coordination has been solved using optimization algorithms; generally metaheuristic methods present an adequate result due to their good ability to find solutions. The performance of metaheuristic algorithms is adequate in much operative conditions of MG, obtaining adequate solutions for protection purposes. However, the performance of the algorithms is different depending on the topology and the operating conditions of the MG since different operating conditions represent different problems for the coordination algorithms. Thus the results obtained can be very diverse, from violation of restrictions to inappropriate times for protection. It is important that a formulation is used that offers an acceptable solution for protection, for various operating conditions such as the presence of RES and topological changes can result in excessive protection operating times or loss of sensitivity. This is due to because the formulation for relay coordination is based on a functional dependency, where a slow relay affects neighboring relays or a group of relays.

The performance of the overcurrent protection in the MG must be evaluated depending on its connection with the Utility. The connection represents an operational support to the MG through the contribution of reactive power and the contribution of fault current. Thus the intermittent behavior of the MG is less significant, benefiting the performance of the overcurrent protection scheme. Likewise, if the fault is in the Utility, it is possible that the MG overcurrent protection schemes have no sensitivity, presenting a weak contribution to the fault. This weak source scenario can go undetected, maintaining the contribution to the fault and avoiding the activation reclosing schemes.

6.3.1 Weak source contribution

Interconnection relays prevent damage during sustained operation under unfavorable conditions of the MG. The island operation protection criterion detects the electrical signals that are modified by the energy imbalance in the MG, such as loss of synchronism, frequency variation, and voltage and current variations. However, when the fault is in the Utility, the weak and sustained contribution of the DGs represents a weak source if R_1 fails to operate. These fault currents are difficult to detect by DOCRs (Fig. 6.12).

The weak source represents a loss of sensitivity of the protection system, depending on the type of generation source. In the case of MT, the contribution to the fault is weak, but constant, which allows the implementation of detection schemes with high sensitivity and the activation of adjustment groups for fault detection. When the contribution of the fault is by RES, the fault current is limited in magnitude and time, in this case it is necessary to enable transferred trips or breaker failure scheme (BF) to trip the relay R_2 of the MG.

Under certain operating conditions, the contribution of sources may be limited, either by the type of the source or by the high impedance of the interconnection. This low contribution may not be detected by the relays, losing sensitivity, and maintaining a fault contribution that can cause sustained low voltages and the blocking fault line reclosing actions. The impact of relays coordination due to low-fault current contribution of weak source must be evaluated because the relay coordination must be done where the I_{pickup} of the primary

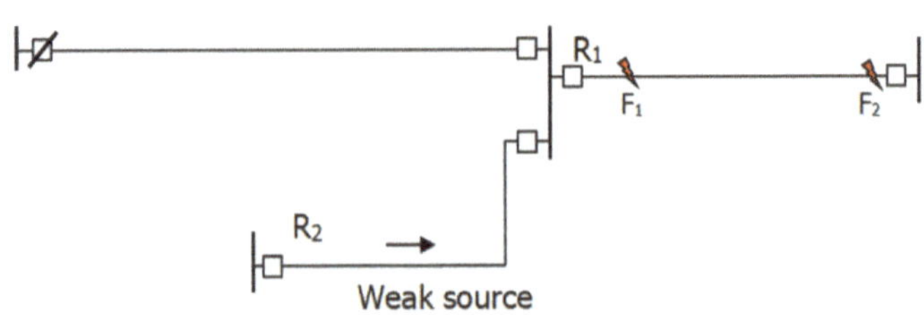

FIG. 6.12 Weak source contribution.

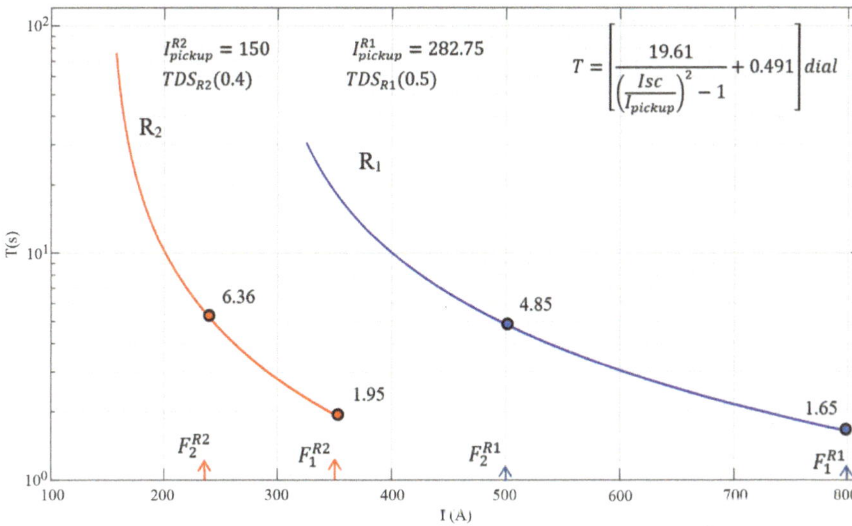

FIG. 6.13 Relay coordination with weak source.

relay (R_1) is greater than the I_{pickup} of the backup relay (R_2). This operating condition is shown in Fig. 6.13, where the comparison of operating time between a relay R_2 and relay R_1 is presented and the relay coordination is obtained. The implementation of transfer trips improves the protection relay significantly because the time operation can be reduced and the weak source contribution is detected.

6.3.2 Islanding detection

Islanding occurs when the capacity of the MG is insufficient to supply the local loads without the support of the Utility; the result of islanding is the loss of energy balance, presenting frequency variations, loss of reactive power support, and loss of inertia, causing damage to RES and sensitive loads. The MG must be disconnected after the loss of connection with the Utility [13]. The IEEE 929-1988 standard establishes the criteria for immediate disconnection of the island, whereas the IEEE 1547-2003 standard proposes a maximum delay of 2 s [14,15].

The antiisland protection systems have been implemented using relays in base frequency and voltage variation for the separation of the MG. The island affects users and owners of the MG by the energy losses; another scheme is to enable control systems in the RES for frequency and voltage control, thus having an island saving scheme. Based on the duration of the short island operation, the control systems could maintain the island without the need for load shedding.

Islanding detection. The main function of the island detection scheme after confirming the power imbalance condition is to disconnect at the point of common coupling (PCC) the MG from the Utility. The objective of this remedial action is to protect the MG from abnormal frequency variations that can cause the unwanted tripping of converters and the damage of small generators such as turbo diesel. The frequency ranges tolerable by converters are shown in detail in [14,15]. For the reconnection of the MG to the Utility, the synchronism of both systems must be verified.

There are two groups of methods for island detection (Fig. 6.14). The methods based on remote schemes use communication channels to establish the link between the Utility and the MG generators. The most widely used methods for simplicity and less use of resources are the local schemes; they can be active or passive depending on the interaction they make with the electrical network. Thus the active methods act to affect the operation of the system by injecting disturbances into the MG. For example, in [16] the method consists of a sliding frequency change, by injecting a disturbance to cause a phase change. Likewise, in [17] the frequency deviation is performed by modifying the waveform, through the injection of a zero amplitude in a short period.

In contrast, passive schemes are based on the supervision of voltage signals to determine the frequency at the MG terminals, with the aim of detecting the island condition. These methods are the most widely used because of their simplicity based on the direct measurement of frequency dynamics. In many proposals, the derivative of the frequency is analyzed through the ROCOF to detect the variation and severity of an abnormal frequency condition [18].

Implementation. A model illustrated in Fig. 6.15 is used to implement the passive method of island detection through ROCOF. The MG is made up of a wind farm with a nominal power of 4 MW, the connection to the distribution network is made through a 13.8/0.48-kV distribution transformer, and the lines

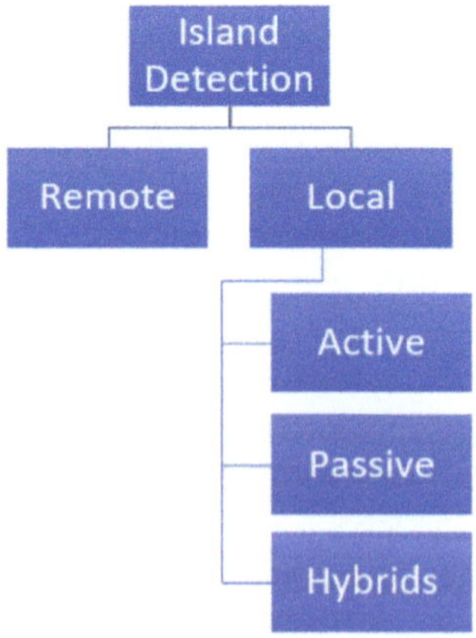

FIG. 6.14 Classification of island detection methods.

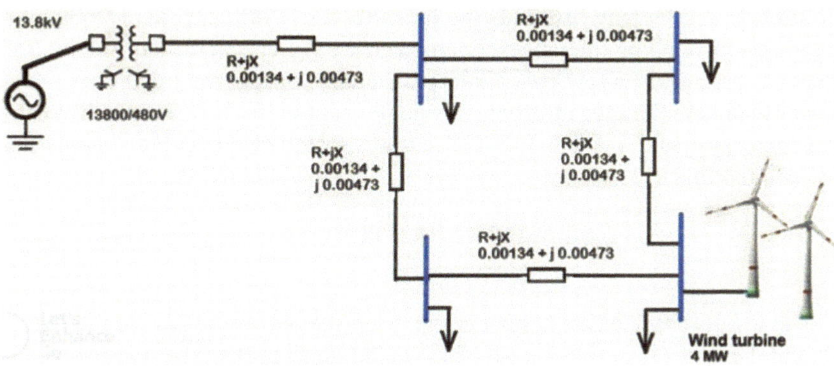

FIG. 6.15 Model of the MG system.

are connected at four nodes. The impedances of the lines and the voltage level of the distribution network are shown in Fig. 6.15. It is common in many works to consider a microgrid composed of a single node for the connection of all the RES. More complex topologies are included in this evaluation to represent MG losses and their impact on voltage and frequency magnitudes.

Frequency change rate method (ROCOF). The method consists of determining the magnitude of the disturbance based on the derivative of the MG frequency. In Fig. 6.16, the MG composed of the connection of an equivalent generator feeding a local load is illustrated. The line with the Utility is controlled by a CB switch, which receives the signal from the detection scheme based on the ROCOF to disconnect the MG. In case the MG is disconnected from the Utility and the energy balance is lost, the frequency will experience a magnitude change and the voltage a phase change. The frequency variation rate df/dt is measured and verified, in case of violating the tolerances the island condition is declared and the MG is disconnected.

The frequency variation measurement is weighted onto the voltage signal over a specified time, typically in a window of two to forty cycles. The number of cycles used for the weighting affects the magnitude of the frequency

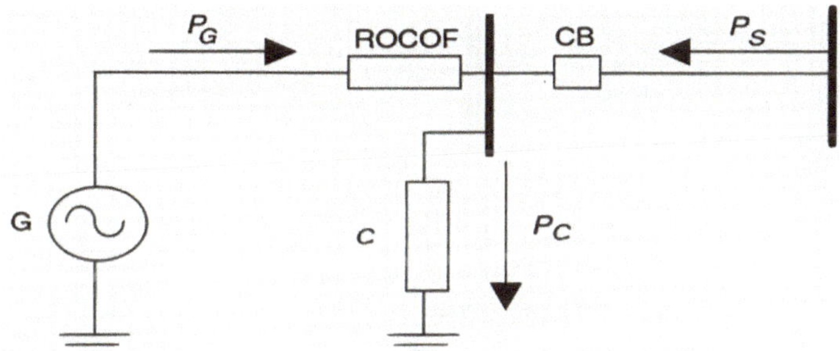

FIG. 6.16 ROCOF scheme.

variation, which is why care must be taken to verify the magnitudes obtained and the corresponding tolerances. Even when the voltage signal has limited frequency components, a low-pass filter is necessary. In case the frequency rate is greater than the tolerance threshold, a signal is sent to trip the CB switch to separate both systems. Typical ROCOF settings installed on 60-Hz systems are between 0.1 and 1.2 Hz/s for a five-cycle average value (6.17).

$$\frac{df}{dt} = \frac{1}{5}\sum_{i=1}^{5}\frac{\Delta f_i}{\Delta t_i} \tag{6.17}$$

The results obtained are shown in Fig. 6.17, the duration of the simulation is 60s and the ROCOF activation for island detection is at 30s. The operation consists of the following stages:

1. The estimation of the phasor of the voltage signal is continuous; it is possible to use a detector to start the estimation, but it does not represent important advantages.
2. The A/D conversion of the voltage signal is performed; a frequency estimation algorithm is used based on the orthogonal sine and cosine components. Algorithms based on the zero crossing of the voltage signal can also be used.
3. Having the estimate of the frequency, its derivative is determined, calculating its rate of change for every five values of the signal (Eq. 6.17). In the event that the disturbance determined by the ROCOF is greater than the threshold, a verification is made by the voltage signal, because during the island operation there will be a voltage depression. If the frequency variation is due to a fault or the start of a generator, the trip signal is blocked.

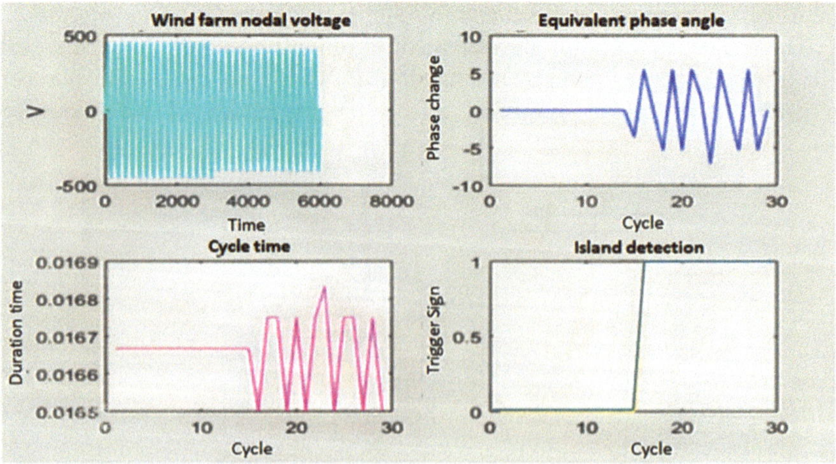

FIG. 6.17 ROCOF for antiisland protection.

```
ROCOF pseudocode
% df/dt (ROCOF)
dt=1/length(data);
ndata=length(data);
in = 1; fin = 5; N = 5;
m = [zeros(1,(ndata-N+1))];
while fin<=(ndata)
  nm = data(1,in:fin); sum=0;
  for j=1:(length(nm)-1)
    sum=sum+((nm(1,(j+1))-nm(1,(j)))/(dt));
  end
  dfdt=(1/2)*sum;
  m(1,in)=dfdt; in = in+1; fin = fin+1;
end

% df/dt trip
disp= [zeros(1,(ndata-N+1))];
for i=1:length(m)-1
    difr=m(1,i); ss=abs(difr);
    if ss<=0.3
       disp(1,i)=0;
    elseif ss>0.3
       disp(1,i:length(m))=1;
       break
    end
end
```

6.3.3 Island saving scheme

The successful operation of an electrical island must maintain an energy balance between generation and load, in case of having an imbalance due to disconnection with the Utility, remedial schemes must be activated to activate the generation control systems and storage systems. Likewise, load control schemes through trigger actions or consumption control must be activated. Island saving schemes are activated in short periods of time, while the MG is reconnecting, or they can be more durable to keep the island operating for long periods of time or permanently.

Intermittent generation sources, in general, are not capable of giving that natural inertial response or frequency regulation response. The frequency response tends to decrease with the increasing penetration of RES in electrical systems and even more so an isolated system fed by intermittent sources, such as wind and photovoltaic plants will be unable to maintain frequency stability. Frequency control in the island saving scheme is based on monitoring the flow power at the *PCC* and the output power at the MG or DG terminals (Fig. 6.18).

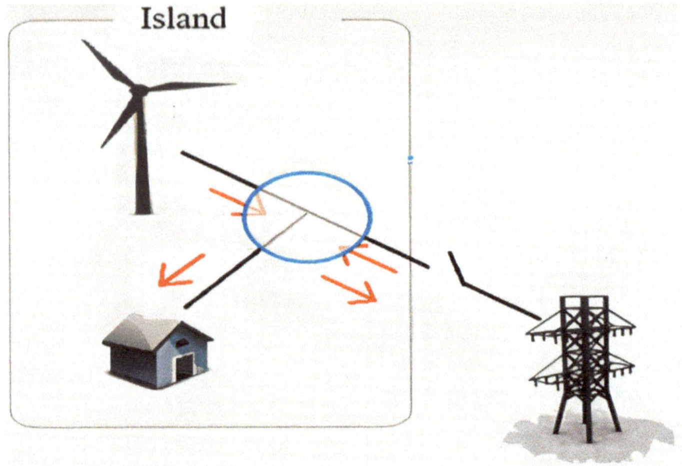

FIG. 6.18 Monitoring of powers in the DG, in PCC, and loads.

Once the island detection signal is received, the registered power value is verified and the amount of excess or missing power in the isolated network after losing the support of the main network. Then the algorithm makes the decision based on the Remedial Action Scheme (RAS) based on the information on power deficit or surplus in the isolated network (Fig. 6.19).

Lack of generation. The action corresponds to eliminating load; it is necessary to carry out a load sensitivity study to trigger the loads that have a greater impact on restoring the frequency.

Excess of generation. The action corresponds to decreasing the amount of power output of the MG. Depending on the magnitude of the imbalance, the RAS must consider the triggering of generation units or activate control actions to regulate the power delivered by the MG's RES or DG. In the case of PV, the control must establish an operating condition outside the maximum power generation point and for the PWs, the pitch control can be activated to regulate the delivered power. In the case of controlled sources such as MV and BESS, the control is simpler because only the island rescue scheme commands the order of the reduction of energy necessary to achieve energy balance.

For the example shown in this section, a wind power plant is considered; the power generated is decreased by controlling the speed of rotation of the blades in the turbine by rotating the speed control angle. The value of the angle of rotation of the blades is estimated based on the characteristic curve of the generator; this curve is a vector of the power extracted from the generator at different wind speeds.

Wind generators are designed to produce electrical power efficiently generally at speeds between 9 and 15 m/s. Wind turbines are designed with a protection that controls the pitch angle of the blades in the turbine and consequently the generated power is controlled. The electronic turbine controller checks the generated power several times per second, which is normally reflected in the

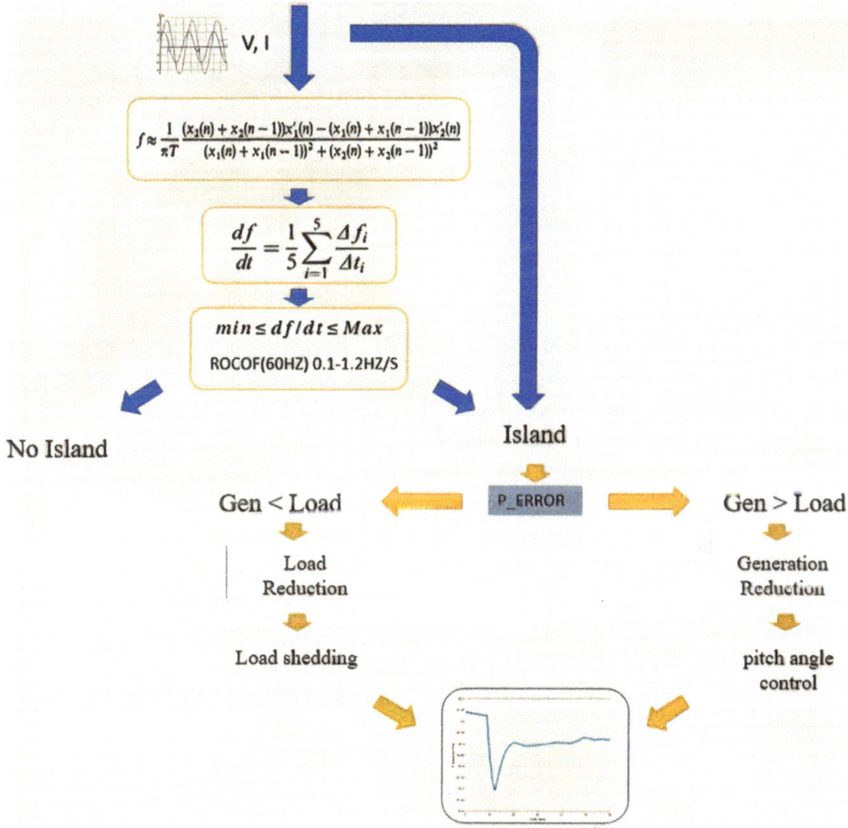

FIG. 6.19 Island detection and island saving scheme.

rotor speed (Fig. 6.20). When it reaches too high a value, the controller sends a command to the pitch angle change mechanism, which immediately turns the rotor blades. The time constant of generating torque is T_g and K_s is the transmission system spring constant.

The power delivered by a turbine depends directly on the speed of the wind and the pitch angle of the blades. The C_p is a function of the pitch angle (β) and the tip speed ratio (λ).

$$C_p(\lambda, \beta) = C_1\left(\frac{C_2}{\lambda_i} - C_3\beta - C_4\beta^{C_5} - C_6\right)e^{\frac{-c^7}{\lambda_i}} \tag{6.18}$$

where λ_i is represented by Eq. (6.19):

$$\lambda_i = \frac{1}{\frac{1}{\lambda + C_8\beta} - \frac{C_9}{\beta^3 + 1}} \tag{6.19}$$

Taking the values of the constants C_1 to C_9 applicable to variable speed wind turbines from [19], the mechanical power extracted from a wind turbine varies

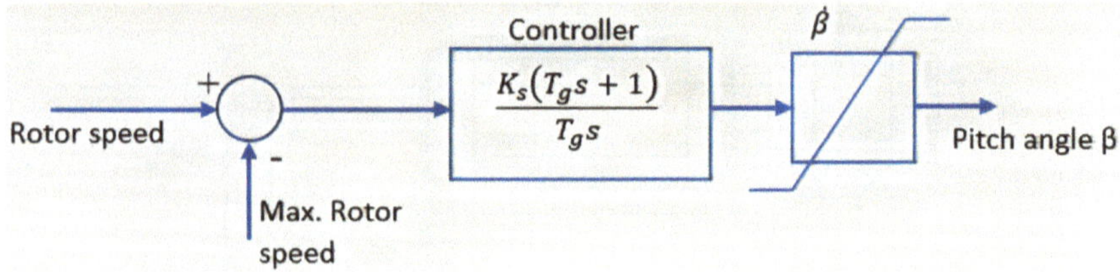

FIG. 6.20 Step angle control chart.

with the value of the angle of the blades. The estimated power values are recorded in the island saving algorithm; these are used once the error power is determined to obtain the corresponding angle. To observe the reduction of the extracted power with the increase of the step angle, a simulation of an algebraic model is carried out to obtain the mechanical power through the wind speed (Fig. 6.21).

The results obtained are reprogrammed in the system and the simulation is carried out with the position of the blades. In Figs. 6.22 and 6.23, the results of the low and high frequency case are presented, respectively. In the low-frequency RAS, the loads closer to the PCC are triggered. In the RAS due to low frequency, the pitch control is enabled. In both cases, the recovery of the frequency can be observed once the actions calculated to save the system are carried out.

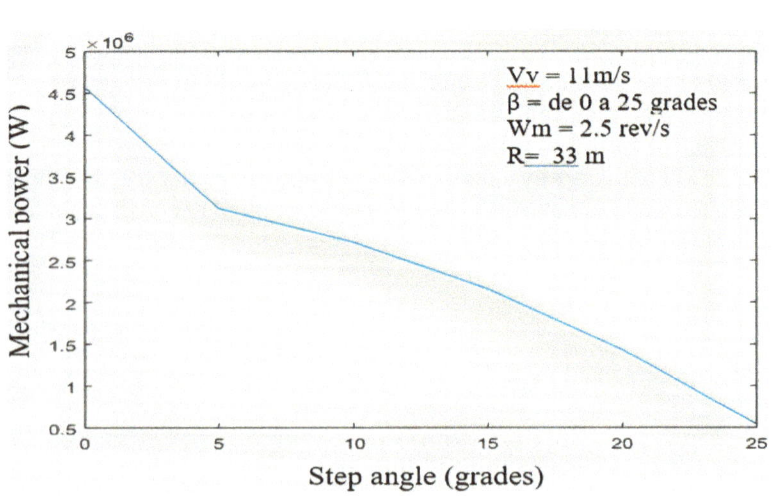

FIG. 6.21 Mechanical power extracted from a wind turbine by varying the angle β.

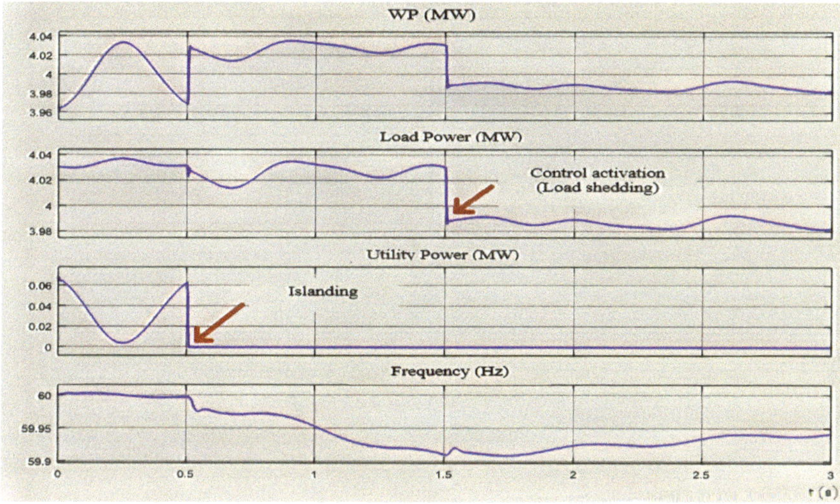

FIG. 6.22 Island frequency control. Low frequency.

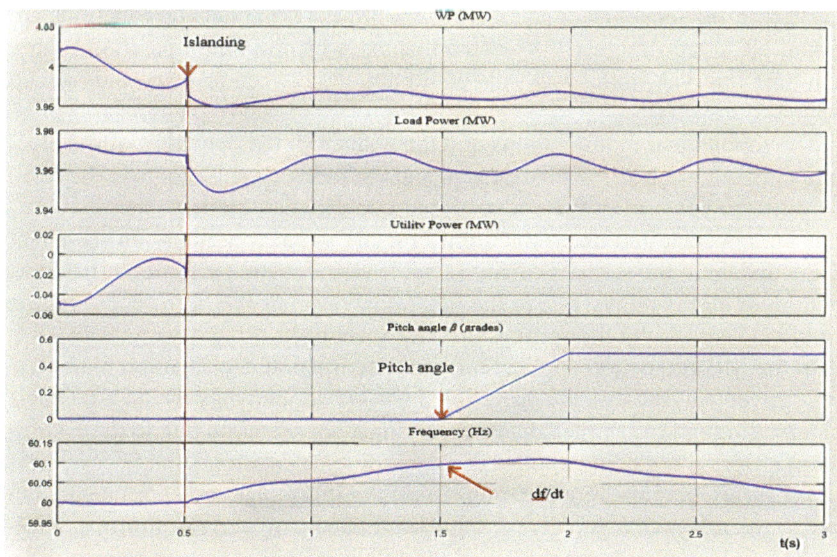

FIG. 6.23 Island frequency control. High frequency.

6.4 DOCRs coordination proposal for MG protection

The MG coordination is solved with optimization algorithms. The proposed formulation, described in Chapter 5, is based on time constraints to obtain relay coordination. The proposed method is illustrated in Fig. 6.24, where the time

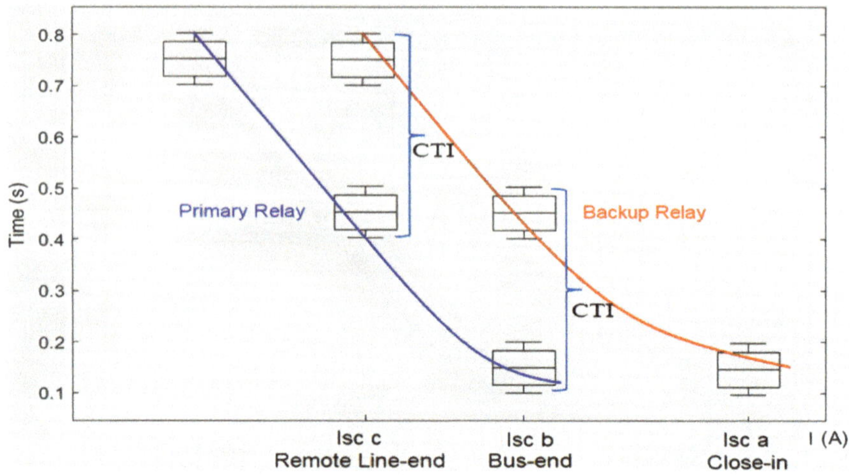

FIG. 6.24 Time constraints for relay coordination.

intervals in the locations of interest for the primary and backup functions are shown. NCC are designed for accomplishing with the time operation within each time interval, in this way, the coordination between primary and backup relays is done.

The topological modifications and the change in the operation of the electrical network, such as the load current and the fault currents, degrade the relay coordination because in the conventional coordination method, the relay settings are constant and determined only for the worst conditions. In the event of any update in the settings of any relay, it would mean performing the coordination of all the relays, due to the functional dependence of the coordination criterion. That is why the settings are rarely changed.

In the MG protection application, the time interval coordination method is very convenient, because the operation times are defined only by the time interval, resulting in lower times than conventional coordination due to the coordination criterion not being applied for relay time calculation. And because it is not necessary to make modifications to all the relays that are coordinated, but only to the relays that perceive a significant change in current. For this, a tolerance can be defined to establish if the relays require a setting change. This allows that in the face of the different operating conditions of the MG, it is possible to maintain a selective and sensitive protection with a change of setting of few relays.

The concept of autonomous protection systems with a centralized adjustment update system, which modifies the settings and coordination according to the diverse dynamic operating conditions of the MG, offers an adaptable alternative to improve the quality of protection.

6.4.1 Academic test system

The evaluation of the proposed method is carried out in the IEEE 9-bus test system; since there is no test system that includes renewable sources, the necessary modifications have been made in this test system. The topology of the system is highly interconnected to obtain more complex test scenarios in the evaluation of optimization methods as the number of coordination pairs increases and so does the number of constraints. The renewable sources that are incorporated into the test system to have different capacities and dynamics are WPP, PV, Turbo diesel, and BESS. The 34.5-kV system has 22 relays with 32 coordination pairs (Fig. 6.25).

Three-phase close-in and phase-to-phase line-end fault currents with the remote-end open are determined. In the tests presented, in case of considering any topology change scenario or during generation changes, the fault currents are updated to weight their impact more adequately on protection. Table 6.2 presents the branch impedances and Table 6.3 the values of I_{sc} and I_{load}.

To solve the coordination, the DE algorithm is used, considering 1000 generations with a population of 80 individuals. The minimum limit of the *TDS* was 0.1 to maintain the security in the operation of the relays avoiding having extremely low times. The algorithm was programmed in Matlab, and Digsilent was used to generate the signals for the test cases. The contingency analysis was carried out to determine the I_{pickup} and obtain the border values of the dynamic scenarios evaluated.

Case 1 *Microgrid connected to the Utility.*

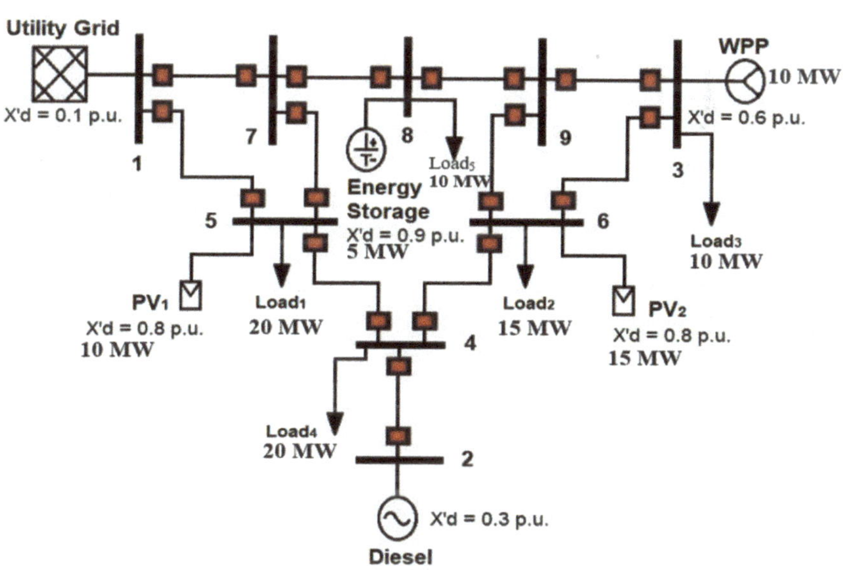

FIG. 6.25 The analyzed microgrid system.

TABLE 6.2 Branch reactance of the test system.

Bus	Bus	X (p.u)
1	5	0.371
1	7	0.292
2	4	0.365
3	6	0.312
3	9	0.381
4	5	0.335
4	6	0.292
5	7	0.255
6	9	0.377
7	8	0.356
8	9	0.352

In this scenario, the MG has a weak connection with the Utility, either because the capacity of the transformer is less than the load demanded by the MG or because there is also a high-impedance link from the transformer to the MG. Under this condition, a contribution to the combined fault between the Utility and the MG is evaluated. A comparison is made between the conventional coordination methods using *NCC* and the proposed method of time intervals. The distinction between both methodologies is that while in conventional coordination there is a functional relationship between relays, the rate of change between the resulting operation times is greater than with the proposed method of time intervals. In the conventional method, the OF used in [20] was used.

The result of the relay operation times is presented in Table 6.4; it is observed that the use of the NCC using the conventional method results in coordinated times, both for the primary and backup relays. However, using the proposed method, the operation times are further reduced because the optimization algorithm only evaluates the time constraints of each relay. The sum of the times for t_p in the conventional method is more than twice the sum of the times of the second interval in the proposed method.

In the time interval coordination method, there are no coordination pairs, since the commitment of each relay is that its time curve complies with the time intervals. In contrast, in the conventional coordination method, coordination pairs must be formed to meet the coordination criterion. Furthermore, in the case of relay 39, it is visible that the backup time obtained with the proposed method (0.408) is much lower than with the conventional method (0.546 and 0.818).

TABLE 6.3 Power flow and fault currents (normal operation).

Relay		I_{sc} (A) Normal operation		I_{sc} (A) Island mode		I_{load} (A)	
Primary	*Backup*	*Primary*	*Backup*	*Primary*	*Backup*	*Primary*	*Backup*
39	63	5454	2729	5343	2606	499	314
45	64	5144	2679	5112	2646	398	162
46	54	5400	2941	4553	2085	162	398
57	45	7914	2428	5347	2321	139	398
69	36	6902	2256	6517	2153	247	314
69	46	6902	2621	6517	2319	247	162
78	57	6945	3090	2929	2096	88	139
89	78	4508	2711	3582	1762	208	88
51	45	7246	2340	5689	2274	1109	398
51	75	7246	2901	5689	1372	1109	139
71	57	5284	3305	4399	2446	1252	139
71	87	5284	1998	4399	1962	1252	88
63	46	6576	2594	6091	2264	314	162
63	96	6576	1959	6091	1783	314	247
93	69	4695	2509	4254	2397	499	247
93	89	4695	2195	4254	1866	499	208
54	75	7826	2660	3722	1190	398	139
64	36	5331	1788	5222	1756	162	314
64	96	5331	1483	5222	1400	162	247
75	87	6210	2074	3481	2013	139	88
96	39	4503	2290	4152	2236	247	499
96	89	4503	2223	4152	1925	247	208
87	98	3924	2104	3904	2081	88	208
98	39	3941	1847	3886	1832	208	499
98	69	3941	2103	3886	2062	208	247

TABLE 6.4 Comparison between the conventional coordination method with NCC and the interval method.

Conventional method (NCC)				Proposed interval method			
Relay Primary-backup	*tp*	*tb*	*CTI_c*	*Relay*	*t (Isc_a)*	*t (Isc_b)*	*t (Isc_c)*
39–63	0.188	0.494	0.306	36	0.132	0.402	0.704
45–64	0.355	0.684	0.329	**39**	0.162	**0.408**	0.732
46–54	0.315	0.697	0.382	45	0.174	0.401	0.706
57–45	0.332	0.648	0.315	46	0.178	0.404	0.700
69–36	0.238	0.541	0.302	57	0.187	0.403	0.711
69–46	0.238	0.541	0.303	69	0.147	0.414	0.703
78–57	0.331	0.642	0.310	78	0.189	0.400	0.760
89–78	0.301	0.618	0.316	89	0.171	0.436	0.704
51–45	0.203	0.670	0.467	63	0.144	0.402	0.702
51–75	0.203	0.506	0.302	54	0.172	0.426	0.707
71–57	0.226	0.618	0.392	64	0.193	0.412	0.707
71–87	0.226	0.623	0.398	75	0.172	0.420	0.714
63–46	0.234	0.546	0.312	96	0.178	0.400	0.769
63–96	0.234	0.537	0.303	87	0.194	0.400	0.727
93–69	0.058	0.512	0.454	98	0.188	0.468	0.702
93–89	0.058	0.568	0.510	Total:	∑2.58	**∑6.19**	∑10.74
54–75	0.226	0.540	0.314	Violations (NV)			0
64–36	0.365	0.734	0.368	Computation time (s)			19.36
64–96	0.365	0.748	0.383				
75–87	0.285	0.604	0.319				
96–**39**	0.246	**0.546**	0.300				
96–89	0.246	0.561	0.315				
87–98	0.340	0.642	0.302				
98–**39**	0.313	**0.818**	0.505				
98–69	0.313	0.619	0.306				
Total:	∑6.43	**∑15.25**	∑8.81				
Violations (NV)			0				
Computation time (s)			24.71				

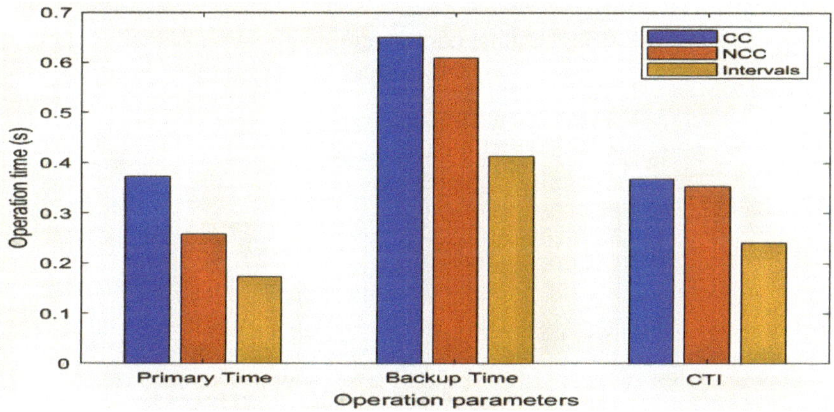

FIG. 6.26 Average relay times of coordination methods.

The times resulting from the interval method are within the times defined in each interval, guaranteeing that the relays are coordinated without using the coordination criterion, which forces a functional dependency.

In Fig. 6.26, the average times of the analyzed coordination methods are shown, the conventional coordination method is evaluated with conventional time curves (CC) and with NCC, and the proposed method of time intervals is included. The graph shows that the proposed method of intervals obtains the shortest times for both primary and backup protection, in addition, the *CTI* is closer to 0.2 s, obtaining better coordination.

Due to the lack of flexibility achieved with increasing degrees of freedom, the conventional coordination with CCs presents many violations, the results are unfavorable, and it does not represent an alternative solution for the proposed method of coordination; the results are presented in Table 6.5.

The proposed method of coordination of time intervals determines the necessary time curves to satisfy the operation times with the fault currents in each interval. In the previous results, interval times were analyzed from 0.1 to 0.2 in the first interval, in the second from 0.4 to 0.5, and the third used to guarantee sensitivity is from 0.7 to 0.8. The results obtained were satisfactory, obtaining the times within each interval and having zero violations. It is possible to use different intervals with the proposed methodology, thus the evaluation at different intervals will allow to reduce the operation times even more. It is important to maintain a *CTI* between intervals to ensure coordination. However, there is a risk of violations when the intervals are very close, and it is not possible to find a suitable time curve. In the electrical network studied, different time intervals were evaluated, there were no violations, and operating times were reduced (Fig. 6.27). The operating times obtained for the intervals from 0.1 to 0.15, 0.35 to 0.4, and 0.6 to 0.65 are presented in Table 6.6.

Case 2 *Microgrid connected to the Utility.*

TABLE 6.5 Conventional coordination with conventional curves.

Relay Primary-Backup	tp	tb	CTI_c	Relay Primary-Backup	tp	tb	CTI_c
39 - 63	0.208	0.513	0.305	63 - 96	0.267	0.609	0.342
45 - 64	0.368	0.670	0.301	93 - 69	0.239	0.678	0.438
46 - 54	0.550	0.906	0.355	93 - 89	0.239	0.638	0.398
57 - 45	0.505	0.807	0.303	54 - 75	0.440	0.247	-0.193
69 - 36	0.379	0.683	0.304	64 - 36	0.561	0.970	0.409
69 - 46	0.379	0.679	0.301	64 - 96	0.561	0.871	0.310
78 - 57	0.123	0.560	0.438	75 - 87	0.214	0.562	0.348
89 - 78	0.430	0.133	-0.297	96 - 39	0.322	0.627	0.305
51 - 45	0.455	0.844	0.389	96 - 89	0.322	0.632	0.310
51 - 75	0.455	0.241	-0.214	87 - 98	0.499	0.804	0.305
71 - 57	0.251	0.554	0.303	98 - 39	0.520	0.943	0.422
71 - 87	0.251	0.558	0.308	98 - 69	0.520	0.834	0.313
63 - 46	0.267	0.683	0.416				
Number of Violations (NV)					3		
Computation Time DE (s)					17.35		

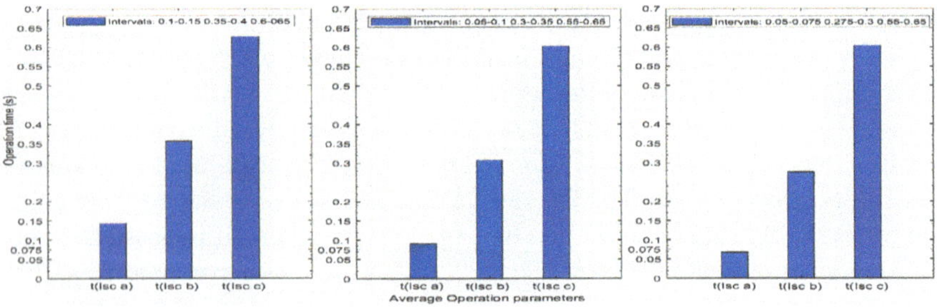

FIG. 6.27 Different time intervals for the proposed coordination method.

Different contingencies were studied to evaluate the performance of the coordination methods under different operating conditions of the electrical network. The robustness of the coordination methods is determined since different coordination problems arise under different operating conditions. In each contingency the coordination was formulated, and the results shown are obtained in

TABLE 6.6 Operation times defining different preestablished intervals.

Proposed interval method							
Relay	t *(Isc_a)*	t *(Isc_b)*	t *(Isc_c)*	Relay	t *(Isc_a)*	t *(Isc_b)*	t *(Isc_c)*
36	0.149	0.367	0.604	63	0.121	0.361	0.643
39	0.143	0.351	0.649	54	0.146	0.358	0.601
45	0.150	0.350	0.629	64	0.141	0.353	0.635
46	0.148	0.350	0.605	75	0.148	0.365	0.602
57	0.147	0.350	0.619	96	0.146	0.363	0.645
69	0.125	0.351	0.616	87	0.150	0.351	0.606
78	0.149	0.368	0.607	98	0.146	0.383	0.601
89	0.140	0.360	0.602				
Violations (NV)				0			
Computation time (s)				17.83			

each specific condition. The contingencies considered are shown in Fig. 6.28; the generated output is analyzed with the tripping of the turbo diesel to affect the magnitudes of the short-circuit current; likewise, the island operation of the MG is evaluated.

To evaluate the effect of coordination under different current flow conditions, the operating times are determined by triggering lines 4–7 and 1–7. Criteria such as the average operating time of the relays, the number of coordination pairs for the conventional method, and the number of relays adjusted for the proposed method are evaluated. Finally, the execution time of both algorithms is weighted. In all cases, the proposed method of intervals was better, obtaining shorter operation and execution times for all the studied contingencies.

Case 3 *Microgrid disconnected from the Utility.*

In this case study, the island operation of the MG is evaluated. During this operating condition the short-circuit currents are reduced in magnitude and duration, since the contributing sources are generally RES and small sources such as turbo diesel (Fig. 6.29). It is to be expected that with the reduction of short-circuit currents, the operating time of the relays will increase. It is also possible that the sensitivity of relays is lost because they keep their I_{pickup} constant. In these cases, it is very convenient to activate the adjustment groups when the island operation of the MG is detected.

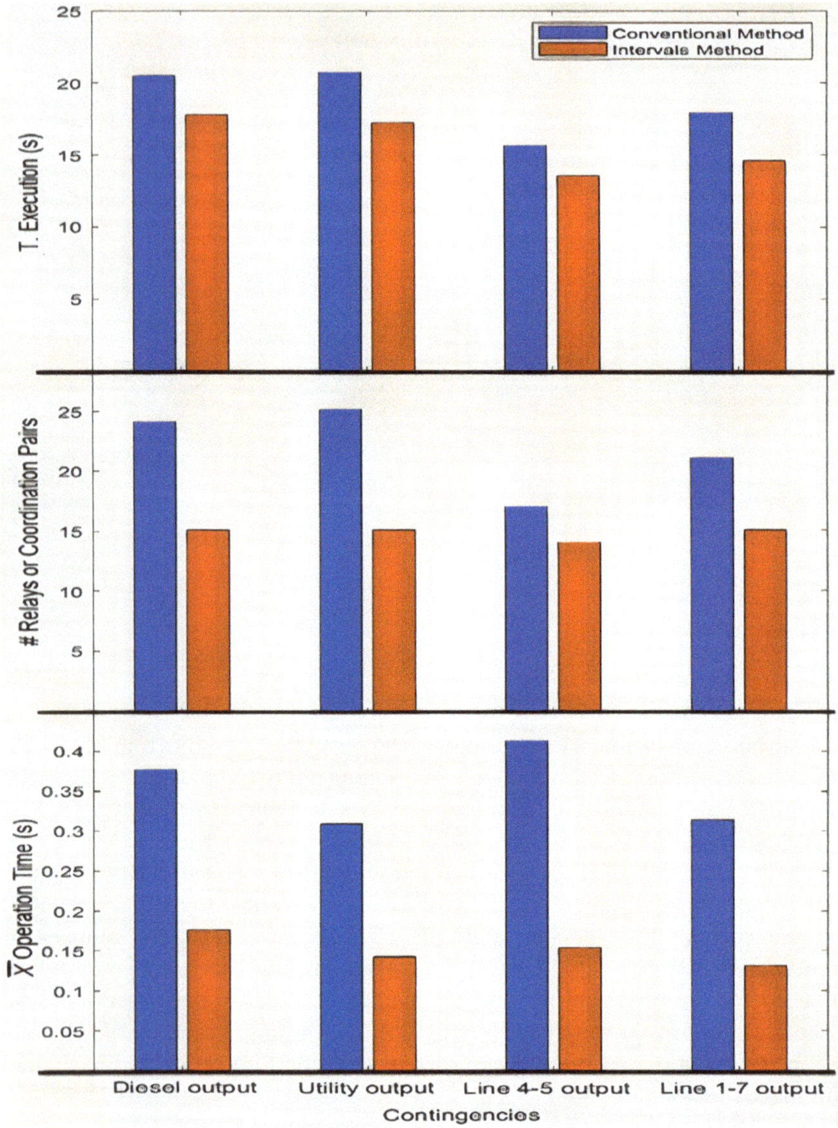

FIG. 6.28 Comparison between coordination methods under contingency analysis.

The evaluation of the coordination in these operating conditions consists in determining the operating times. It is a condition that represents a greater coordination requirement, because with a reduction in the fault current it is likely that coordination will be lost or that the operating times obtained from the

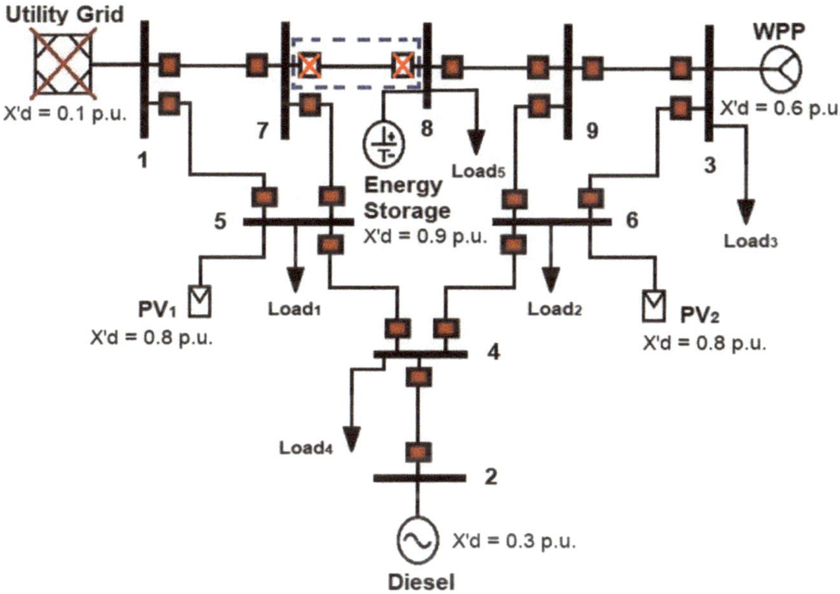

FIG. 6.29 The analyzed microgrid operating in island mode.

algorithms will be long. This condition occurs in the conventional coordination method, as shown in Fig. 6.30. Large operation times are observed for both primary and backup protection, a very large difference is observed between the operation times of the relays. In the case of the method proposed by intervals, much shorter times are observed and because the algorithm tries to determine the time curve that complies with the intervals, very similar operating times between relays are observed. This is important as fault release times must meet coordination time constraints for both equipment protection and power quality.

During the island operation, the output of line 7–8 is simulated to evaluate the coordination obtained with both coordination methods during topological changes. The adjustment of the coordination algorithm is made for this contingency, and under the resulting adjustments the operation times of the relays are weighted (see Table 6.7). It is observed in the table that the operation times of the NCC present a violation.

On the other hand, in the proposed method of intervals, it is only required to make the change of adjustment and coordination of 4 relays that are affected by the variation of the load current and the fault current; a tolerance of ±10% was considered to change the configuration of the relays. Thus the proposed interval algorithm only requires processing 4 relays, thus its computation time is significantly reduced (Fig. 6.31). The relay settings for the interval coordination method are presented in Table 6.8.

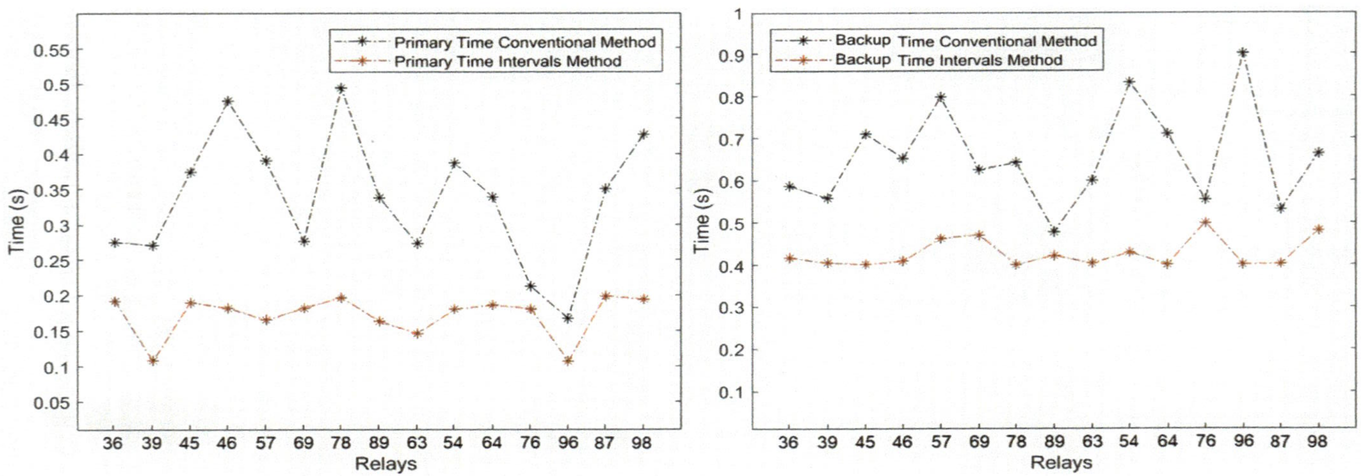

FIG. 6.30 Islanding condition: (A) primary relay times and (B) backup relay times.

TABLE 6.7 Conventional coordination with NCC, line 7–8 output.

Relay Primary-Backup	*tp*	*tb*	*CTI_c*	*Relay Primary-Backup*	*tp*	*tb*	*CTI_c*
39 - 63	0.147	0.460	0.313	71 - 57	0.674	0.021	-0.654
45 - 64	0.175	0.476	0.301	63 - 46	0.226	0.526	0.300
46 - 54	0.385	0.693	0.307	63 - 96	0.226	0.526	0.300
57 - 45	0.020	0.432	0.413	93 - 69	0.159	0.484	0.324
69 - 36	0.188	0.494	0.306	64 - 36	0.321	0.625	0.304
69 - 46	0.188	0.526	0.338	64 - 96	0.321	0.746	0.425
51 - 45	0.130	0.432	0.302	96 - 39	0.345	0.654	0.309
Number of Violations (NV)					1		
Computation Time DE (s)					13.27		

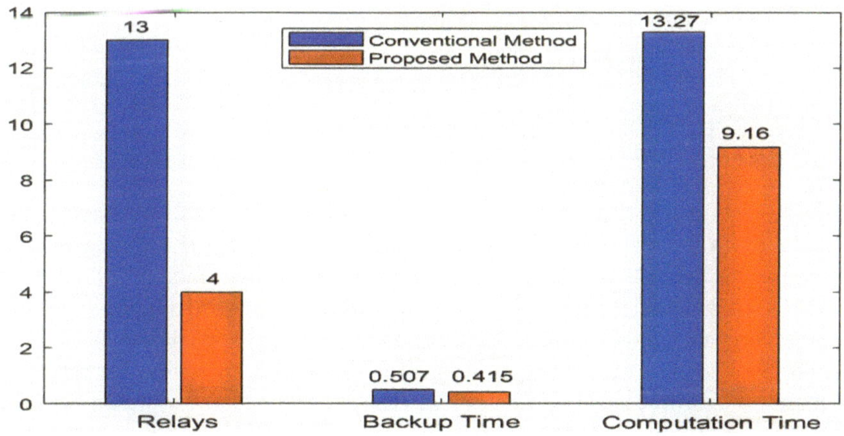

FIG. 6.31 Comparison between the conventional and proposed methods under output of line 7–8.

6.5 Summary

The dynamic condition of MGs makes it necessary to find new strategies so that the electricity grid provides reliable and secure power users. In this study, an interval coordination method was proposed. The presented formulation guarantees the coordination of the DOCR through preestablished time intervals for the primary and backup functions. The functional independence between relays is obtained because each time curve is computed to comply with the time intervals. The nonconventional time curves obtained with the proposed algorithm

TABLE 6.8 Settings for the time curve. Relays operating in island mode.

Relay	(TDS)	(PSM)	(A)	(B)	(p)
36	0.699	1.571	6.876	0.199	1.861
39	0.149	1.536	12.037	0.17	1.66
45	0.189	1.462	5.901	0.339	1.041
46	0.664	1.403	25.246	0.139	1.649
57	0.835	1.464	26.176	0.131	1.686
69	0.420	1.470	14.248	0.235	1.519
78	0.442	1.549	22.072	0.124	1.424
89	0.360	1.507	27.077	0.153	1.839
63	0.251	1.403	12.761	0.155	1.246
54	0.240	1.589	11.567	0.273	1.751
64	0.245	1.403	24.009	0.178	1.127
75	0.702	1.522	20.716	0.163	1.751
96	0.147	1.407	12.435	0.113	1.247
87	0.445	1.592	25.472	0.138	1.318
98	0.684	1.481	23.575	0.116	1.918

provide the flexibility necessary to comply with the preestablished operating ranges. From the results, the proposed coordination method shows less operation time for the relays than the conventional coordination method in the MG systems studied, and the resulting operating times between the relays have a flat profile, indicating similar operating times for the primary and backup protection functions. The proposed coordination method is very advantageous for network dynamic operations such as topological changes and island formation, as it is necessary to update the setting only in the affected relays, obtaining lower operating times than conventional methods of coordination in the cases studied. The proposed coordination method was applied to MG networks because of their highly dynamic operating conditions, but it can be applied to other electrical systems, such as subtransmission or distribution systems.

References

[1] IEEE Application Guide for IEEE Std 1547™, IEEE Standard for Interconnecting Distributed Resources with Electric Power Systems, 2008, pp. 60–61.

[2] Standard, U.L. "1741", Inverters, Converters, and Controllers for Independent Power Systems, 2002.

[3] G.L. Johnson, Wind Energy Systems, Electronic Edition, Manhattan, KS, 2010, pp. 4-1–4-54.
[4] H. Chen, T.N. Cong, W. Yang, C. Tan, Y. Li, Y. Ding, Progress in electrical energy storage system: a critical review, Prog. Nat. Sci. 19 (2009) 291–312.
[5] H. Louie, Off-Grid Electrical Systems in Developing Countries, Springer, Electrical and Computer Engineering Seattle University, 2018, p. 140, ISBN: 978-3-319-91889-1.
[6] H. Laaksonen, K. Kauhaniemi, Voltage and Frequency Control of Low Voltage Microgrid with Converter-Based DG Units, vol. 1, Published by International Science Press, 2009. No. 1. January-June.
[7] H. Wu, X. Liu, M. Ding, Dynamic economic dispatch of a microgrid: mathematical models and solution algorithm, Electr. Power Energy Syst. [Online] 63 (2014) 336–346. Available: https://www.sciencedirect.com/science/article/abs/pii/S0142061514003482.
[8] H.A. Kiehne, Battery Technology Handbook, second ed., Marcel Dekker Inc, 2003, ISBN: 0-8247-4249-4.
[9] K.C. Divya, J. Østergaard, Battery energy storage technology for power systems—an overview, Electr. Power Syst. Res. 79 (2009) 511–520.
[10] A. Poullikkas, A comparative overview of large-scale battery systems for electricity storage, Renew. Sust. Energ. Rev. 27 (2013) 778–788.
[11] R. Bussar, et al., Battery energy storage for smart grid applications, EUROBAT (2013).
[12] A. Castillo, X. Jiang, D.F. Gayme, Lossy DCOPF for Optimizing Congested Grids with Renewable Energy and Storage, in: American Control Conference (ACC) June 4–6, 2014. Portland, Oregon, USA, 2014.
[13] S. Chowdhury, C.F. Ten, P. Crossley, Islanding protection of distribution systems with distributed generators: a comprehensive survey report, in: IEEE Power and Energy Society General Meeting, IEEE, 2008, pp. 1–8.
[14] S.I. Jang, K.H. Kim, An islanding detection method for distributed generations using voltage unbalance and total harmonic distortion of current, IEEE Trans. Power Deliv. 19 (2004) 745–752.
[15] 929-2000 - IEEE Recommended Practice for Utility Interface of Photovoltaic (PV) Systems, (N.d.) (2000).
[16] G. Smith, P. Onions, D. Infield, Predicting islanding operation of grid connected PV inverters, IEE Proc. Electr. Power Appl. 147 (1) (2000) 1.
[17] M. Ropp, M. Begovic, A. Rohatgi, Analysis and performance assessment of the active frequency drift method of islanding prevention, IEEE Trans. Energy Convers. 14 (3) (1999) 810–816.
[18] J. Vieira, W. Freitas, W. Xu, A. Morelato, Efficient coordination of ROCOF and frequency relays for distributed generation protection by using the application region, IEEE Trans. Power Deliv. 21 (2006) 1878–1884.
[19] J. Slootweg, H. Polinder, W. Kling, Representing wind turbine electrical generating systems in fundamental frequency simulations, IEEE Trans. Energy Convers. 18 (2003) 516–524.
[20] M. Shih, A. Conde, T. Hsiao, L. Torres, Enhanced differential evolution algorithm for coordination of directional overcurrent relays, Electr. Power Syst. Res. 143 (2017) 365–375, https://doi.org/10.1016/j.epsr.2016.09.011.

Chapter 7

Sensitivity enhancement via adaptive methods for phase DOCR in active electrical networks with high CIGRs connection

7.1 Sensitivity of directional overcurrent relays overview

The protection of medium- and low-voltage electrical networks has been carried out mainly with the use of directional overcurrent relays (DOCR) due to the operational advantages, tolerance to temporary overloads, and easy coordination with primary equipment damage curves. Sensible solutions can focus on the pickup current (I_{pickup}) since modifications in the time curve or coordination improve the operating times, but it does not increase sensitivity. The lack of sensitivity of the relays is very likely and can cause current faults (I_{sc}) in the electrical network to be undetected, or that due to the reduced values of the fault current, the operating times are very long, causing long-lasting voltage sags. High impedance fault (HIF) schemes can be included if selectivity and security of the protection operation are guaranteed [1]. Also, the use and coordination of distance relays to substitute overcurrent relays with coordination conflicts is proposed [2,3].

In case of microgrids (MG), a minimal fault can be very common under poor utility connection or under isolated operation [4]. The connection of MG with the utility does not represent sensitivity problems generally, but the similar fault currents in all buses of MG can be frequent. The use of change of current polarity when the fault occurs gives an alternative for fault detection with inverter-based microgrid [5,6]. The benefits of having micro-phasor measurement units (μPMUs) to determine faults rates to determine their location [7,8] and the use of traveling waves to estimate the faulty area in a wide area protection scheme [9] allow to improve protection functions.

Diverse objective functions have been proposed in optimization algorithms for DOCRs coordination where the sum of the operating times of all the primary relays is the priority; however, the backup function can also be considered to

Overcurrent Relay Advances for Modern Electricity Networks. https://doi.org/10.1016/B978-0-323-91777-3.00007-6

evaluate the backup operation time. Also, the sensitivity of the relays must be considered, because depending on the type, operating condition, and topology of the electrical network, some relays may have a lack of sensitivity [10,11]. These relays can be excluded from the formulation of the coordination algorithms to avoid violations and increase in solution times. But it is clear, that these solutions hide a serious protection problem.

The activation schemes for setting groups of relays must be evaluated, verifying that all the different operating conditions of the electrical network have been considered. The setting groups have the advantage that they are incorporated in the control logic of the current relays; however, they can hardly cover all the operating scenarios of the electrical network. For this alternative, it is necessary to have a central computer that calculates and sends the new settings through communication channels. Also, the use of auxiliary relays with communication links to perform backup functions has been proposed as a substitute for time slow relays resulting from the coordination process to speed up the protection scheme [12].

The change of the I_{pickup} of the relay either by activation of the setting group or by load current (I_{load}) will affect the coordination because the time curve is shifted horizontally. In this way, the change in the relay setting will affect the neighboring relays having a domino effect, and requiring the coordination of all or most of the relays. This problem is more complex as the network becomes more meshed, but similarly the problem occurs in radial networks. Therefore it is necessary that before any change in the I_{pickup}, the coordination of the relays must be carried out. The relay coordination can be done automatically using optimization algorithms [13]. The optimization algorithm for online execution must show the robustness necessary for protection systems to carry out the correct operation of the protections. Chapter 5 presents a coordination method based on time intervals to avoid the coordination of all relays, only the setting of the relay(s) to which the setting change was made is performed.

Overcurrent relay sensitivity. The operation of DOCRs is very simple which makes this protection scheme to be widely used, especially on interconnected systems. For proper relay operation, the relay setting must be greater than the maximum I_{load} and below the minimum fault current on the protected line. The operation of the relay is ensured by normally setting its I_{pickup} greater than the maximum possible I_{load} by using a plug setting multiplier (*PSM*) as shown in:

$$I_{pickup} = I_{load} \times PSM \tag{7.1}$$

The *PSM* is a factor that prevents operations under normal conditions caused by temporary overloads, power transfer, and errors in current transducers. The *PSM* values should ensure the reliability of the relay operation and at the same time prevent this current from exceeding the minimum fault current that the relay can detect, 1.5–2.0 is suggested in [14].

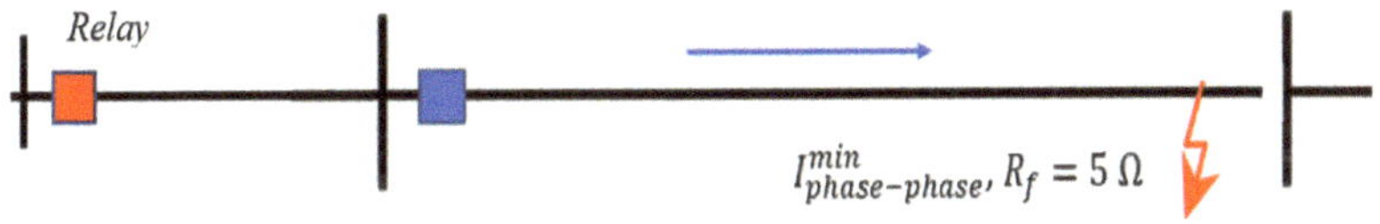

FIG. 7.1 Fault current evaluated for relay sensitivity.

To determine the I_{load}, the operating scenarios that increase the current during line or generator operations and that may result in power transfer must be considered. Therefore an analysis of contingencies $n-1$ is carried out.

The relay must be able to detect minimal faults at the adjacent remote line-end. Phase-to-phase faults must be evaluated and a resistive fault of 5 Ω can be considered (Fig. 7.1).

As a result of the disconnection of a section of the electrical network due to the presence of a fault or abnormal operating conditions, the isolated section may be in an unintentional island mode. It is common for the electrical island to present unsustainable energy imbalances, and that the short-circuit current profiles are very low due to the contribution of weak sources or due to renewable energy sources (RES). The most critical scenarios of loss of sensitivity occur in medium- and low-voltage networks, especially in MG or active networks with high CIGRs connection.

7.2 The lack of sensitivity in active networks with CIGRs connection

The connection of smaller networks with the utility is beneficial because the deficiencies of active and reactive power will be provided by the company maintaining a control of frequency and voltage. Likewise, the main contribution of the fault current will be from the utility; it can result in flat fault current profiles for the network buses. This does not represent a problem of coordination of the relays, but will result in high operating times. In networks with a high presence of RES, the operating time during islanding condition can be short due to the intermittency of the sources and their lack of dispatchability. In case of having controllable energy support such as turbo diesel or battery bank [15,16], the operating times on the island can be increased; to achieve this, energy resource management systems may be required [17–19]. Due to the high connection of CIGRs, it can be very common that the protections do not have sufficient sensibility for fault detection because of low-fault currents.

The test system used (Fig. 7.2A) is derived from the IEEE 9 bus with a selected voltage of 34.5 kV, to which DER, loads, and lines are incorporated, seeking to evaluate the proposal and the coordination method in a microgrid with a higher degree of complexity. The system has 12 coordination pairs; the fault analysis of the maximum three-phase current seen by the relays and the minimum fault for each bus considering the open remote end are determined

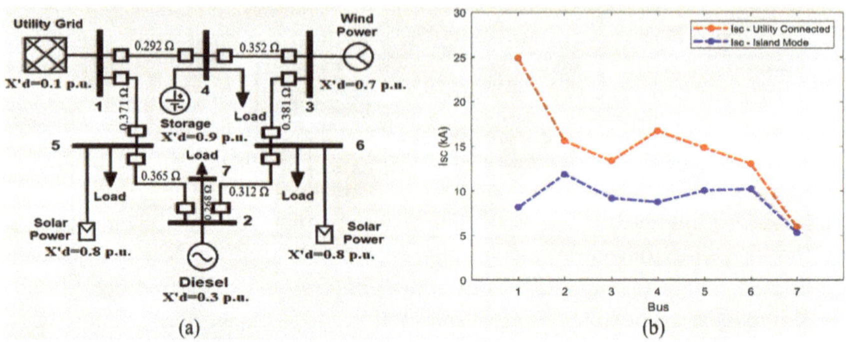

FIG. 7.2 Active networks with CIGRs connection. (A) electrical network and (B) fault current profiles.

(Fig. 7.2B). A minimum *TDS* of 0.1 is used in each case, which allows obtaining the time ranges required. Matlab software is used for simulation and DigSilent is used for $n-1$ contingency for the calculation of I_{pickup} and islanding conditions.

Each relay is assigned a name based on the busbar where it is located and the line to be protected. Different operation scenarios are analyzed to define the setting parameters of each relay, as well as the minimum contribution to the faults presented at the remote end of the line to be protected as a backup; likewise, the I_{pickup} is established and the sensitivity is determined; the values are presented in Table 7.1. The MG is analyzed for two operative conditions, utility connection and islanding connection. The lack of sensitivity relays is flagged with red numbers; we can see that of the 12 pairs of relays, 5 relays have a lack of sensitivity with utility connected operation, and for islanding 7 relays are affected.

Sensitivity is most affected when fault current values are reduced during minimum demand. The results of the evaluation of the sensitivity of the relays during the minimum fault current contribution of 0.7 times of maximum demand are presented in Table 7.2.

7.2.1 Negative sequence relays

Negative sequence relays (67Q) are widely used for the protection of electrical networks by solving sensitivity problems in electrical networks. It detects unbalances in the network and provides greater sensitivity in scenarios where it is lost by phase overcurrent relays. The imbalance during normal operative conditions of electrical systems must be considered to avoid wrong operations. Its configuration is a function of the contribution to the minimum phase-to-phase fault current presented at the remote end of the line to be protected.

The negative sequence overcurrent function can be enabled with the following setting criteria. The setting criterion is shown in Fig. 7.3.

TABLE 7.1 Sensitivity evaluation results.

Backup relay	I_{pickup} (A)	Utility connected		Island mode	
		$Isc_{2\phi}$ (A)	Sensitivity	$Isc_{2\phi}$ (A)	Sensitivity
51	1718	2750	1.60	3296	1.92
41	1992	2144	1.08	2909	1.46
62	2428	3380	1.39	2920	1.20
52	2916	4016	1.38	2758	0.95
63	909	3345	3.68	3131	3.44
43	800	4186	5.23	2676	3.34
34	800	3729	4.66	3170	3.96
25	2916	4122	1.41	3914	1.34
15	1718	3850	2.24	2554	1.49
36	909	3496	3.85	2855	3.14
14	1992	4948	2.48	2813	1.41
26	2428	3122	1.29	2775	1.14

TABLE 7.2 Relay sensitivity during minimal current contribution.

Backup relay	I_{pickup} (A)	Utility connected		Island mode	
		$Isc_{2\phi}$ (A)	Sensitivity	$Isc_{2\phi}$ (A)	Sensitivity
51	1925	2750	1.120	3296	0.584
41	1500	2144	0.753	2909	0.516
62	2366	3380	0.974	2920	0.810
52	2811.2	4016	0.964	2758	1.019
63	2341.5	3345	2.576	3131	0.748
43	2930.2	4186	3.663	2676	1.095
34	2610.3	3729	3.263	3170	0.823
25	2885.4	4122	0.990	3914	0.737
15	2695	3850	1.569	2554	1.055
36	2447.2	3496	2.692	2855	0.857
14	3463.6	4948	1.739	2813	1.231
26	2185.4	3122	0.900	2775	0.788

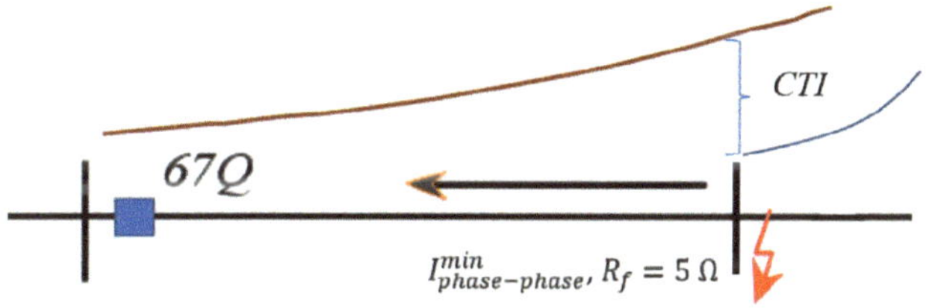

FIG. 7.3 Negative sequence relay.

(a) The I_{pickup} of negative sequence relay must be the 50% of the minimal close in phase-to-phase faults.

$$I^{Q}_{pickup} = 0.5{*}I^{min}_{phase-phase} \tag{7.2}$$

(b) It must backup the primary relay of the protected line for phase-to-phase and phase-to-ground faults. Verify that a coordination margin is maintained with the remote bus line protection, for phase-to-phase "close-in" faults.
(c) Its application should be evaluated in networks with high imbalances and by asymmetric contributions of the RES.

67Q application limitation MG networks. Dynamic studies on the contribution of faults by RES should be evaluated, since the high-frequency contamination of the currents can affect the operation and detection of the negative sequence relays. It can have complications in their adjustments due to a high imbalance in the electrical network, causing unwanted tripping during load current or external fault. In some network codes, it is established that the converters must deliver negative sequence current in order to achieve fault detection.

In addition, the complexity of the coordination of DOCRs is greater between positive and negative sequence relays, being necessary to carry out the coordination for each type of relay. Also, a phase-to-phase fault can cause nonselective operation because both relays exhibit different dynamics for the same fault. Due to configuration, a negative sequence backup relay may overreach a positive sequence relay and a loss of selectivity occurs.

7.3 Sensitivity detection system

The protection of active networks is mainly through directional overcurrent relays, due to the multiterminal contribution to the fault, the tolerance to overload conditions, and the simplicity in coordination with fuses and reclosers, which are generally used in the protection of feeders. It is possible to use other protection principles such as differential protection; however, this principle does not offer the backup function. Another alternative is to use distance relays in specific locations of the active network to replace overcurrent relays where it is not possible to improve their sensitivity.

The online determination system is proposed for vulnerable zone detection; the system executes routines at specific time intervals for fault calculation under topological changes. Power contributions, whether it is the utility, batteries, microturbines, or RES is actualized in each predefined time interval. In addition, the proposed system identifies the different states of demand based on the load profiles, already known in the electrical network for updating the

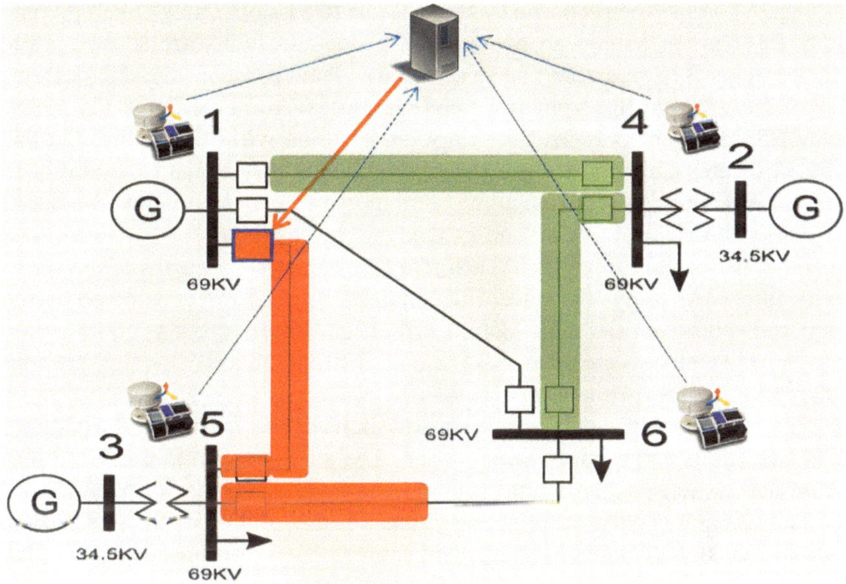

FIG. 7.4 Online determination of vulnerable zones for loss of sensitivity.

I_{pickup}, through previously defined setting groups of the nonsensitivity relays. This routine is very important for the proposed system, since in addition to increasing the sensitivity of the relays, the operating times are reduced. Therefore, in each defined demand state it is necessary to execute relay coordination routines because the setting change of any relay affects the relay coordination of the remaining relays. The online determination of the vulnerable zones will be carried out through the execution of routines for the estimated calculation of faults, refreshing before topological changes and the I_{pickup} of relays, according to Eq. (7.1). The intermediate source effect, unlike distance relays, will not be considered because it does not affect the reach of the DOCR.

The topological detection systems must have the position of auxiliary contacts (52a) and the current measured in that branch (Fig. 7.4). This information verified at both ends will confirm the current topology and will allow to represent the topological modifications in the nodal arrays. The voltage phasor and the sample time will allow to obtain the directionality of the measured current and the latency of communication available in the electricity grid must be considered. The time updating before any topological change must be almost in real time, and the update by variation of the demand can be carried out at intervals of no more than 15 min, similar to the times used for the calculation of the demand. The I_{pickup} is updated with the measured load current within the time interval considered, and the I_{sc} will be modified in case of a topology change or the connection of generation sources. When changing the I_{pickup} or I_{sc}, it will be necessary to coordinate the relays.

Fig. 7.4 shows the proposed system and the relay's operating coverage. The relay R14 (green zone) can detect remote line-end faults are shown, while another relay R15 (red zone) lacks sensitivity, leaving a considerable percentage of the adjacent line vulnerable and without backup protection. If primary relay R56 does not operate, the utility contribution will be sustained. In this way, with measurement at each bus and having a system with the ability to check areas with a lack of sensitivity, actions can be carried out on these relays to guarantee backup function and enhance their sensitivity.

First, the backup function of each relay is evaluated by means of the sensitivity filter. The relays that meet the condition $S_b \geq 1.5$ are considered by the proposed coordination. The settings that guarantee correct coordination between the elements are obtained. The nonsensitivity relays ($S_b < 1.5$) are considered by the following approaches.

The main goal of dynamic setting is to increase the sensitivity during poor load conditions and reduced fault current. The alternative to improve the sensitivity of the nonsensitivity relay is to modify the I_{pickup}, since the magnitude of the fault current is not controllable because it is determined by the intermittent contributions of RES and by controlled sources. To accomplish this, the relay should have a dynamic setting of the I_{pickup} that reflects the operational state of the electrical grid [20]. It is important to weigh the action of the system when detecting vulnerable zones, determining if it is temporary and depending on its duration, the modification in the relay settings can be established. However, in conditions where the loss of sensitivity is permanent, the consideration of including negative sequence relays or the change of protection principle, such as distance relays, must be evaluated in the design of the protection scheme.

7.4 Adaptive pickup current

The dynamic setting of overcurrent relays breaks with the conventional protective relay adjustment paradigm, where the value is calculated for the most critical condition and remains constant even when there are dynamic operating conditions of the electrical network. The advantage is the simplicity in the operation of the relay, but the disadvantage is the loss of sensitivity for fault detection, which is the main function of the relay.

When the I_{pickup} is variable, in addition to requiring online coordination of the protections, there is the complexity that the relay must identify the different operating states of the electrical network to avoid false operations, such as load connection, topological changes, cold load, and continuous variation of load current. There are different degrees of complexity, compromising the infrastructure requirement and the increase in the sensitivity of the relay. A low-demand alternative can be based solely on topological changes of the electrical network, updating the relay setting through breaker monitoring and branch measurement validation. Even though this alternative is simple, its impact on sensitivity is important; however, it is necessary to carry out the contingency analysis.

The I_{pickup} can be updated in various ways depending on the dynamics of the electrical network and the availability of infrastructure for communication and adjustment of the relays. Due to the dynamics of the electrical network, the adjustment can be made at intervals similar to the calculation of the demand, in subtransmission and distribution networks, for example, using the instantaneous value acquired every 15 mins or using the demand current with rolled integration with 5-min subintervals. But in microgrids or industrial systems where the dynamics of the current are greater, the updating of the settings can be done more frequently, for example, every minute. As the protection is based on the activation of adjustable thresholds to determine its operating criteria, it is for this reason that it is not necessary for the OCR or the DOCR to have synchronized phasor measurements.

7.4.1 Relay setting groups

The main goal of dynamic setting is to increase the sensitivity during poor load conditions and by minimal fault currents. The alternative to improve the sensitivity of the overcurrent relay is to modify the I_{pickup}, since the magnitude of the fault current is determined by the intermittent contributions of RES and by controlled sources. To accomplish this, the relay should have a dynamic setting of the I_{pickup} that reflects the operational state of the electrical grid. It is proposed to make a modification of the settings by means of setting groups defined in each relay; in this alternative, although the sensitivity is improved, it is insensitive to topological changes in the electrical network.

The activation of the relay setting groups is nowadays feasible; it is possible to activate the setting groups depending on the level of demand. By distinguishing between different demand operating states by activating sensors, it is possible to improve the sensitivity of the relays. Defining I_{pickup} settings to improve sensitivity and time dial settings (TDS) to achieve coordination for each demand state, relays that meet sensitivity are held without setting change. The activation of the groups can be carried out following the historical demand forecasts.

Fig. 7.5 shows the discretization of the demand states for the assignment of setting groups; in this way when an increase in demand exceeds the setting of a group, the verification of the fault condition is carried out through a control logic. If there is no fault present, it switches to the new setting group. In the maximum demand is when there is a greater number of relays with a lack of sensitivity; therefore, the detection of these cases can activate distance relays temporarily or permanently to guarantee sensitivity. The operation of the MG mode connected to the utility or in isolated operation affects the sensitivity of the protections, so the mode of operation must be detected. The activation of the groups will be defined for both conditions.

The system can determine the relays with low sensitivity and the setting groups for these relays will be assigned. Relays with no sensitivity issues remain

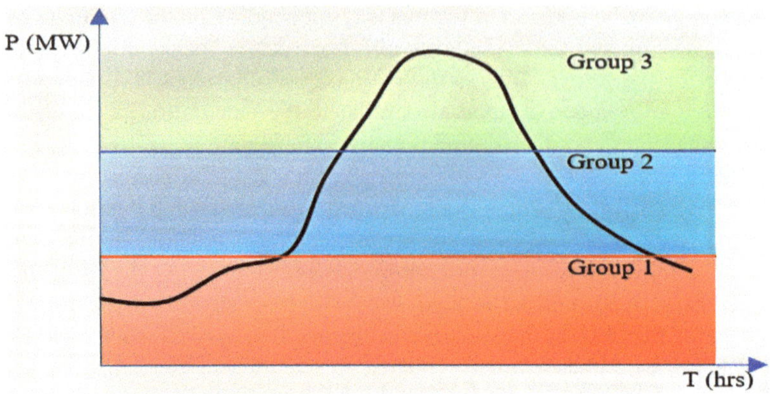

FIG. 7.5 Relay setting groups.

unchanged and their settings will be assigned in the conventional way. For non-sensitivity relays, a control logic that distinguishes between a fault condition and an operational change in the network is required. Switching between setting groups of a relay is done by discriminating between a fault condition and a load increase by a fault detector.

7.4.2 Test of microgrid connected to the utility

In this case, the proposal is evaluated with the microgrid connected to the utility (Fig. 7.2) and several demand scenarios (maximum, average, and minimum) are taken into account in Table 7.3. The performance of the adaptive I_{pickup} methodology is analyzed for each scenario (Fig. 7.6), as well as its influence on the sensitivity of the relays. The results obtained for this case are shown. During

TABLE 7.3 Sensitivity of the relay setting groups.

Relay	Sensitivity		
	D_{max}	D_{mean}	D_{min}
51	1.01	1.41	2.12
41	0.47	0.66	0.99
62	1.29	1.81	2.71
52	1.32	1.85	2.77
63	3.51	4.92	7.38
43	4.69	6.56	9.84
34	4.38	6.13	9.19
25	1.08	1.51	2.26
15	2.98	4.17	6.25
36	3.86	5.41	8.11
14	3.53	4.94	7.42
26	1.55	2.16	3.25

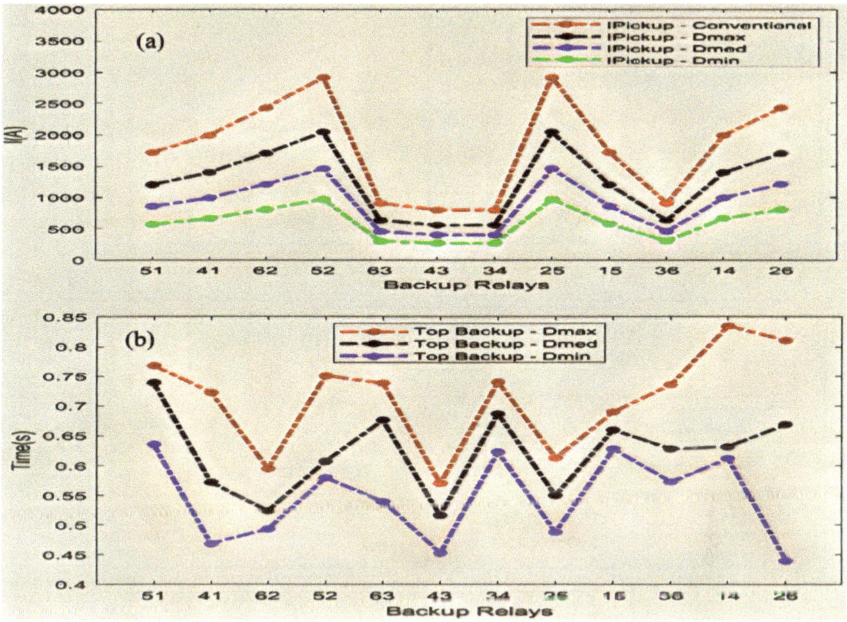

FIG. 7.6 Adaptive I_{pickup} to different variations of the I_{load} during MG connected. (A) I_{pickup} and (B) operation time.

peak demand (D_{max}) there will be less sensitivity since the historical peak demand is not considered and the calculation of $n-1$ contingencies is not included. The group setting relays of the I_{pickup} will be lower than the conventional setting.

The I_{pickup} is affected by the group setting, so that in minimum demand, greater sensitivity is obtained and less operating time of the relay. In case all relays have variable I_{pickup} setting groups, the operation time reduction effect is obtained in all relays. This behavior is shown in Fig. 7.6 for the different operating relay setting groups. For example, the operation time of Relay 26 is reduced from 0.81 during D_{max} to 0.43 during D_{min}.

Fig. 7.7 shows the improvement in sensitivity of each relay by different relay setting groups, in this case with the MG connected to the utility.

Consequences of dynamic pickup current. With the increase in sensitivity through the reduction of the I_{pickup}, the operation time of the OCR is reduced too. The effect is illustrated in Fig. 7.8. The relay operating time depends on the I_{pickup}, then each change of I_{pickup} provokes that the time operation also changes, so the loss of relay coordination may be present. It will be necessary to update the coordination before changes in the I_{pickup} groups because the operating times will be affected.

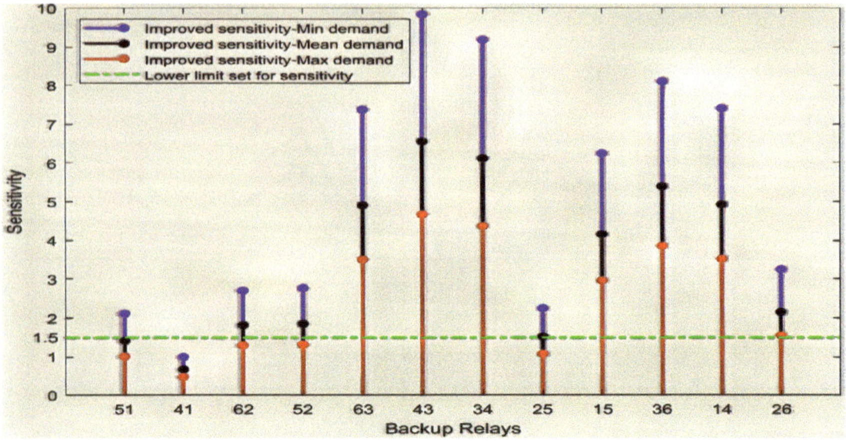

FIG. 7.7 Sensitivity improvement achieved by adaptive I_{pickup}.

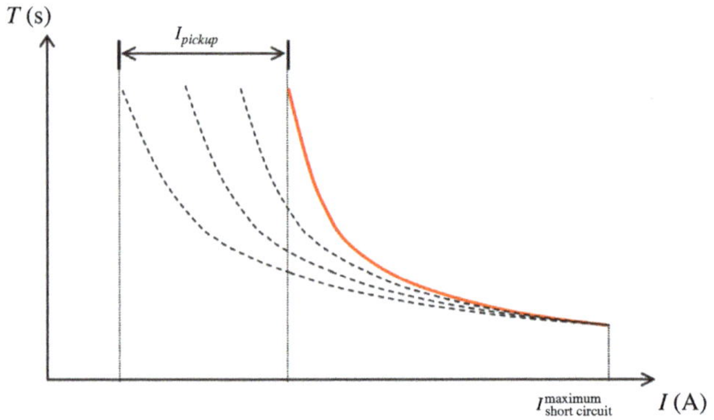

FIG. 7.8 Relay operating times for I_{pickup} changes.

The alternatives of dynamic variation of the I_{pickup} using the I_{load} allow to further increase the sensitivity of the relays. For this reason, a control logic is required depending on the way in which the I_{pickup} is updated.

In the operation of active networks with the presence of RES, the forecast of demand can have a high value of uncertainty, this complicates the activation of relay settings since it is not possible to distinguish between states of minimum and maximum demand. For example, at overnight, batteries can be charged by wind generation. Under these conditions, it is proposed that the I_{pickup} be continuously adaptable depending on the current demand. Through the continuous monitoring of the operation of the electrical network, there are more benefits in the operation of the protection scheme since it is possible to modify the settings and coordination of the relays in the face of topological changes and conditions of formation of islands or isolated operation of the utility.

7.5 Continuous adaptive overcurrent

A new adaptive current function for overcurrent relays is proposed. This criterion can be applied to phase time overcurrent relays and can be applied to both power and industrial systems. The main goal of a time overcurrent adaptive relay is to increase the sensitivity to minimum fault currents in poor load conditions. To accomplish this, the relay should have a dynamic adjustment of the I_{pickup} that reflects the operational state of the electrical grid. This function requires a control logic that distinguishes between a fault condition and an operational change in the network; in addition, operating and backup times are reduced. Finally, the proposed relay is obtained with a minimal change in the relay's firmware and without an increase in cost. However, as adaptive relay has an additional dynamic of the I_{pickup}, the operating time of the relay will be affected; it is, therefore, the adaptation of the operating time or online relay coordination system must be incorporated.

Eq. (7.3) ensures that the OCR always has the I_{pickup} necessary to avoid incorrect operation due to the effect of the load. This gives it great sensitivity, since under minimum demand conditions the I_{pickup} value is also small (Fig. 7.9).

$$I_{pickup}^{Adp} = I_{load} + \Delta I \tag{7.3}$$

where ΔI represents a safety margin, with a value of 15% of the maximum I_{load}, considering a 5% error in the protection system and a 10% safety factor.

It is necessary to establish a maximum limit of the value of the adaptive I_{pickup} of the relay for coordination purposes; this maximum pickup value is that assigned as follows:

$$I_{pickup\_max} = k\, I_{load\_mÆx} \tag{7.4}$$

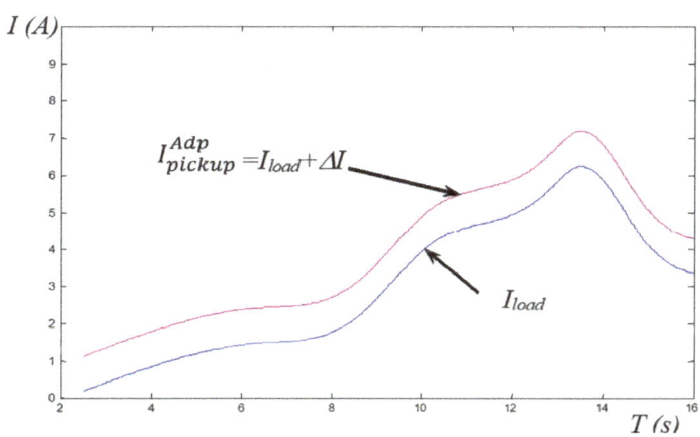

FIG. 7.9 Adaptive pickup current.

where k is a factor that considers the safety margin and future load growth (typical values 1.5–2) and I_{load_max} is the maximum load current. The maximum limit of the I_{pickup} Eq. (7.4) is different from the conventional setting of a conventional relay because in the adaptive function the contingency analysis and the power transfer are not considered.

The adaptive overcurrent relay logic must ensure that the I_{pickup} does not exceed the maximum value given by Eq. (7.4):

$$I_{pickup}^{Adp} \leq I_{pickup\_max} \tag{7.5}$$

7.5.1 Adaptive characteristic of the overcurrent relay

The adaptability of the I_{pickup} described in this chapter makes the operating time of the adaptive relay dependent, not only on the electrical distance to fault (which can be expressed in terms of impedance from source to a fault), but also of the actual I_{pickup} at the time of fault. Therefore it is convenient to represent the operation of the adaptive inverse time element by an operation characteristic $T = f(Z, I_{pickup})$. This characteristic, unlike the case of the conventional relay, has two independent variables, so it must be represented in three-dimensional form only to visualize the operation of the adaptive relay. In its coordination with other relays, it is not necessary to have this three-dimensional representation since the variation of the pickup current is limited to its intervals.

In Fig. 7.10, the characteristic $T = f(Z)$ for relay n is represented, considering that it is conventional. The characteristic given by IEC [20] is used, with $\eta = 2$ and $K = 80$, and a I_{pickup} of 50.2 A. The representation is three-dimensional, as a basis for comparison with the adaptive case. The plane is

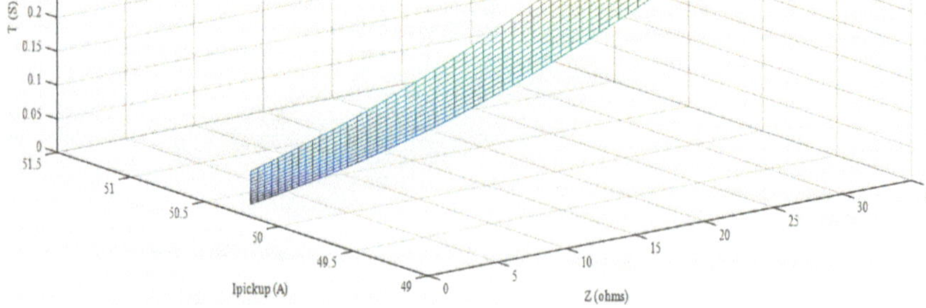

FIG. 7.10 Time curve of a conventional relay.

delimited by the time curve corresponding to maximum short-circuit current at the bottom and by the curve for minimum short-circuit current at the top. That is, for maximum fault currents the minimum operation time is obtained and for minimum faults, the protection operation time is longer.

In Fig. 7.11, the characteristic $T = f\,(Z, I_{pickup})$ considered as adaptive of OCR is shown. The two surfaces are delimited by the interval variation of $I_{pickup\_max} = 50.2$ A to $I_{pickup\_min} = 25$ A. It is observed that in minimum generation there is greater sensitivity (lower I_{pickup}) and minor operating time of the relay for the different operating states of the system. This behavior is manifested in the graph as a reduction in the space between the two surfaces as I_{pickup} is reduced. This increased sensitivity eliminates the compromise in conventional relays. An operating state on the lower curve of minimum demand (25 A) would represent a condition in which the source impedance is very small compared to the total impedance to the point of fault. This implies that the level of fault is relatively independent of the generation level of the system.

If a linear dependence between the source short-circuits level and the load demand is assumed, the operating zone illustrated in Fig. 7.12 will be obtained. This surface constitutes a transition between the two surfaces in Fig. 7.11. For example, in maximum demand there is a generation impedance corresponding to that operating state (lower curve), a maximum I_{pickup} (50.2 A), and a shorter operating time (ψ in Fig. 7.11). From the earlier mentioned considerations, a curve is obtained that diagonalizes the space between the two surfaces.

Sensitivity analysis was carried out in the radial power system shown in Fig. 7.13. It is not necessary to consider a more complex configuration of the power system, as use of a complex power system does not lead to

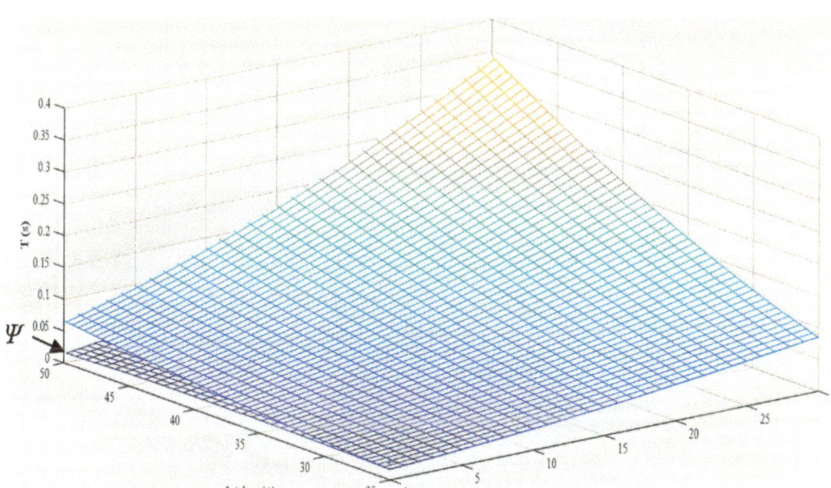

FIG. 7.11 Time characteristic of adaptive relay.

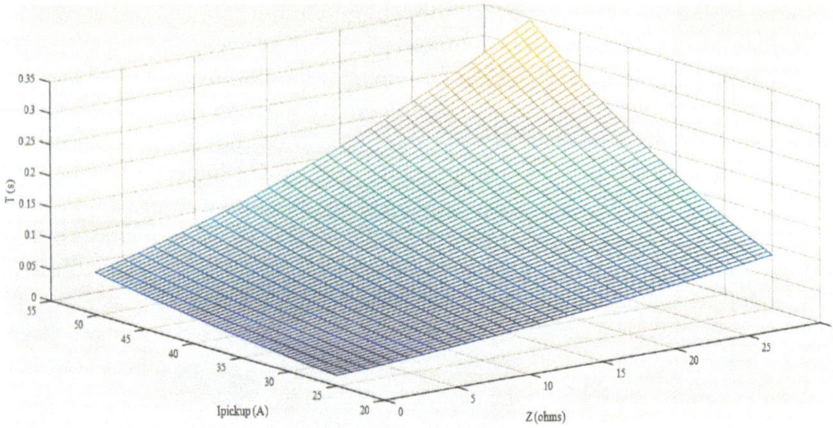

FIG. 7.12 Time characteristic with linear dependence criterion.

unexpected values. The variable load impedances Z_2 and Z_3 are used for obtaining a diverse operative condition of Relay B, also the minimum phase-to-phase fault is simulated in Bus 3. Fig. 7.13 shows the relay sensitivities for different I_{load} and source contributions (P_{source}). The proposed dynamic setting (plane 3), conventional (plane 1), and negative sequence (plane 2) relays are compared under minimum load and poor fault contribution. The conventional and negative sequence relays have a pickup setting that is independent of the dynamic

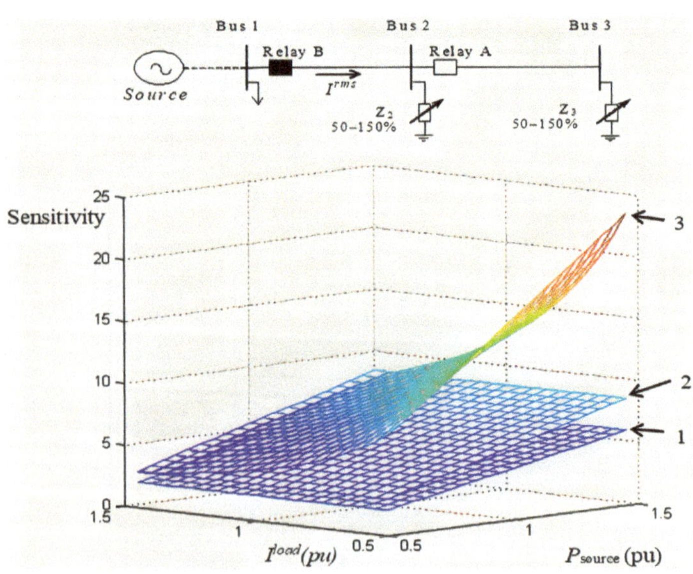

FIG. 7.13 Sensitivity analysis for OCRs.

load having a low sensitivity compared with dynamic setting under minimal load current. Under high power conditions, the sensitivity of the three relays is higher (larger short-circuit currents), but under low contribution the sensitivity of adaptive relay setting is better.

7.5.2 Control logic for adaptive relay

The control logic of the I_{pickup} (Fig. 7.14) has the task of maintaining constancy of I_{pickup} during a fault to avoid changes in pickup settings. If the line is deenergized the control logic assigns a maximum value, which can be the setting of a conventional relay.

When a large load is added to a feeder, the relay has a detection fault logic to supervise the I_{pickup}. This logic includes a negative sequence verification and a high-level current detector, both combined with *OR* logic. The negative sequence was proposed to detect phase-to-phase faults and its setting has been described previously. On low-voltage networks, the negative sequence setting detects unbalance faults and a conventional setting detects a three-phase fault. Then the logic discriminates between a large change in load current and a fault current (symmetrical or asymmetrical).

When the I_{load} is less than the I_{pickup}, the dynamic I_{pickup} is updated. Instead, if the I_{load} is greater than the dynamic I_{pickup}, the control logic must confirm the existence of a fault by combining fault detectors. In case of confirming the fault, the actual value of the I_{pickup} is used to determine the operating time by means of the time curve of the relay.

The adaptation of the I_{pickup} of an inverse time relay can be done in two ways. *RMS* of the current that includes an instantaneous actualization; or

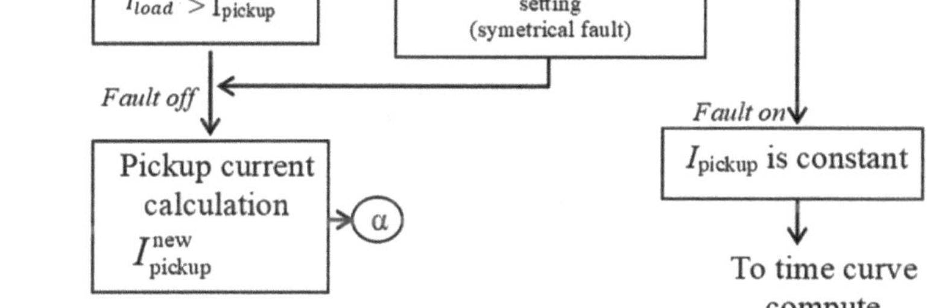

FIG. 7.14 Control logic of an adaptive pickup current.

using the demand concept, having the I_{pickup} update when the demand is calculated. The demand can be calculated like billing, with integration rolled every 5 min and calculation of the demand every 15 min. The difference between these alternatives lies in the requirement of the relay or the centralized protection system to update its settings. In this sense, the concept of demand offers advantages as it has low computational and communication demands.

7.6 Adaptation of the pickup current with RMS

Control logic is required to discriminate between load variations and system faults. This logic must keep the adaptive I_{pickup} constant when a short circuit occurs and update during normal states of load on the protected line. The logic for the adaptive control includes a fault detection function, to identify the load and fault states, and a set of functions aimed at controlling the value of the adaptive I_{pickup}.

Fault detection. The detection of the instant in which the fault occurs is the basis for the adaptive control of the adaptive OCR; detection of the onset of the fault can be made for the abrupt change in the current signal. A fault detector must have two fundamental characteristics: (a) sufficient sensitivity to detect faults with reduced currents and (b) immunity to noise present in the fault current. The transient detector [21] was used to detect the discontinuity caused by the fault of the input signals to the relay.

The transient detector applies a least-squared estimation algorithm to obtain the parameters of the components of the input current to the relay; the estimate will have an appreciable error when the data window is crossing the discontinuity that the fault causes. This is the data containing information on the prefault and fault states, which deteriorates the estimate. The transient detector uses this feature to detect the presence of the fault. The basic idea is to calculate the values of the dummy samples that result from the signal estimate and to compare these values with those of the real samples.

The input signal $y(t)$ to a digital relay can, in general, be expressed by:

$$y(t) = \sum_{n=1}^{M} Y_n S_n(t) + \varepsilon(t) \quad (7.6)$$

where $S_n(t)$ represents the set of elementary signals that are assumed to be present in $y(t)$, Y_n are their coefficients, and $\varepsilon(t)$ represents the errors. In protection applications, $s_n(t)$ typically includes the fundamental frequency component, an exponential aperiodic component, and a set of harmonics [21,22]. In its discreet way, the parameter estimation problem consists of estimating the M values of Y_n from a set of N_m samples of the measurements $y[k]$, in such a way that $\varepsilon[k]$ is minimal. That is, we must solve (7.7) for the minimum error:

$$\begin{bmatrix} y[1] \\ y[2] \\ \vdots \\ y[N_m] \end{bmatrix} = \begin{bmatrix} s_1(\Delta t) & s_2(\Delta t) & \cdots & s_M(\Delta t) \\ s_1(2\Delta t) & s_2(2\Delta t) & \cdots & s_M(2\Delta t) \\ \vdots & \vdots & \ddots & \vdots \\ s_1(N_m\,\Delta t) & s_2(N_m\,\Delta t) & \cdots & s_M(N_m\,\Delta t) \end{bmatrix} \begin{bmatrix} Y_1 \\ Y_2 \\ \vdots \\ Y_{N_m} \end{bmatrix} + \begin{bmatrix} \varepsilon[1] \\ \varepsilon[2] \\ \vdots \\ \varepsilon[N_m] \end{bmatrix} \tag{7.7}$$

Eq. (7.7) can be written in matrix form as follows:

$$y = SY + \varepsilon \tag{7.8}$$

where the matrix $\boldsymbol{S}$ represents the set of elementary signals that are assumed to be present in $\boldsymbol{y}$, $\boldsymbol{Y}$ are their coefficients, and $\boldsymbol{\varepsilon}$ represents the errors. If it is considered that the errors corresponding to different samples are not correlated, and that the covariance matrix is a multiple of the unit matrix, the least-squared solution [23,24] is as follows:

$$\widehat{Y} = \left(S^T S\right)^{-1} S^T y = S^+ y \tag{7.9}$$

where $\widehat{\boldsymbol{Y}}$ is the vector of the estimated coefficients of $s_n(t)$ and $\boldsymbol{S}^+$ is the pseudo-inverse matrix of $\boldsymbol{S}$.

The elements of the matrix $\boldsymbol{S}$ depend on the elementary signals that are selected to represent the signal $y(t)$, as well as on the sampling frequency. The length of the window and the time reference adopted, that is, the instant that is designated as $t = 0$. When the data window passes through the fault insertion time, the estimation results show error because the data window contains prefault and fault samples. The estimator is trying to fit into a single signal model information that belongs to two different models.

Using the least squares solution given by Eq. (7.9), it is possible to estimate the values of the fictitious samples, according to:

$$\sim y = S\widehat{Y} = S\left(S^T S\right)^{-1} S^T y \tag{7.10}$$

Vectors $\sim \boldsymbol{y}$ and $\boldsymbol{y}$ differ due to two factors: one of them is the set of measurement errors that takes place in the process. The other factor is the estimated error, which is influenced by the noise present in the signal, and the crossing of the data window by the magnitude and/or phase discontinuity caused by the signal fault. The estimation error caused by the crossing of the data window is much greater than the rest, so if $\sim \boldsymbol{y}$has values significantly different from those of $\boldsymbol{y}$, it is logical to assume that the transition from the prefault state of the fault state is occurring.

An example of the difference of the vectors $\sim \boldsymbol{y}$ and $\boldsymbol{y}$ before a discontinuity is illustrated in Fig. 7.15, in which the input signal is represented by a solid line, with the real samples $y[k]$ indicated by circles (the sampling frequency is 16 samples per cycle). Reconstructed samples $\sim y[k]$ are represented by asterisks. In this case where there is no noise in the fault signal, the mismatch between the real and reconstructed samples is present only during the 16 sampling instants corresponding to the crossing of the data window due to a discontinuity caused by the start of fault.

A vector of residuals can be calculated:

$$\boldsymbol{r} = \sim \boldsymbol{y} - \boldsymbol{y} = \left[\boldsymbol{S}\left(\boldsymbol{S}^T\boldsymbol{S}\right)^{-1}\boldsymbol{S}^T - \boldsymbol{I}\right]\boldsymbol{y} \tag{7.11}$$

The function of the transient detector can be formed in different variants from the vector of residuals $\boldsymbol{r}$. In (7.12) the recommended detector model is presented. This model, in addition to having excellent sensitivity, has a reducing effect on the errors of the estimate before and after passing the data window through the discontinuity.

$$t[k] = \left[\sum\nolimits_{j=k-N}^{k} |r[j]|\right]^2 \tag{7.12}$$

In (7.12), $t[k]$ is the function of the transient detector corresponding to sample k, $r[j]$ is the residual calculated according to (7.11) for sample j, and N is the number of samples in the data window. To detect the fault, the $t[k]$ value of the transient detector compares with a certain threshold value, which must be high enough so that $t[k]$ does not exceed it due to the noise present in the fault current.

An important aspect in the design of the transient detector is the selection of the signal components to be included in the least-squared estimator model. The model must meet two basic requirements:

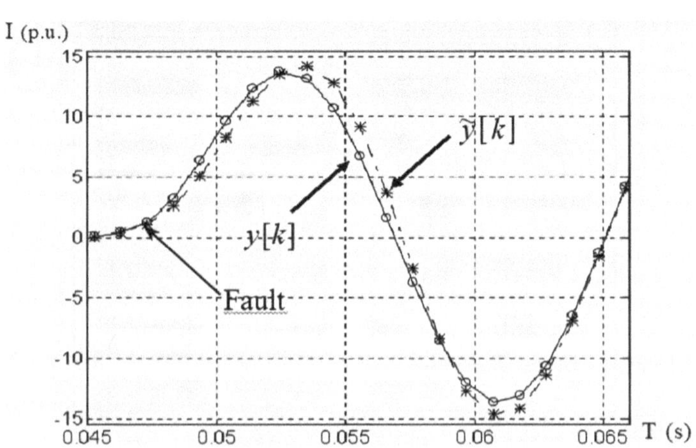

FIG. 7.15 Estimation of the current signal before a discontinuity.

- Represent with fidelity the waveform of the fault current, to avoid errors in the estimator due to an inadequate representation in your model. With this, it is achieved that the output of the transient detector falls to values close to zero at the conclusion of the crossing of the data window due to the discontinuity.
- Do not reduce the sensitivity of the detector, that is, offer high values in its output when the window is crossing the discontinuity.

To meet the first requirement, the type of contamination that may be present in the fault current must be considered. In the input signals to the relays during short circuits, the following components can be expected, among others [22]: fundamental component, exponential aperiodic component, and high-frequency damped oscillatory components. However, in the current signal, the high-frequency components generally have low values. Additionally, the antialiasing filter also tends to suppress high-frequency components when their cutoff frequency is low.

The model for the transient detector must be selected to comply with the requirement of faithful representation of the fault current, and with the sensitivity of the transient detector to detect the crossing of the data window by the discontinuity. The best signal model for the least squares estimator of the transient detector for this application is the one formed by the fundamental component and the exponential aperiodic component.

The transient detector depends on the model provided to estimate the signal, greater number of components results better the estimate will be, but the computational load will be increased. For fault detection purposes, it is important that the expected signal components are available in steady state to achieve a good estimate. The fault detection is carried out by the change of signal magnitudes due to the presence of a fault; the magnitudes of the estimated errors in steady state and fault are generally very different, which allows an adequate detection of the fault. Following is an application example of Fig. 7.16.

```
The modeled signals for both steady state and fault state are:
State stable:   y(k)=4*sin(wo*j+0.5)

Fault   y(k)=(7*sin(wo*j+0.5)+1.1666*sin(3*wo*j+0.5)+
            0.05833*sin(5*wo*j+0.5)+7*exp(-3.5*j))*1.1*exp(-3.5*j);

The signal model of the estimator is:
        s1(a)=cos(wo*i);
        s2(a)=sin(wo*i);
        s3(a)=cos(3*wo*i);
        s4(a)=sin(3*wo*i);
        s5(a)=cos(5*wo*i);
        s6(a)=sin(5*wo*i);
        s7(a)=1;
        s8(a)=i;
```

FIG. 7.16 Model of signals for transient detector example.

Fig. 7.17 shows an example of the elementary signals, matrix S and the pseudo-inverse matrix S^{+}.

The input signal and the estimated signal and the difference between them are shown in Fig. 7.18.

S

16x8 double

	1	2	3	4	5	6	7	8
1	1	0	1	0	1	0	1	0
2	0.9239	0.3827	0.3827	0.9239	-0.3827	0.9239	1	0.0010
3	0.7071	0.7071	-0.7071	0.7071	-0.7070	-0.7072	1	0.0021
4	0.3827	0.9239	-0.9238	-0.3828	0.9239	-0.3826	1	0.0031
5	-3.7007e-05	1.0000	1.1102e-04	-1.0000	-1.8503e-04	1.0000	1	0.0042
6	-0.3827	0.9239	0.9239	-0.3826	-0.9238	-0.3829	1	0.0052
7	-0.7071	0.7071	0.7070	0.7072	0.7073	-0.7069	1	0.0063
8	-0.9239	0.3826	-0.3829	0.9238	0.3824	0.9240	1	0.0073
9	-1.0000	-7.4013e-05	-1.0000	-2.2204e-04	-1.0000	-3.7007e-04	1	0.0083
10	-0.9238	-0.3828	-0.3825	-0.9240	0.3831	-0.9237	1	0.0094
11	-0.7070	-0.7072	0.7073	-0.7069	0.7068	0.7074	1	0.0104
12	-0.3826	-0.9239	0.9238	0.3830	-0.9241	0.3822	1	0.0115
13	1.1102e-04	-1.0000	-3.3306e-04	1.0000	5.5510e-04	-1.0000	1	0.0125
14	0.3828	-0.9238	-0.9240	0.3824	0.9236	0.3832	1	0.0135
15	0.7072	-0.7070	-0.7068	-0.7074	-0.7076	0.7066	1	0.0146
16	0.9239	-0.3826	0.3831	-0.9237	-0.3820	-0.9241	1	0.0156

SI

8x16 double

	1	2	3	4	5	6	7	8
1	0.0763	0.0976	0.0660	0.0446	0.0076	-0.0377	-0.0522	-0.0884
2	-0.2451	-0.0421	-0.0241	0.0991	0.1631	0.1663	0.2707	0.1840
3	0.0763	0.0299	-0.1108	-0.1188	0.0076	0.1256	0.1246	-0.0208
4	-0.0730	0.0887	0.0549	-0.0527	-0.1137	-0.0327	0.1427	0.1560
5	0.0763	-0.0657	-0.1108	0.1122	0.0075	-0.1054	0.1247	0.0749
6	-0.0326	0.1035	-0.1034	-0.0500	0.1301	-0.0411	-0.0641	0.1336
7	0.4281	0.1967	0.2304	0.0870	0.0057	-0.0133	-0.2094	-0.1406
8	-46.7969	-17.1776	-21.4865	-3.1326	7.2671	9.7045	34.8069	25.9943

	9	10	11	12	13	14	15	16
1	-0.1521	-0.1517	-0.0985	-0.0554	0.0033	0.0702	0.1063	0.1642
2	-0.1361	-0.2301	-0.1392	-0.1535	-0.1086	-0.0030	0.0016	0.1972
3	-0.1521	-0.0841	0.0783	0.1079	0.0032	-0.0931	-0.0705	0.0966
4	-0.0406	-0.1698	-0.1035	0.0365	0.1299	0.0813	-0.0616	-0.0425
5	-0.1521	0.0116	0.0782	-0.1231	0.0033	0.1378	-0.0706	9.6646e-04
6	-0.0182	-0.1397	0.0817	0.0427	-0.1228	0.0629	0.1003	-0.0829
7	0.2656	0.3344	0.1383	0.1193	0.0380	-0.1054	-0.0717	-0.3031
8	-25.9943	-34.8069	-9.7045	-7.2671	3.1326	21.4865	17.1776	46.7969

FIG. 7.17 Matrix of data.

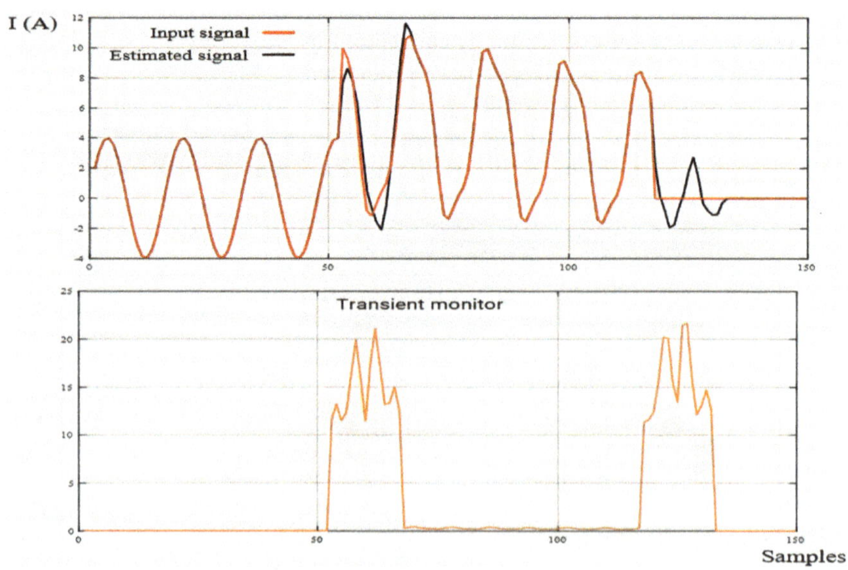

FIG. 7.18 Signal obtained from transient monitor.

7.6.1 Control logic

The I_{pickup} control logic of the adaptive overcurrent relay is summarized in the block diagram in Fig. 7.19. These functions must consider the following operating states: (a) stable state, in which there is a normal operating regime of the power system; (b) transitory states caused by the occurrence of faults; and (c) conditions in which the protected line is disconnected.

The adaptive relay logic in a normal operating state must: (a) allow the I_{pickup} to vary with the operating regime, according to (7.3); (b) continuously supervise that the I_{pickup} of the relay does not violate the maximum limits established.

Transient states associated with faults or line closing and opening operations are detected by the transient detector. This should activate a logic with the following functions: (a) set the I_{pickup} at its current value at the time of the fault, to allow the current due to the fault to exceed the setting value of the relay; (b) return to the I_{pickup} variation regime if the fault is eliminated by another protection or if it self-extinguishes; (c) set the I_{pickup} to its maximum value in case the fault is eliminated by the tripping of the adaptive relay itself (opening of the protected line).

7.6.2 Evaluation of adaptive pickup current

Based on these considerations, a program was created that simulates the logic of the adaptive relay. To carry out the tests, a simulation program was developed, also using Matlab. The program allows the generation of I_{load} signals (including

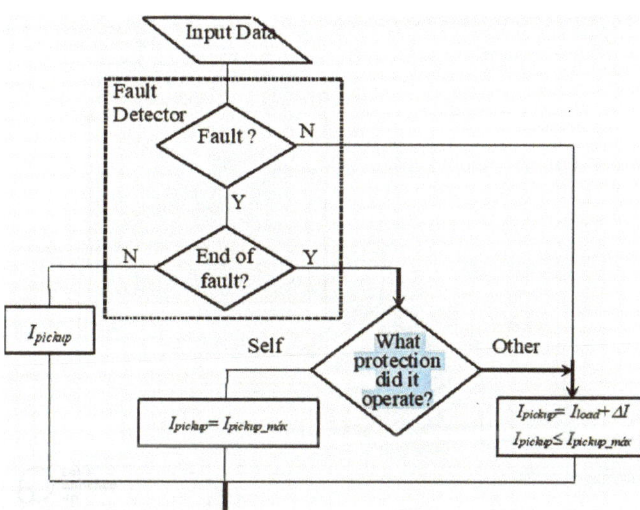

FIG. 7.19 Control logic block diagram.

variable load regimes over time), as well as the appearance of faults and their subsequent trip by opening the own switch or some other switch. The generated signals are applied to the relay as a sequence of samples representing the instantaneous values of the current.

In Fig. 7.20, the behavior of the logic is shown in the case in which a fault occurs in the line, which is tripped by the adaptive relay and subsequently a successful reclosing of the circuit breaker takes place. These are graphs showing the variation with time of the effective value of the I_{load} and I_{sc}, as well as the adaptive I_{pickup} of the adaptive relay.

During the fault state (Fig. 7.21) the value of adaptive I_{pickup} remains constant and the short-circuit current value is greater, allowing the start of the relay operation. The fault is eliminated by the opening of another breaker. In this case the adaptive I_{pickup} is fixed at its prefault value during the fault and returns to the float regime when the fault current disappears.

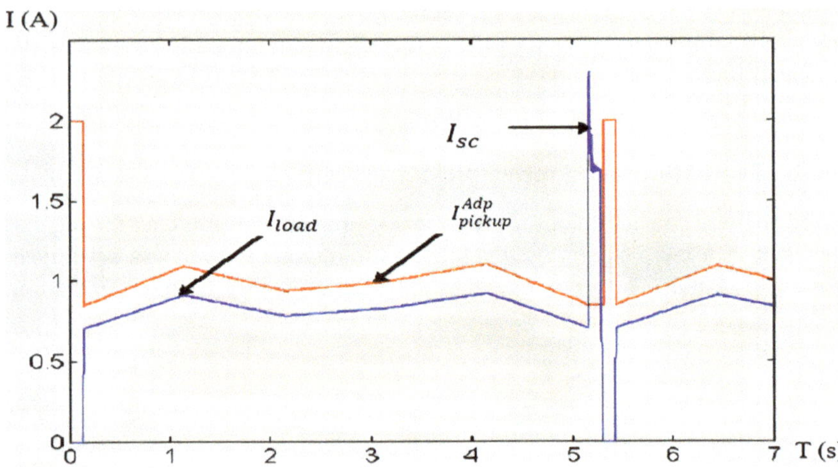

FIG. 7.20 Behavior of the control logic.

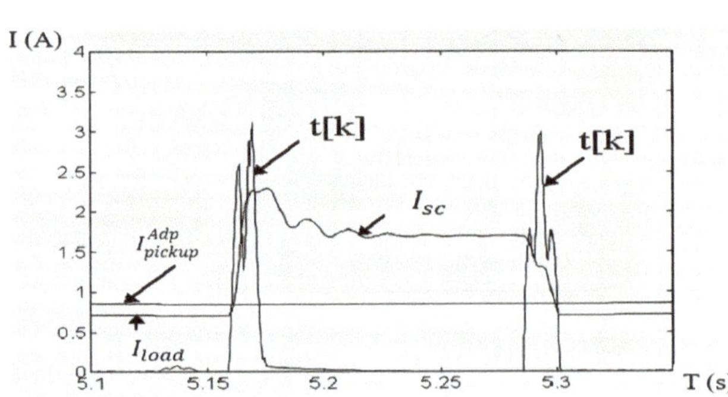

FIG. 7.21 Fault detection of adaptive relay.

Fig. 7.22 shows the behavior of the logic of the case in which an evolutionary fault occurs on the line, which is released by the adaptive relay. Note that the detector activates a second time with the sudden increase in current caused by the evolution of the fault. In this state, the adaptive logic makes a comparison between the magnitude of the fault current and the last value of the prefault current; if the fault value is greater, the adaptive logic maintains the constant value of the adaptive I_{pickup}, with which the relay maintains its operating condition activated.

In general, when the protected line is out of service, the adaptive I_{pickup} must be set to the maximum value, from there to start the floating regime on the load current once the service is restored. This gives the relay a certain immunity to the high values that the load current (cold load) can have during the transient state associated with the restoration of electrical service after a prolonged interruption of the same. During long periods of out of service in an electrical system where recovery currents of several times the nominal current are expected, it is necessary to make temporary modifications to the protection setting values so that it does not trip under this operating condition.

Traditionally, the operation of the protected system is avoided, blocking the protection elements that may be sensitive or can operate in the face of this increase in current, the first zone is blocked for distance relays, and the instantaneous element and/or inverse time element is blocked for overcurrent relays. In digital relays it is possible to assign a setting group, consisting of several stages of inrush current and time values. The setting values are programmed in such a way that they adapt to the cold load current and more adequately protect the electrical system during faults in this condition. This cold load restoration tolerance function is implemented in the adaptive relay.

The behavior of the adaptive relay when a group of settings is activated to tolerate cold load reset effects is shown in Fig. 7.23. The adaptive relay has a cold load tolerance function, this function when activated assigns a group of settings in a few stages defined by the user. By default, the setting group consists

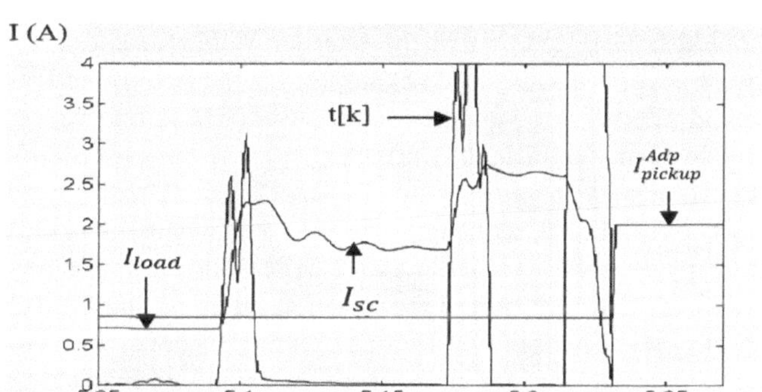

FIG. 7.22 Behavior of the control logic before an evolutionary fault.

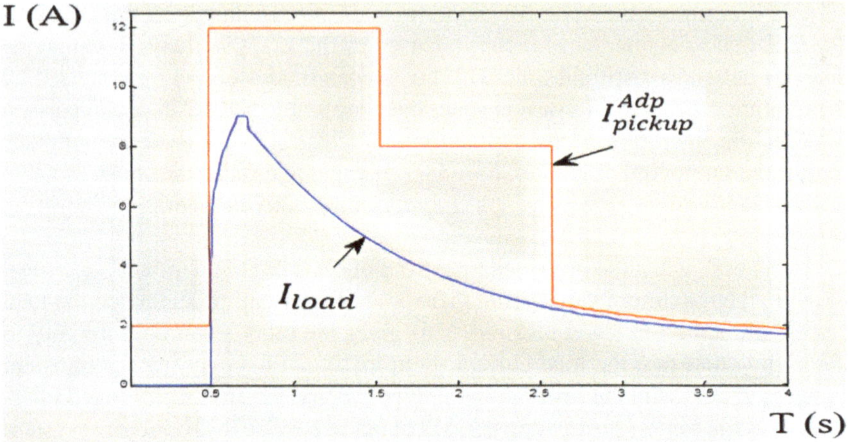

FIG. 7.23 Behavior of the control logic in the face of cold load restoration currents.

of two I_{pickup} setting steps with an effective time of each step. The test signal represents starting a 250-HP motor at 13.8 kV with full connection.

7.7 Adaptation of the pickup current with load demand

The adaptive I_{pickup} criterion based on the current demand concept performs a process of updating in time intervals defined for the calculation of the demand. The I_{pickup} is calculated based on the demand value calculated in the previous interval; therefore, the use of fault detection algorithms is not necessary, which simplifies the control logic of the adaptive relay. The current demand is the average value of the current in a certain time interval:

$$I_{a\_pickup} = \frac{1}{N_d}\sum_{j=1}^{N_d} I_j + \Delta I \tag{7.13}$$

where N_d should be selected so that the interval N_d (Δt) has a duration of the order of one or several minutes. For example, for the determination of power demand, typical intervals of 5 and 15 min are used. These intervals ensure that the demand value is not significantly affected by short-term transient variations in power or current, such as the starting of a large capacity motor or the appearance of a short circuit. For the calculation of the current demand in the relay, it is recommended to use intervals smaller than the previous ones to achieve a closer follow-up to the normal variation of the load. On the basis that the duration of a typical short circuit in distribution networks is a few seconds, the recommended interval for calculating current demand is 1–5 min; with this interval, the demand value calculated in the interval containing the fault is not significantly affected.

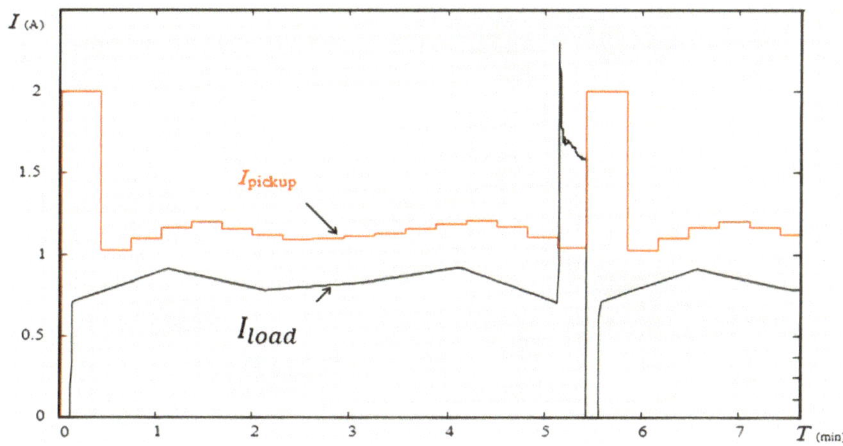

FIG. 7.24 Control logic of an adaptive pickup current using a signal test.

Eq. (7.13) represents a sliding average filter; that is, during the interval N_d (Δt) the summation is being formed, and the average value is obtained when $j = N_d$. During the entire interval, the value of I_{pickup} calculated at the end of the previous interval is in force in the relay. That is, it varies in a stepwise manner following the feeder load curve. The low-pass filtering action inherent in the current demand concept simplifies the adaptive relay logic.

The dynamic pickup setting can be shown in Fig. 7.24. The setting of I_{pickup} is constant in each demand interval, but lower than conventional setting, thus the increase of sensitivity is obtained.

7.7.1 Demand control logic

The adaptive relay pickup current calculation procedure is illustrated in Fig. 7.25. The process consists of calculating I_{pickup} according to (7.13), where S_d represents the accumulated value of the current samples in the interval N_d. At the end of each demand interval ($j = N_d$), the value of S_d is used to calculate I_{pickup}. The value of the resulting I_{pickup} remains in force until it changes again, that is, always the current value is the one obtained for the previous period of demand calculation.

In Fig. 7.26, the block diagram of the control logic is shown. The basic function is to discriminate between different possible operating states to suit the relay setting to those conditions.

The absence of current in the protected line is evaluated by comparing each current sample I_i (where $I_i = I_k$) with a certain small threshold value ε and checking that the condition $I_i \leq \varepsilon$ is maintained for a certain number of samples N_m; this ensures a reliable identification of the absence of current. In that case, I_{pickup} is set at its maximum value and remains at that value until, once the

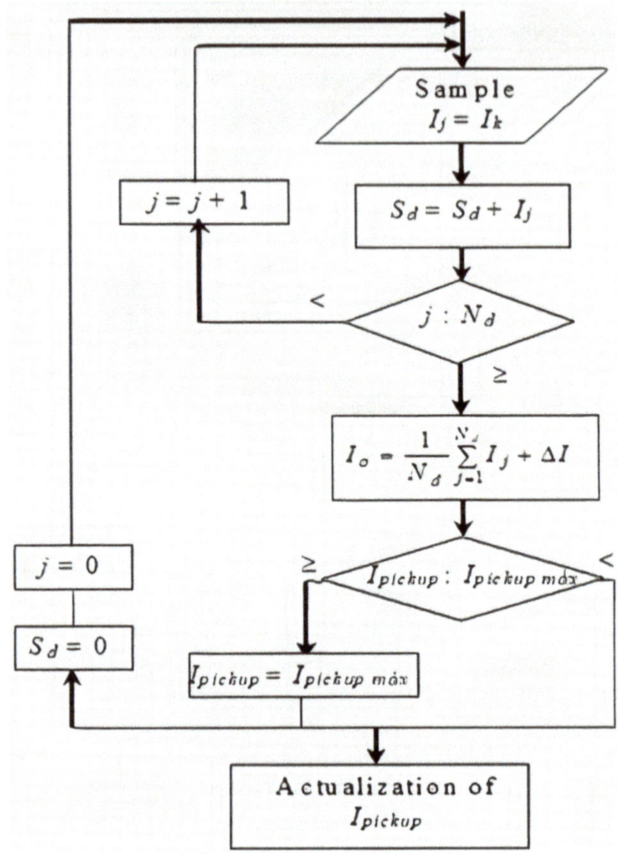

FIG. 7.25 Block diagram for the calculation of I_{pickup}.

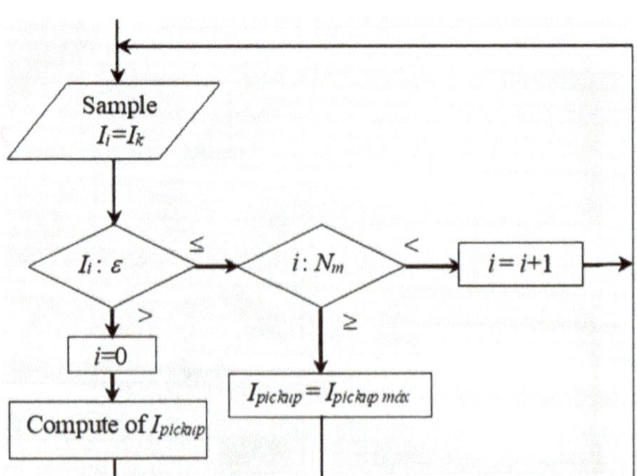

FIG. 7.26 Control logic of I_{pickup}.

service is restored, a new calculation is completed. That is, when the service is restored, $I_{pickup} = I_{pickup\ max}$ during an entire current demand calculation interval. On the other hand, if there is current in the line ($I_i > \varepsilon$), the value of I_{pickup} is being calculated according to the procedure in Fig. 7.25.

7.7.2 Evaluation

In the tests carried out, the operation of the control logic was evaluated; it was verified that it operated adequately before the different operating states of the system, such as energization, load current variation, fault, and trip. The tests carried out on logic were satisfactory in all cases. As an example, graphs are shown later that illustrate the behavior of the I_{pickup} of the relay applying the concept of current demand.

In the simulations, the waiting time after the line is energized is 2 cycles. This is observed in Fig. 7.27 where the I_{pickup} starts its floating regime after energization. During the fault state the relay operation logic behaves satisfactorily (Fig. 7.28).

In Fig. 7.29, an evolutionary fault in the line is simulated; the operation logic based on demand is not affected and does not require additional algorithms to tolerate this type of fault.

Fig. 7.30 shows the behavior of the relay logic when the setting group is activated to tolerate cold load effects. The test signal from Fig. 7.23 was used. The performance of demand current I_d is shown, also the behavior of the adaptive logic of the I_{pickup} is also satisfactory.

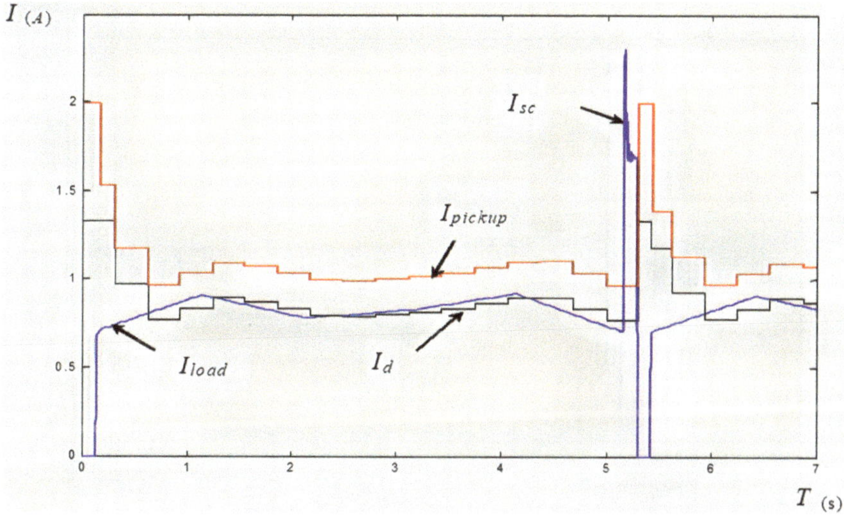

FIG. 7.27 Performance of demand I_{pickup} during line fault and reclosing operation.

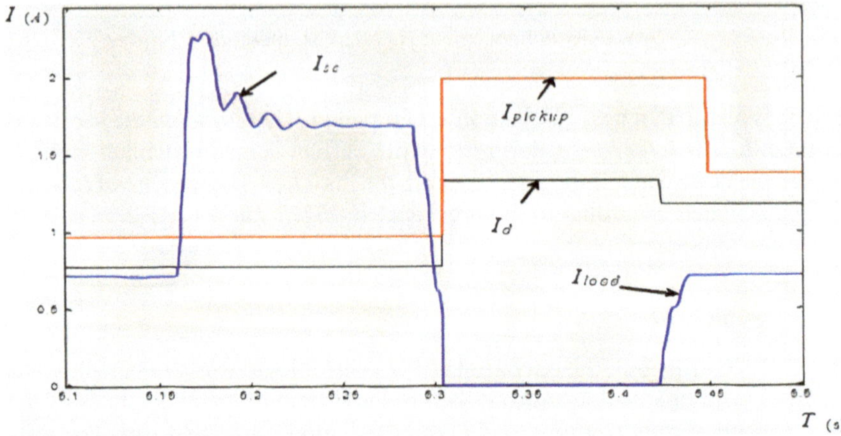

FIG. 7.28 Fault detection, trip line, and reclose line.

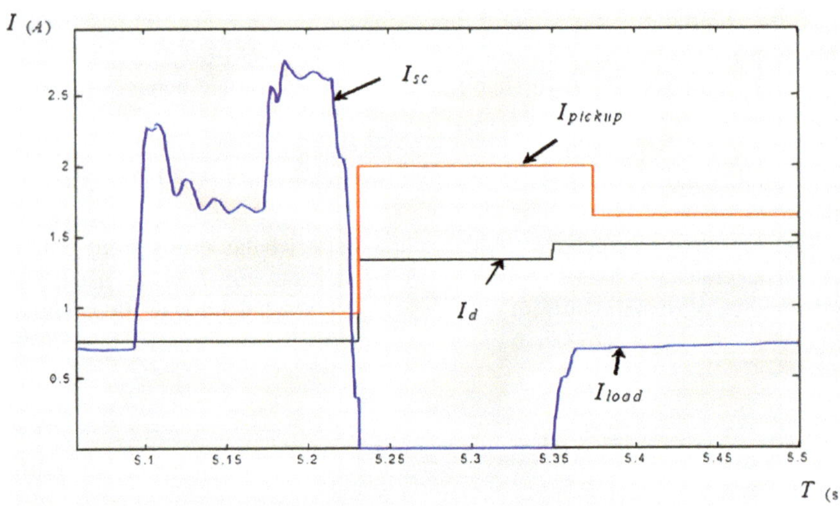

FIG. 7.29 Behavior of the control logic before an evolutionary fault.

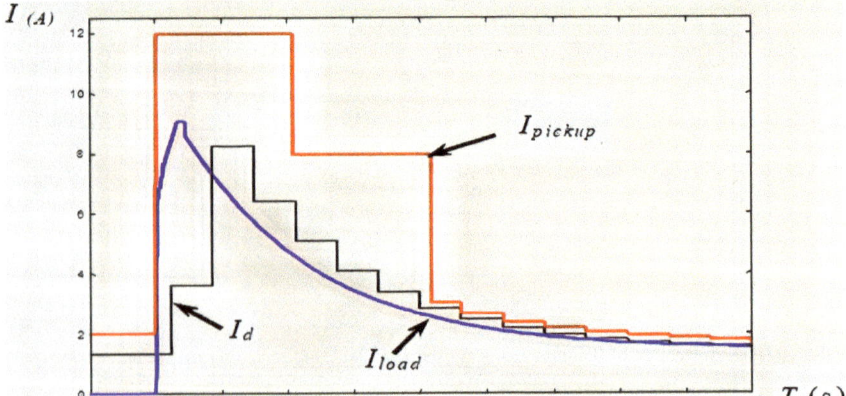

FIG. 7.30 Behavior of the control logic in the face of cold load restoration currents.

7.8 Comparison between the two current values used in adaptive logic

The principle of adaptation of the I_{pickup} is based on increasing the sensitivity of the relay, mainly during the regimes of minimum demand, where there is more probability that the conventional relay does not have sufficient sensitivity for the detection of the fault. The I_{pickup} adaptation process includes the necessary operating logic so that the relay can identify the different operating situations of the system and can adapt to these operating states. Based on the above, the criteria of sensitivity, dependability, safety, and "software" requirements were considered as the basis for the comparison between the two alternatives for adapting the adaptive relay pickup current.

The algorithm based on fault detection performs an adaptation process with greater sensitivity to changes in the system variables than the algorithm based on demand current. In practice, this situation is not important, because the slow dynamics of the demand current makes the difference in sensitivity between these two algorithms practically nil. Based on the results presented, it is observed that for the different operational scenarios that were used to evaluate the operation logics, both algorithms present satisfactory results.

Reliability is the degree of certainty that the protection system operates correctly when a fault occurs. The algorithm based on a fault detector has a greater risk of not detecting high impedance faults than the algorithm based on demand current. This possible lack of dependability of the fault detector algorithm is basically due to the tolerance that must be given to the fault detector and that has a direct impact on the sensitivity of the relay.

Safety is the degree of certainty that the protection system will not operate incorrectly. The demand-based algorithm presents a greater risk of incorrect operation in the face of substantial and rapid increases in load current, although from the tests carried out it is concluded that the margin defined in Eq. (7.13) is adequate.

About the software requirements, the demand current variant, by dispensing with a fault detection algorithm, presents clear advantages over the variant that uses the effective value of the fundamental component of the current; since in addition to requiring the fault detection algorithm, the relay control logic is more complex in algorithmic requirements than the logic of RMS variant.

From the above, it can be concluded that the demand algorithm is the most convenient to implement and is the one that is taken as the basis for the conception of the relay.

7.9 Summary

The adaptation of the I_{pickup} of the inverse time relay is obtained by varying as a function of the I_{load}. For the calculation of the adaptive I_{pickup} of the inverse time element, two current values are proposed and analyzed: effective value of the fundamental component of the phase current, and current demand value.

It is concluded that the variant that uses the current demand value, because it performs a process of updating the relay adjustment value in defined intervals for the calculation of the demand value and does not require the use of fault detection algorithms, is the one that most simplifies the operating logic of the adaptive overcurrent relay.

References

[1] F. Costa, A. Monti, S. Paiva, Overcurrent protection in distribution systems with distributed generation based on the real-time boundary wavelet transform, IEEE Trans. Power Deliv. 32 (1) (2015) 462–473.

[2] C. Castillo, A. Conde, E. Fernandez, Mitigation of DOCR miscoordination through distance relays and non- standard overcurrent curves, Electr. Power Syst. Res. 163 (2018) 242–251.

[3] M. Singh, V. Telukunta, S. Srivani, Enhanced real time coordination of distance and user defined over current relays, Int. J. Electr. Power Energy Syst. 98 (2018) 430–441.

[4] T. Adefarati, R. Bansal, Reliability, economic and environmental analysis of a microgrid system in the presence of renewable energy resources, Appl. Energy 236 (2019) 1089–1114.

[5] S. Razavi, E. Rahimib, M. Sadegh, A. Esmaeel, M. Lotfi, M. Shafie-khah, J. Catalão, Impact of distributed generation on protection and voltage regulation of distribution systems: a review, Renew. Sust. Energ. Rev. 105 (2019) 157–167.

[6] B. Wang, L. Jing, A protection method for inverter-based microgrid using current-only polarity comparison, J. Mod. Power Syst. Clean Energy 8 (3) (2019) 446–453.

[7] A. Memon, K. Kauhaniemi, A critical review of AC microgrid protection issues and available solutions, Electr. Power Syst. Res. 25 (2015) 23–31.

[8] M. Elbana, N. Abbasy, A. Meghed, N. Shaker, μPMU-based smart adaptive protection scheme for microgrids, J. Mod. Power Syst. Clean Energy 7 (4) (2019) 887–898.

[9] J. Lee, W. Kim, J. Han, W. Jang, C. Kim, Fault area estimation using traveling wave for wide area protection, J. Mod. Power Syst. Clean Energy 4 (3) (2016) 478–486.

[10] M. Yousaf, K. Muttaqi, D. Sutanto, Assessment of protective device sensitivity with increasing penetration of distributed energy resources, in: Presented at 2018 Australasian Universities Power Engineering Conference (AUPEC), 2019.

[11] M. Yousaf, K. Muttaqi, D. Sutanto, A control strategy to mitigate the sensitivity deterioration of overcurrent protection in distribution networks with the higher concentration of the synchronous and inverter-based DG units, IEEE Trans. Ind. Appl. 57 (3) (2021) 2298–2306.

[12] A. Yazdaninejadi, D. Nazarpour, V. Talavat, Fast protection scheme for active distribution networks: breaking chains by utilizing auxiliary relays, J. Mod. Power Syst. Clean Energy 8 (3) (2020) 540–548.

[13] S. Allamsetty, R. Thangaraj, T. Chelliah, M. Pant, Sensitivity analysis on inverse characteristics of directional over current relays using differential evolution algorithm, Int. J. Syst. Assur. Eng. Manag. 9 (3) (2018) 620–629.

[14] IEEE Recommended Practice for Protection and Coordination of Industrial and Commercial Power Systems (IEEE Buff Book), IEEE Std 242-2001 (Revision of IEEE Std 242-1986) [IEEE Buff Book], 17 December 2001, pp. 1–710, https://doi.org/10.1109/IEEESTD.2001.93369.

[15] A. Sharma, B. Panigrahi, Optimal relay coordination suitable for grid-connected and islanded operational modes of microgrid, in: Presented at IEEE Conference. (INDICON), 2015.

[16] S. Burger, J. Jenkins, S. Huntington, I. Perez-Arriaga, Why distributed?: a critical review of the tradeoffs between centralized and decentralized resources, IEEE Power Energy Mag. 17 (2) (2019) 16–24.

[17] K. Peterson, R. Ahmadiahangar, N. Shabbir, T. Vinnal, Analysis of microgrid configuration effects on energy efficiency, in: Presented at IEEE 60th International Scientific Conference on Power and Electrical Engineering of Riga Technical University (RTUCON), 2019.

[18] F. Mehmood, B. Khan, S. Muhammad, J. Anthony, Distributed MPC for economic dispatch and intermittence control of renewable based autonomous microgrid, Electr. Power Syst. Res. 195 (2021) 107131.

[19] 1547-2018. IEEE Standard for Interconnection and Interoperability of Distributed Energy Resources with Associated Electric Power Systems Interfaces, SASB/SCC21 - SCC21 -, Fuel Cells, Photovoltaics, Dispersed Generation, and Energy Storage. N.d. (2018).

[20] IEC Standard 255-4, Single Input Energizing Measuring Relays with Dependent Specified Time, first ed., IEC Publication 255-4, 1976.

[21] A.G. Phadke, J.S. Thorp, Computer Relaying for Power Systems, Taunton, Somerset, Research Studies Press Ltd, England, 1988.

[22] E.O. Schwietzer, D. Hou, Filtering for protective relays, in: 19th Annual Western Protective Relay Conference, Spokane, Washington, October 1992.

[23] M.S. Sachdev, M.A. Baribeau, A new algorithm for digital impedance relays, IEEE Trans. Power Appar. Syst. 98 (6) (1979) 2232–2240.

[24] M.S. Sachdev, M. Nagpal, A recursive least error squares algorithm for power system relaying and measurement applications, IEEE Trans. Power Deliv. 6 (3) (1991) 1008–1015.

Chapter 8

Distance and overcurrent relay coordination

8.1 Overcurrent and distance relay coordination

The coordination between different protection principles has become an important topic explored by different researchers in recent years. Since the protection principles' combination increases the problem variables and restrictions, the coordination problem complexity grows. The coordination process increases its complexity in real-life applications where multiple adjacent lines with different lengths and bilateral generation are common factors. For example, the line lengths will cause the second zone to under or overreach the desired coverage in distance relays—depending on which line percentage is used to compute the second zone distance—while the infeed effect will cause the protection to detect an impedance greater than the real, underreaching the fault and causing an increased tripping time. In case of DOCRs, the dynamic performance of fault currents determines the operation time, for that, the coordination with other principles of protective relays must be reviewed. In case of distance relay coordination, few points of coordination must be evaluated to avoid loss of coordination between relays.

There have been several attempts to automate the coordination process between directional overcurrent and distance relays [1–4]. The main objectives of these algorithms are to reduce the burden to the protection engineers while obtaining adequate solutions to this problem. In a more robust and still used work, Ramaswami et al. [5] present enhanced analytical techniques for coordinating three zones of distance relays with overcurrent ones. Pérez and Urdaneta [6] improve the work presented in [7] by including the timing of the second zone of distance relays to the coordination process. An important conclusion of this work is the importance of the selected coordination time interval, in some scenarios the second zone tripping time should be greater than the classical and widely used 0.3 s. In a different approach, Khederzadeh [8] proposes a fixed second zone operation time while the shape of the OCR would change in accordance with the fault location. This work is based on the use of universal protection devices—which are shaped by the combination of inverse-time and definite-time overcurrent tripping shapes [9]—instead of inverse definite minimum time relays. Sadeh et al. [10,11] work with the coordination of DOCR

Overcurrent Relay Advances for Modern Electricity Networks. https://doi.org/10.1016/B978-0-323-91777-3.00008-8

with the second zones of distance ones; these works include the second zone tripping time as a variable on their PSO formulation. Singh et al. [12] implement a DE algorithm with hybrid mutation with the objective to optimize the time dial setting (*TDS*) and the I_{pickup} of a DOCR, while the tripping times of the distance relay zones are fixed. Four coordination points are considered in this work.

The coordination is carried out offline considering the upper limits of the load current and the fault current. However, the distribution and subtransmission networks have the highest topological dynamics of the entire electric network. As a result, these dynamic changes in the electrical network affect the sensitivity and selectivity of the DOCRs, causing inappropriate operations manifesting in the lack of fault detection capacity and in nonpermissible operation times of the protections, affecting the voltage quality of the network and increasing the damage of the primary equipment. In the coordination solution, it is common to find operation times greater than those required by the operating criteria of the electrical network and that several relays are not coordinated. The difficulty in adjusting relays and performing the coordination increases rapidly as the system grows, particularly in highly interconnected power systems. These functional limitations are accentuated by increasing the diversity of operation of the electricity network and, in recent years, the diversity of dynamics of nonregulated generation elements.

The degradation of coordination is often due to the limitation by the overcurrent principle; although the use of nonstandard curves offers an alternative, not in all cases it is possible to solve the problem of violation of restrictions, mainly in highly interconnected networks where the lack of sensitivity of relays can be very common. The electric companies have chosen to incorporate different protection principles to improve trip times. However, the problem of lack of sensitivity, mainly in its backup function, is not solved.

In several works the coordination between overcurrent relays and distance relays has been studied. GA has been used frequently in first literatures due to its simplicity, robustness, and ease of implementation. In other works, alternative formulations are proposed to improve the coordination between distance and overcurrent relays. Nair and Reshma [13] use a small six-bus system to obtain the optimal curve selection and time multiplier settings; the genetic algorithm (GA) considers a fixed tripping time for the second zone. The adaptation of a new multiple embedded crossover PSO to optimize the TDS, the I_{pickup}, and the second zone's operating time at three critical coordination points is presented in Farzinfar et al. [14].

Fig. 8.1 represents a radial system with mixed distance and overcurrent protection principles; distance relay 23 backups OCR 34 and it is backed up by OCR 12. This scenario represents the problem of substituting a nonsensitive DOCR for a distance relay. Then, the coordination between the distance relay with its primary and its backup is evaluated and represented in a time-impedance plane. The coordination results in four verification points if the three zones of the distance relay are considered; however, the most critical points are

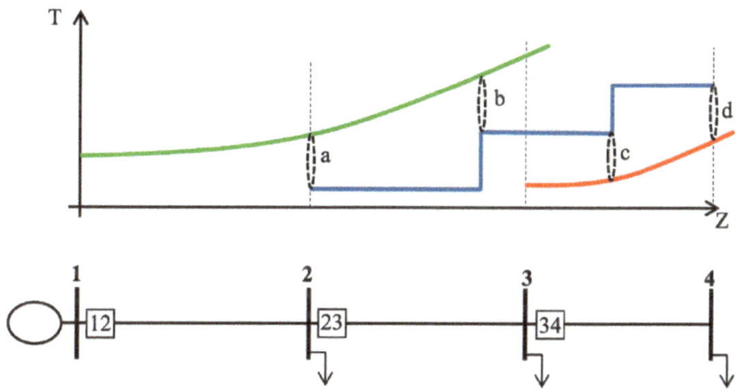

FIG. 8.1 Critical points between distance and overcurrent relays coordination.

points *b* and *d*. The coordination complexity increases even more when a protection principle must be coordinated with different common operative points. In case of considering only two zones of the distance relay, the coordination commitments are simplified, because point *c* disappears.

Because both relays present a different performance, it is important that the critical points are included in the coordination. The protection zone of the overcurrent relay is highly dynamic because it depends only on one parameter, whereas the distance relay has a relatively constant reach. The most representative dynamics of the distance relay is an infeed source, while other dynamics such as polarization and mutual coupling could also be considered, but they are less representative.

8.2 Use of distance relay to replace a DOCR for lack of sensitivity

The distance relay is more sensitive than the overcurrent relay, so where the overcurrent relays do not comply with the sensitivity, they can be replaced by distance relays. Usually, the lack of sensitivity is the result of combination of long lines, large loads, and moderate short-circuit contribution—this condition can be very common in microgrids. Also, the intertie zones represent the coordination between distance and overcurrent relays. To have a solution to the lack of sensitivity, the DOCR coordination that considers distance relays to substitute overcurrent relays with coordination conflicts is proposed. A quantitative example of coordination between DOCR and a distance relay is shown later, where a nonsensitive DOCR is substituted, and the result of the coordination obtained allows the sensitivity of the protection scheme to be maintained.

The radial network of Fig. 8.2 is used and the electrical network data is shown in the same figure. The three-phase fault currents in buses are as follows (Table 8.1):

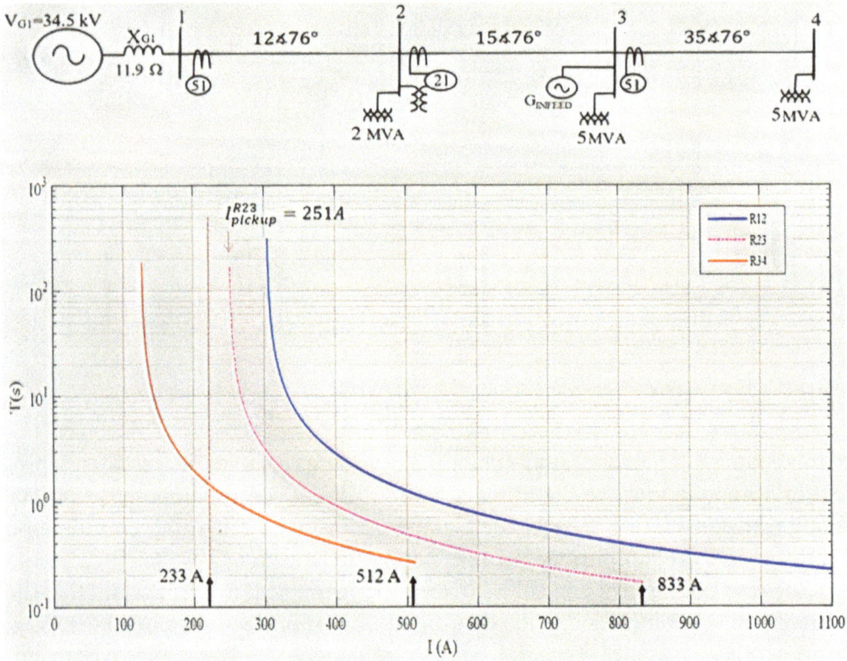

FIG. 8.2 Example of DOCR coordination in radial network when insensitive overcurrent relay R23 is detected.

TABLE 8.1 Fault currents.

Node	I_{sc}
1	1673
2	833
3	512
4	269

Result of the coordination process and the adjustments of the DOCRs using very inverse time for all relays are as follows (Table 8.2):

It is observed that the relay R23 does not comply with the sensitivity because the I_{pickup} is higher than the fault current (233 A); that is, it does not guarantee the backup function for a fault in line 3-4. In case the R34 relay does not operate represents a protection problem and must be resolved. It is possible to incorporate a transferred tripping scheme in case the breaker failure is

TABLE 8.2 Data of relay setting

DOCR	I_{pickup}	TDS	Sensitivity
R12	303.6	0.10	1.46
R23	251.1	0.06	0.92
R34	125.5	0.15	

detected, but if the fault is in another section of the trip circuit, in the transducers or if this 50BF scheme is not available, an alternative solution is to incorporate a distance relay.

8.2.1 Distance relay in DOCR coordination

The replacement of the nonsensitive relay R23 is done by including a distance relay. The coordination can be visualized graphically on a time-current plane or on a time-impedance plane. The setting criterion for distance relay is shown in Fig. 8.3 [15].

Zone I

$$Z1 = 0.9\,Z_{L1}{}^{*}I_{Infeed},\quad T_1 = \text{inst}$$

Zone II

$$Z2 = Z_{L1} + 0.5^{*}Z_{L2}{}^{*}I_{Infeed},\quad T_2 = 0.3\,\text{s}$$

Zone III

$$Z2 = Z_{L1} + Z_{L2}{}^{*}I_{Infeed},\quad T_3 = 0.6\,\text{s}$$

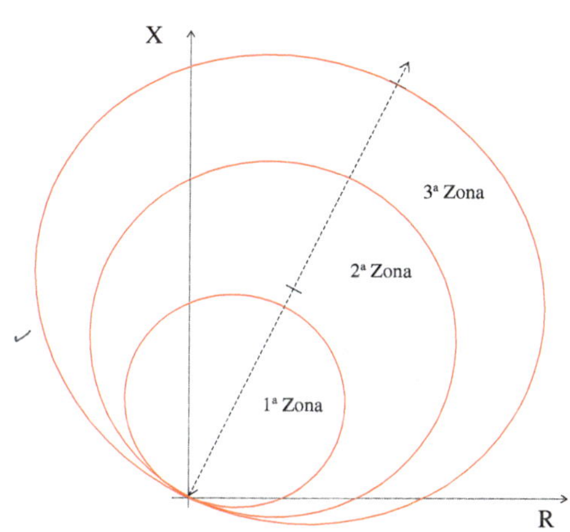

FIG. 8.3 Distance relay operating characteristics.

TABLE 8.3 Distance relay setting.

Distance relay	Time (s)	Impedance (Ω)
Zone I	0.032	13.5
Zone II	0.65	32.5
Zone III	1.12	50

The distance relay replaces the relay 23 for fulfills with the protection coverage. The coordination with the overcurrent relays are like follows (Table 8.3).

It is observed in Figs. 8.4 and 8.5 that the coordination between both relay principles is obtained, guaranteeing the sensitivity of the R23 relay. The coordination between the overcurrent relays and the distance relay is obtained with the *CTI* for both the backup and the primary of the distance relay.

The infeed source shown on bus 3 of Fig. 8.2 only affects the distance relay since it reduces its operating area or range (green line of Fig. 8.5). In contrast, overcurrent relays are not affected in their operation because the contribution of current not seen by the relay results only in a reduction of the primary relay, not affecting the coordination. However, the infeed current does not affect the coordination between relays since the impedance reduction causes the distance relay characteristic to be further away from the critical coordination points.

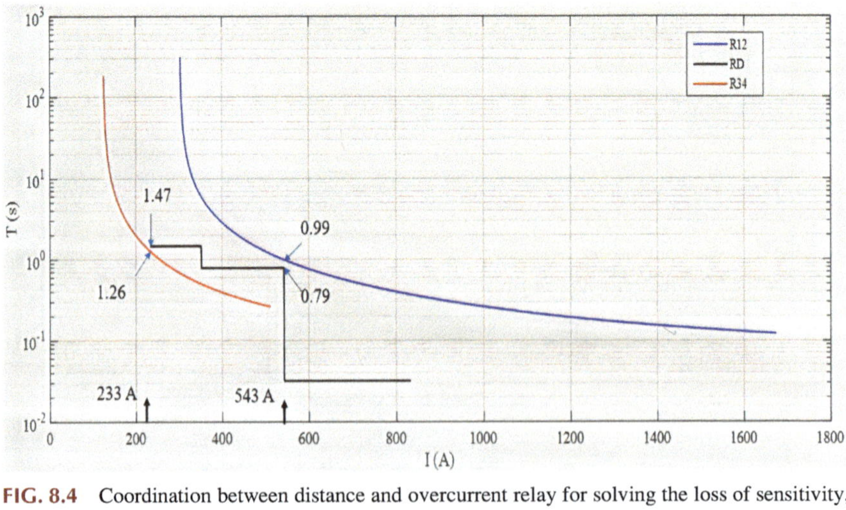

FIG. 8.4 Coordination between distance and overcurrent relay for solving the loss of sensitivity, plane time-current.

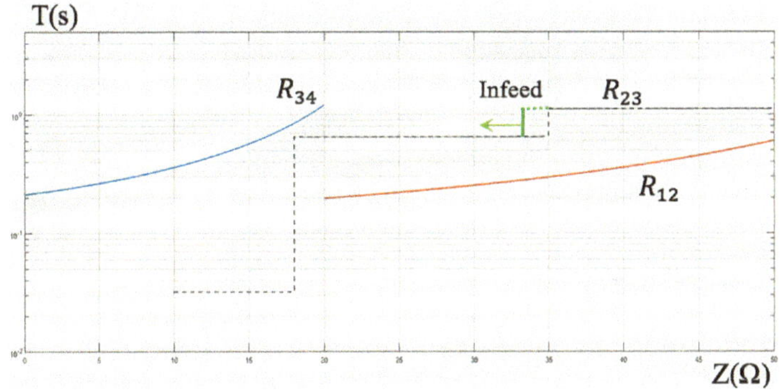

FIG. 8.5 Coordination between distance and overcurrent relay for solving the loss of sensitivity, plane time-impedance.

8.2.2 Nonconventional time curves with distance relay coordination

The use of nonstandard curves in OCRs allows reduced operating times. It is shown in Fig. 8.6 that coordination is improved because the times of the overcurrent and distance relays for the same example of Fig. 8.4 are reduced.

As presented in the previous sections, the coordination between distance relays and overcurrent relays can be obtained by using the curves of each relay. However, it is important to note that in the case of overcurrent relays, the time curve is defined to show the operating time for constant faults, so it does not determine the relay's operating time in the event of a dynamic fault. For the coordination process, the comparison of time curves is sufficient, since all the overcurrent relays will have similar dynamics because they are subjected

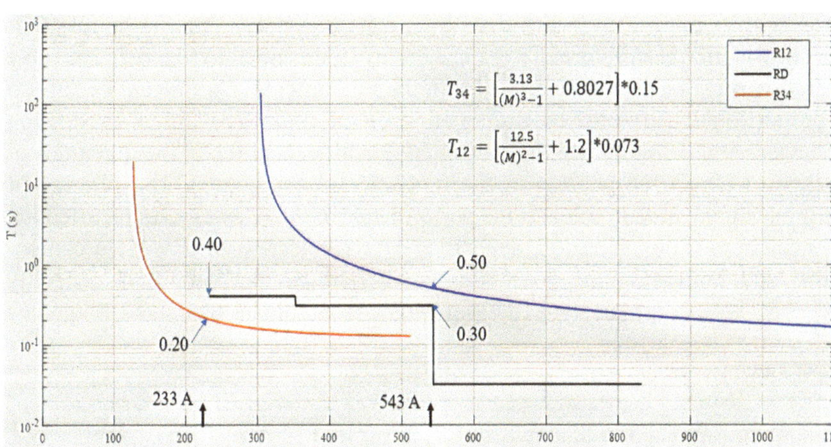

FIG. 8.6 Use of nonstandardized time curves in distance-overcurrent relays coordination.

to the same fault current. But when coordination is carried out between different protection principles such as distance and overcurrent, they will have different dynamics, therefore it is necessary to evaluate the coordination in a dynamic state.

The overcurrent relay model has been presented previously in previous chapters, that is, why in this section only the distance relay model is presented.

8.3 Distance relay

The evaluation of the performance of a distance relay is based on its ability to detect faults within its protection zone. The distance relay detects a fault condition when the apparent impedance is less than the setting and enters its operating characteristics. The trip decision must be confirmed because there are phenomena in the system that can cause penetration of the characteristic without being a fault condition, such as power oscillations.

The relay algorithm must receive the voltage and current phasors to determine the operating condition. The input electrical signals to the protection relay coming from the secondary of the current transformer (CT) and potential transformer (PT) are made up of signals that, although they are representative of the electrical phenomenon, are not required to determine the relay's operating criteria because they have a random character that makes it impossible to determine the location of the fault. The saturation of CT, even when it impacts the performance of the distance relays, is generally present in very specific operating and fault conditions; being directly associated with the dimensioning of the transformer, its effect can be significantly reduced. The application of CCVT (Coupling Capacitor Voltage Transformer) for voltage measurement prints a transient in the voltage signal measured by the relay. Since the nature of the transient depends on the moment of the fault, it is not generally compensated, and its mitigation is achieved by means of a delay time in the relay.

The functional structure of a distance relay is presented in Fig. 8.7. The relay operation must be established only with the fundamental component because it is proportionally affected by the location of the fault. The digital processing of the signal must eliminate the unwanted quantities and retain the quantities of interest to obtain the phasors of the fundamental component. The filtering of the input signal directly impacts the performance of the relay algorithm, affecting the accuracy of the operation, depending on the rejection of unwanted components, and increasing the processing load. For this purpose, it is generally necessary to incorporate two filtering stages to eliminate unwanted frequency

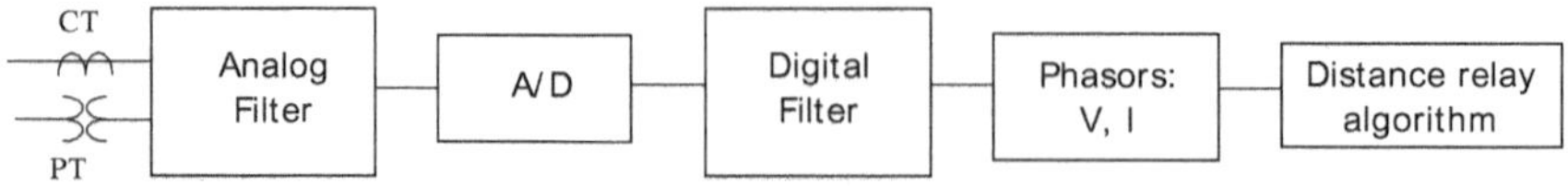

FIG. 8.7 Signal processing for protection purposes.

components, such as noise, harmonics, and direct current components, since the magnitude or polarity is not representative of the location of the fault. Thus these random components are considered as a source of noise that can affect the selectivity of the relay.

There are analog and digital filtering stages to have the configuration with the least delay. Generally, the analog filter used is the Butterworth filter of 2 or 4 order with a cutoff frequency of 360 Hz (Chapter 1)—this filter is preferred for having flat responses in the passband and monotonic decreasing in the stop band.

After the analog filtering stage, the digitalization process of the analog signal is carried out, the increase in the sampling frequency allows to achieve a substantial increase in the resolution of the signal but increases the load of the processor. The reduction of the aliasing effect is obtained through a tuning between the analog filtering and the digital filtering, allowing a frequency overlap. In another alternative it is possible to eliminate the analog filtering by oversampling the signal. The digital filtering is done with finite impulse response (FIR) type filters, since there is no recursion. That is, the output depends only on the input and not on past values of the output, since the memory of the previous conditions of the signal it has no benefit in determining the fault condition by the relay. In addition, infinite impulse response (IIR) filters generally produce phase distortion, contrary to FIRs, which are linear in phase. This condition allows its frequency response to have natural zeros at harmonic frequencies allowing a rejection of these components. The Fourier or Cosine type filter is used regularly.

8.3.1 Distance relay algorithm

The relay generates a discrete output signal that depends on the fulfillment of operating conditions by the electrical input signals. The operating condition is determined by comparing signals; this comparison can be amplitude comparison or phase comparison [16]. The input signals to the comparators depend linearly on the input signals V_r and I_r according to the following expressions:

$$\begin{aligned} S_0 o S_1 &= k_1 \angle \alpha_1 \cdot V_r \angle 0^\circ + Z_{R1} \angle \theta_1 \cdot I_r \angle - \varphi_r \\ S_R o S_2 &= k_2 \angle \alpha_2 \cdot V_r \angle 0^\circ + Z_{R2} \angle \theta_2 \cdot I_r \angle - \varphi_r \end{aligned} \tag{8.1}$$

where k_1, k_2, Z_{R1}, and Z_{R2} are complex constant coefficients; S_0 and S_R represent operating and polarization input signals, respectively, for an amplitude comparator, S_1 and S_2 similarly for a phase comparator. Trying to improve the selectivity and functionality of the distance protection, various operating characteristics have historically been developed. The diversity of operating characteristics is obtained by varying the coefficients of Eq. (8.1).

The comparison signals described in (Eq. 8.1) have a directional reference, the operating signal depends on the current direction, so the reference signal must be supplied by S_2 or S_R which is the amount of polarization, for this reason for the formation of the different directional operating characteristics of distance relays the S_2 or S_R signals are not current dependent; on the other hand, the impedance or displaced mho characteristics that are not directional or have a controlled directional if they present the current term.

In the comparators, the amplitude or phase of the input signals is evaluated; this represents a rectification of the analog signals and a transformation into representative voltage values. It is observed that the factors of the currents in (Eq. 8.1) are always impedances; this impedance is known as the replica impedance and represents the impedance of the line for the operating signal. In the polarization signal, the impedance will represent the offset of the characteristic away from the origin (mho offset). The comparison signals are a sum of voltages and this is due to the legacy of analog relays; however, they can also be currents terms. Fig. 8.8 shows the amplitude and phase comparators and their operating conditions.

The angles β_1 and β_2, called coincident angles, establish the angular limit for the phase comparison; these values are defined for each operating characteristic. Instead, the angle β is defined by the comparison signals: $\beta = \arg(S_1/S_2)$. In Fig. 8.9, an example of visualization of the characteristic of a distance relay is presented, where the variation of the shape of the characteristic is obtained through the variation of the coincident angles. The structure of the comparison signals is simpler for the phase comparator, which is why this is the comparison scheme used for the design of distance relays.

The distance relay programming algorithm is represented by the flowchart in Fig. 8.10. With the phasors corresponding to the fundamental component of voltage and current, the comparison signals for the desired characteristic are formed; the operating condition is established as described in Fig. 8.8. The verification of the operating condition as well as the formation of signals S_1 and S_2 must be carried out for each sampling period. That is, for each new sample of the signal, the data window is formed by discarding the last sample of the window, new phasors are determined, and the operation of the relay is evaluated. Many manufacturers use algorithm starters to avoid continuous processing and processor overload. These starters can be overcurrent fault detectors for example.

The trip decision must be evaluated. If this condition is sustained for a certain number of samples, the relay contact is closed, energizing the circuit

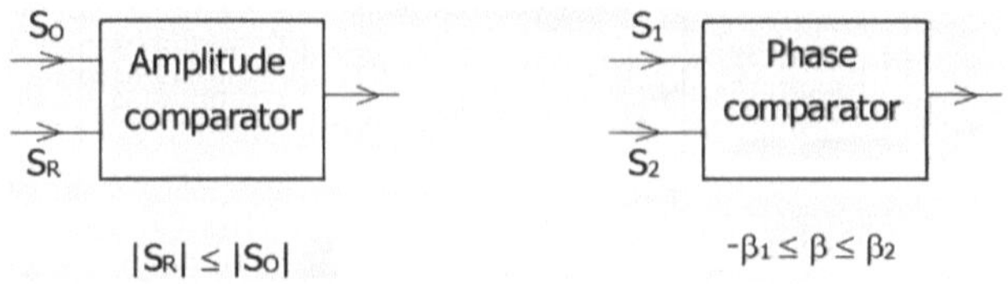

FIG. 8.8 Distance relay comparators.

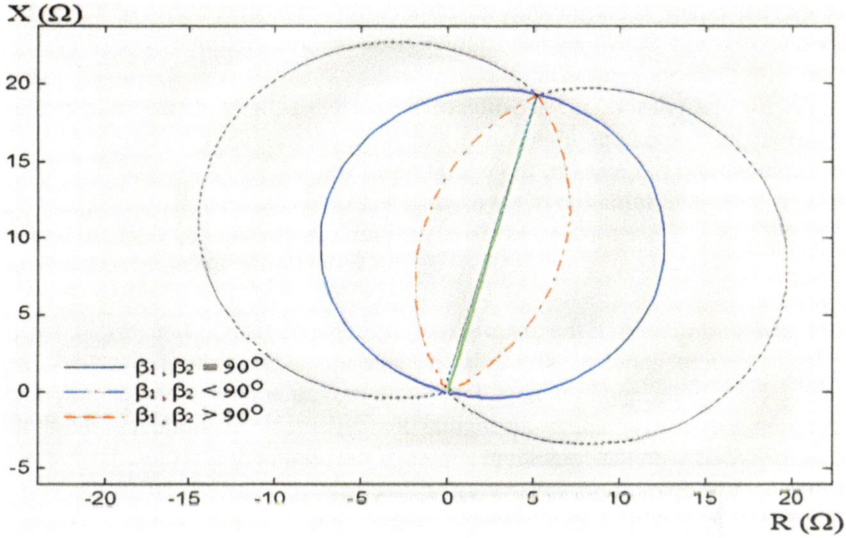

FIG. 8.9 Variation of the coincident angles of a mho characteristic, sensitivity angle $\theta_1 = 75$.

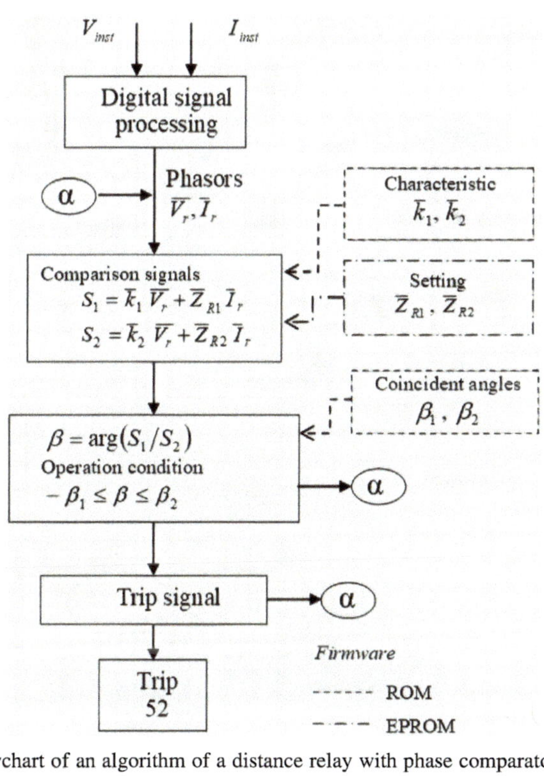

FIG. 8.10 Flowchart of an algorithm of a distance relay with phase comparator.

breaker trip circuit. This prevents false operation of the relay due to power oscillations entering and leaving the characteristic or faults in other zones of operation with temporary penetration.

Having the phasors of the voltage and current signals of the fundamental frequency, the apparent impedance can be represented in a complex plane for visualization purposes in the operation of the relay as shown in Fig. 8.11. However, the determination of the trip is not carried out through the comparison of impedances. The decision is carried out through the angular verification for the phase comparator. The decision of the operating condition is shown in Fig. 8.11.

The determination of the operation of the distance relay is not carried out in the complex impedance plane; this is only a graphical tool to be able to visualize the trajectory of the apparent impedance with the characteristic of the relay. The operation criterion is made by an angular comparison of the quotient of the comparison signals with the coincident angles; if the argument is within the angular tolerance, the operation of the relay is determined. However, it is necessary to enable a timer to be able to emit the trigger signal at the correct time depending on the measured impedance, if it is a second or third zone. In Fig. 8.12,

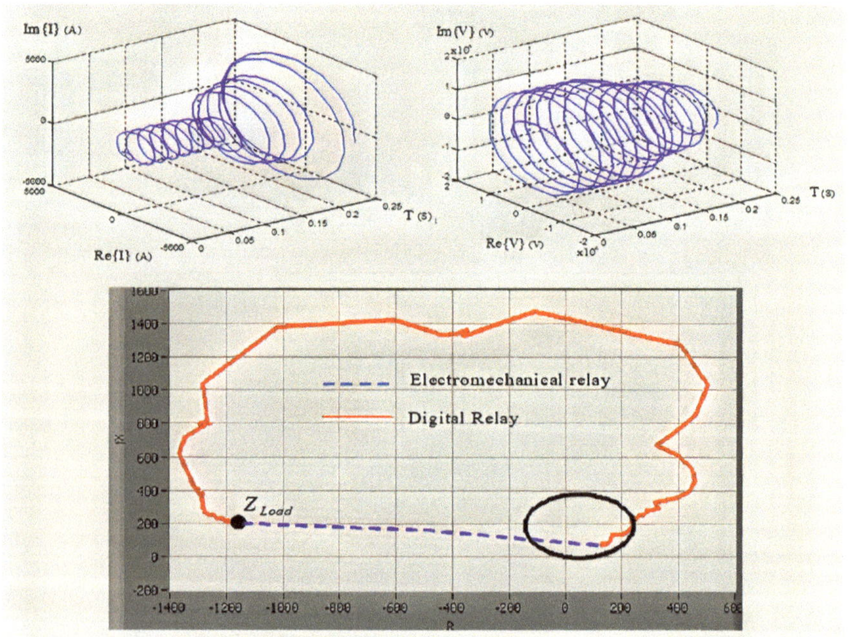

FIG. 8.11 Representation of the apparent impedance in a digital distance relay, (a) current and voltage phasors, (b) apparent impedance in the complex plane.

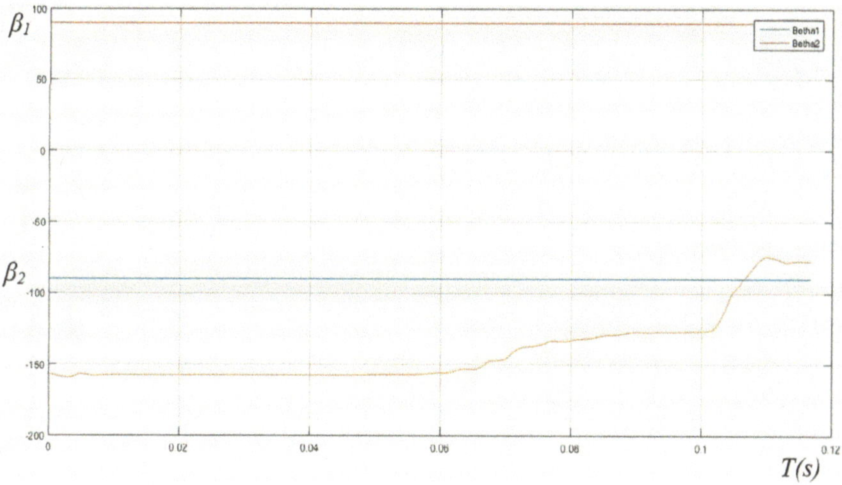

FIG. 8.12 Relay operating condition, $\beta = 90$.

the dynamics of the phase comparator angle during the fault is observed, penetrating within the intervals of the coincident angles to determine the operation of the relay.

To validate the operation of the distance relay, graphic resources are commonly used, where the voltage and current signals obtained from oscillograms are represented in their apparent impedance in a complex plane and are compared with the characteristic of the relay. With this it is possible to determine the operation of the relay indirectly, since the angular comparison is correct.

8.3.2 Validation of relay operation

Event analysis can be performed by recording real or simulated events. In this chapter, a test signal from an oscillograph record of a real event in the 230-kV network and the operation record of a distance relay are used. The test is intended to reproduce the operation of the relay in the field to validate its operation. The record corresponds to a single-phase fault in phase on a 230-kV line with a series impedance of 480 Ω. The field relay has a mho characteristic with a pickup impedance value of $382\angle 75^\circ$ Ω (90% of the line impedance). The event log is shown in Fig. 8.13.

Commonly, the visualization of the operation of relays is done by means of the registration of the time of the actuation of contacts. In the protection scheme used in the fault line, there is an electromechanical relay and a digital relay as double primary protection. It is observed that the operation time of the electromechanical relay is less than the digital relay due to the delay involved for signal processing. As can be seen in the event log, the sequence of operation is given by the detection of the fault in the analog relay 1½ cycles after it occurs,

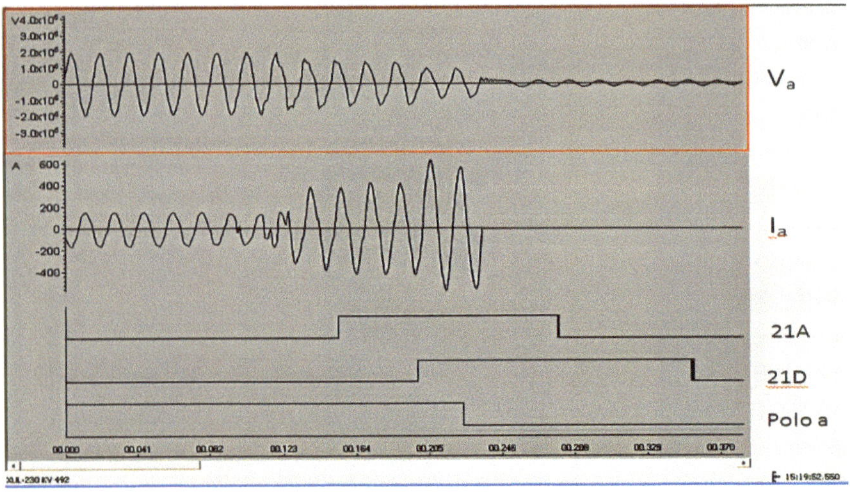

FIG. 8.13 Real event oscillography record.

the digital relay performs the detection 4½ cycles after the fault occurs and 3 cycles after the analog relay detected it, and the opening of pole ***a*** is made 6½ after the fault occurred.

The event registered by the relay is reproduced in a virtual relay developed in Labview shown in Fig. 8.14. From the result obtained, the following observations can be made: (a) The operation time of the virtual relay was like that of the real event relay (21D). The operation time shown in the application

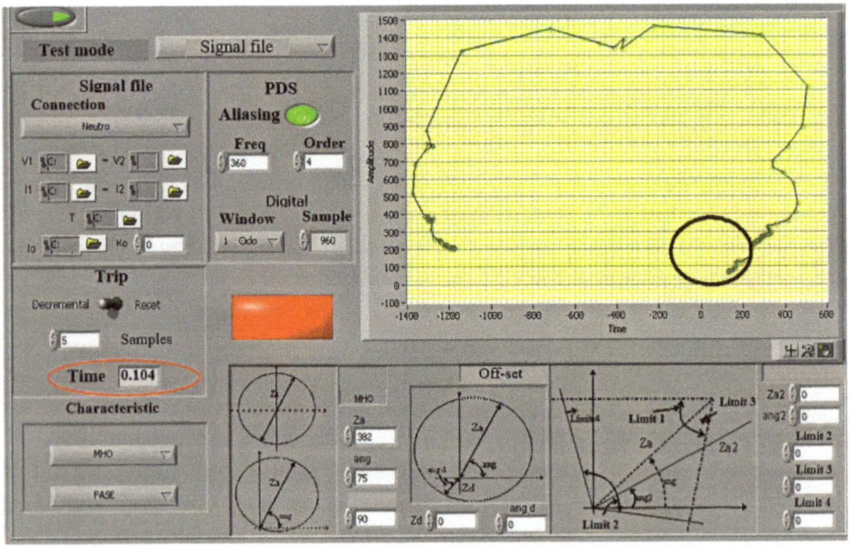

FIG. 8.14 Validation of the operation of a distance relay in a real event through a virtual relay.

corresponds from the beginning of the signal to the trigger output of the relay. (b) As in the real event record, the virtual relay operated seeing the event behind this because the flow was inverted of power by the type of fault. (c) The trajectory that the fault impedance is not a straight line, due to the filtering that is carried out on the input electrical signals. This increases the operating time in the digital relay, which is you can also see in the event log.

8.4 Optimization formulation for overcurrent and distance relay coordination

Overcurrent protection faces limitations regarding the tripping times increase for current magnitudes near the pickup setting. By using nonstandardized inverse time curves, inversion grades can be modified and some of those limitations may be overcome while others can remain and cause miscoordinations. In addition, there are some scenarios where $I_{sc} \leq I_{pickup}$ and consequently the OCR is insensitive. Trying to offer an integrated solution for the OCR coordination problem, a methodology that uses distance relays to substitute overcurrent ones for conflicted coordination pairs is proposed. Three kinds of coordination pairs are expected, overcurrent as main and backup (OC+OC), distance as main and OCR as backup (D+OC), and OCR as main and distance as backup (OC+D).

Optimization problem
The methodology is adapted to be compatible with the already described implementation; a new term that considers distance relay coordination is included to the objective function as shown in Eq. (8.2):

$$\underset{T_{mc}f,g,h,d}{\text{minimize}} \frac{T_{mc}}{n} + \alpha \sum_{i=1}^{m} \frac{f_i(x_i)}{m} + \beta \sum_{j=1}^{n} \frac{g_i^p(y_i)}{n} + \gamma \sum_{j=1}^{n} \frac{h_i^p(x_j, y_i)}{n} + \delta \sum_{k=1}^{o} \frac{d_k(x_k, y_k, Z_k)}{o} \tag{8.2}$$

Eq. (8.2) is a weighted objective function by α, β, γ, and δ, so the first term is used to reduce the number of miscoordinations (T_{mc}), the second term (f) determines the sum of the total times of the m primary relays, in a similar way for the n backups with the function (g). The coordination criterion is established by the difference between the times of the primary and its backups; the function (h) is then the weighting of this difference with respect to the coordination interval (CTI). This objective function is extended to include a fourth term to consider coordination with o distance relays.

The function d_k considers the coordination error of each distance relay k in the critical points of coordination. As a result of tests, it was observed that the critical points were only two, b and d, as described in Fig. 8.15, where o represents the total distance relays, δ is a weighting factor, and Z is a matrix containing the distance relays settings for each one.

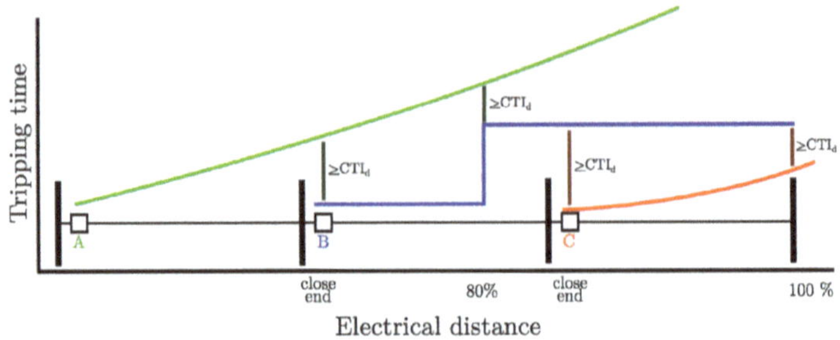

FIG. 8.15 Distance relays and OCR coordination points.

A distance relay is located to replace the operation of OCRs that does not pass the sensitivity filter. As described, there may be some scenarios where an OCR is insensitive for a specific coordination pair and sensitive for others; consequently, the OCR is not completely removed, but its coordination requirements with the appointed coordination pair are discarded. After a distance relay is placed in the selected location, it might have to operate as backup of OCR. Three critical points to be coordinated, a distance relay may be part of as many coordination pairs as adjacent lines are next to its location.

$$d_k(x_k, Z_k) = T_{Zone\ II} - f(x) - CTI_d \tag{8.3}$$

$$d_k(y_k, Z_k) = g(y) - T_{Zone\ II} - CTI_d \tag{8.4}$$

where CTI_d is the coordination time between distance and overcurrent relay, 0.2 s was considered. Three operative scenarios are analyzed under different levels of short-circuit contribution.

The coordination problem formulation is solved through the implementation of metaheuristic methods. Genetic algorithms and invasive weed optimization methods are used in this thesis work; in addition, a sequential quadratic programming method is implemented aiming to improve the convergence process. In this chapter the experimental results obtained by the implemented routines are presented.

The first section presents the preliminary simulations, whose results are used as input by the optimization algorithms. The next section introduces the results of the conducted experiments; those simulations were carried out with the objective of defining adequate tuning parameters for the optimization algorithms. The following sections compare the results of the proposed implementation with a base case that considers two parameters as adjustable settings.

8.4.1 Test systems

The results obtained from the fault and flow analysis for the 14-bus system of Fig. 8.16 are presented in Table 8.4. The two levels of solid short circuit are

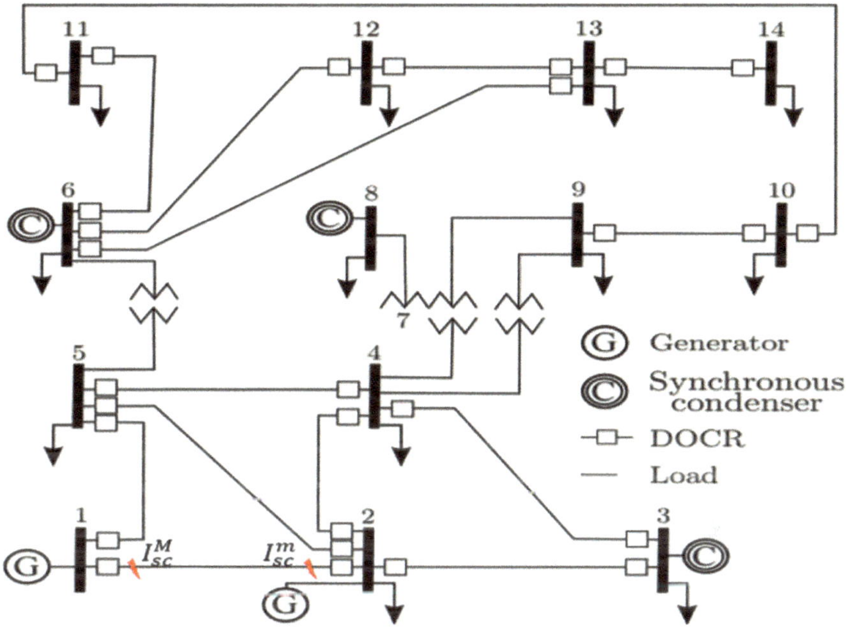

FIG. 8.16 IEEE 14-bus system one-line diagram.

contemplated as coordination current. A maximum fault (I_{sc}^{M}) is contemplated as a three-phase fault located at the close-in of the line and a minimum fault (I_{sc}^{m}) is a phase-to-phase fault located at the line-end.

8.4.2 Results

One of the advantages of metaheuristic methods is their capacity of being adapted to solve several different problems; the parameters involved in the solution process must be previously tuned in order to improve algorithm performance and achieve better results. The parameters to tune in the coordination process include the total iterations, population size, parameter selection range, weighting factors of the objective function, as well as those for the different levels of short-circuit current, the probability of occurrence of each genetic or weed operator.

The coordination pairs are conformed in accordance with the bus connection data of system, after fault and flow analysis coordination pairs pass through the sensitivity filter. The total pair of relays removed from the coordination process for test system is 3. According to reported literature, researchers have implemented algorithms to solve the OCR coordination; nevertheless, nothing has been done about the insensitive relays; furthermore, some works propose distance and OCR coordination offering a redundant solution by placing both

TABLE 8.4 IEEE 14-bus power system faults magnitudes seen by the backup relays.

Primary	Backup	I_{sc}^{m}	I_{sc}^{M}	Primary	Backup	I_{sc}^{m}	I_{sc}^{M}
1-2	5-1	963.745	1606.804	4-5	3-4	1116.421	1328.916
2-1	3-2	1863.462	2252.871	5-4	1-5	1337.088	1601.987
2-1	4-2	1960.285	2475.514	5-4	2-5	950.336	1760.876
2-1	5-2	2002.632	2501.083	6-11	12-6	351.180	443.006
1-5	2-1	1461.318	6972.935	6-11	13-6	285.946	546.542
5-1	2-5	915.267	1852.019	11-6	10-11	324.620	449.118
5-1	4-5	1252.608	3620.568	6-12	11-6	289.486	475.722
2-3	1-2	854.660	7235.935	6-12	13-6	301.282	412.148
2-3	4-2	1012.024	1086.224	12-6	13-12	255.426	354.633
2-3	5-2	903.381	1187.427	6-13	11-6	276.994	490.898
3-2	4-3	1003.620	1179.681	6-13	12-6	313.704	327.655
2-4	1-2	741.773	7235.762	13-6	12-13	324.278	368.262
2-4	3-2	1008.136	1131.460	13-6	14-13	257.591	358.701
2-4	5-2	1082.336	1259.429	9-10	14-9	345.400	428.304

4-2	3-4	636.808	1528.802	10-9	11-10	277.838	418.525
4-2	5-4	838.997	2958.993	9-14	10-9	304.497	600.867
2-5	1-2	1667.953	7235.617	14-9	13-14	249.346	352.457
2-5	3-2	970.057	1249.029	10-11	9-10	325.348	627.134
2-5	4-2	1070.789	1377.241	11-10	6-11	247.838	495.018
5-2	1-5	678.229	1594.613	12-13	6-12	323.019	399.014
5-2	4-5	844.482	3035.837	13-12	6-13	290.277	482.890
3-4	2-3	929.379	1254.278	13-12	14-13	311.048	365.628
4-3	2-4	884.680	1779.245	13-14	6-13	255.256	558.084
4-3	5-4	777.839	2012.752	13-14	12-13	295.784	415.928
4-5	2-4	1371.712	1615.407	14-13	9-14	261.265	445.713

protection principles on every bus. In this work distance relays are used to substitute insensitive overcurrent ones, aiming to offer a better solution to OCR coordination.

Inserted distance relays conform new hybrid coordination pairs with all their related OCR. Fig. 8.15 illustrates the coordination points considered in this section of the work when distance relay is used as main or backup protection. The programmed algorithm is capable to handle different total zones and zone coverage percentage; however for this chapter, distance relays first zone is set to cover 80% of the main line while their second zone is expected to cover the remaining 20% and the 100% of the adjacent one, the third zone is deactivated. As highlighted in Fig. 8.15, each kind of coordination pair—distance as main and overcurrent as backup (D+OC) or vice versa (OC+D)—aims to fulfill coordination requirements for two coordination points. The distance coordination time interval is set to 0.2 s. The total overcurrent and distance coordination pairs for tested system are 11 pairs; since in this approach each pair involves two coordination points, the total points are twice the total pairs.

The objective function is described by Eq. (8.2), and the weighting factors are set to α, β, γ, and δ, which are, respectively, set to 0.5, 0.15, 0.25, and 0.1. Overcurrent and distance relay coordination results are presented in Table 8.5. The proposed methodology obtains lower tripping times and lesser miscoordinations in comparison with DOCR coordination. The fitness, miscoordination percentage (mc%), distance miscoordination percentage (mcd%), average primary (T_p) and backup (T_b) tripping times, and coordination errors are the results to be compared.

It is appreciated that the %mc is drastically reduced with the distance relays. In the coordination process between DOCRs and the distance relays, the difference between the operation curves results in both the primary and

TABLE 8.5 Overcurrent and distance relay coordination results.

	I	DOCRs	D+OC
	Fitness	3.67	0.44
	mc%	11.35	0.67
	mc_d%		22.73
	T_p	0.68	0.91
I_{sc}^m	E_{CTI}	3.70	1.35
	Tb	4.48	2.50
	Tp	0.43	0.75
I_{sc}^M	E_{CTI}	0.35	0.24
	Tb	1.03	1.26

TABLE 8.6 Total pairs and miscoordinations comparing both approaches.

	DOCR		DOCR+DR	
System	9	14	9	14
TCP	36	150	38	163
mc	6	9	0	1
mc%	16.67	6.00	0.00	0.61

backup operation times being able to increase marginally; however, the reduction in E_{CTI} coordination errors is notable.

Table 8.6 presents the total coordination points and total miscoordinations of both approaches. The total and percentage of miscoordinations are then shown in the fourth and fifth rows. The second group of results is obtained by overcurrent and distance relays coordination; the total coordination points (*TCP*) are the sum of OC+OC, OC+D, and D+OC coordination points. As can be appreciated in the last two rows, the inclusion of distance relays decreases the miscoordination percentage if insensitive relays are considered in four out of five cases.

8.5 Dynamic coordination

Although the adjustment criteria can guarantee adequate coordination, as shown in Fig. 8.4, it is important to establish that the times of the overcurrent relays are reflected in their characteristics considering that the fault is constant. That is, it remains at its value maximum throughout the operating time. However, it is important to recognize that DOCRs and distance relays have different dynamics. While the reach of the distance relay is affected by infeed conditions and phasor estimation, the overcurrent relay exhibits much greater dynamics in its reach because it is affected by Thevenin impedance and the type of fault.

The evaluation of the coordination in dynamic state, although it is impractical for the relay adjustment process, is presented in this section to evaluate the coordination protections against variable fault currents and their possible loss of coordination.

The evaluation will be carried out with the DOCR R34 as primary and the distance relay R23 as backup, the simulated fault is in line 34 of Fig. 8.4. From the coordination result of Section 8.2, the adjustments obtained were as follows (Table 8.7).

In Fig. 8.17, the simulated phase-to-phase fault current located at 30% of line 34 is shown; the event was simulated in EMTP with sufficient duration to have the operation of both relays.

In Fig. 8.18, the dynamic behavior of the distance relay is shown; it can be seen that at the moment of the fault 0.1 s the relay performs the detection by

TABLE 8.7 Settings of distance and overcurrent relays

DOCR	I_{pickup}	TDS
R34	125.5	0.15
Zone	**Time**	**Z setting**
Zone II	0.65	32.5

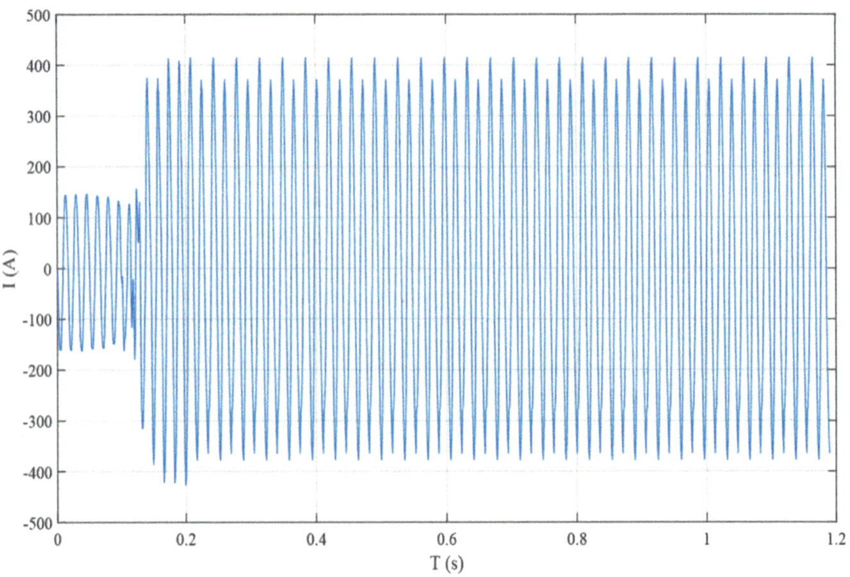

FIG. 8.17 Fault current signal.

means of the angular comparison between the argument of the comparison signals and the coincident angles. In this event, the adjustments were $\beta_1 = 90°$ and $\beta_2 = -90°$ to form a mho-type characteristic. In its coordination with the DOCR, the distance relay will operate in its second zone. From the adjustment obtained in the coordination process, an adjustment of 0.65 s was obtained to coordinate with the DOCRs.

Another more common way of visualizing fault detection is by using a complex impedance plane where the apparent impedance of the fault is shown penetrating the mho characteristic of the relay. Fig. 8.19 shows the path resulting from the apparent impedance formed by the voltage and current phasors and the error in the phasor estimation process. The fault is detected when the apparent impedance stays a few samples within the relay characteristic, commonly four samples or 4 ms, but it depends on the design of each relay.

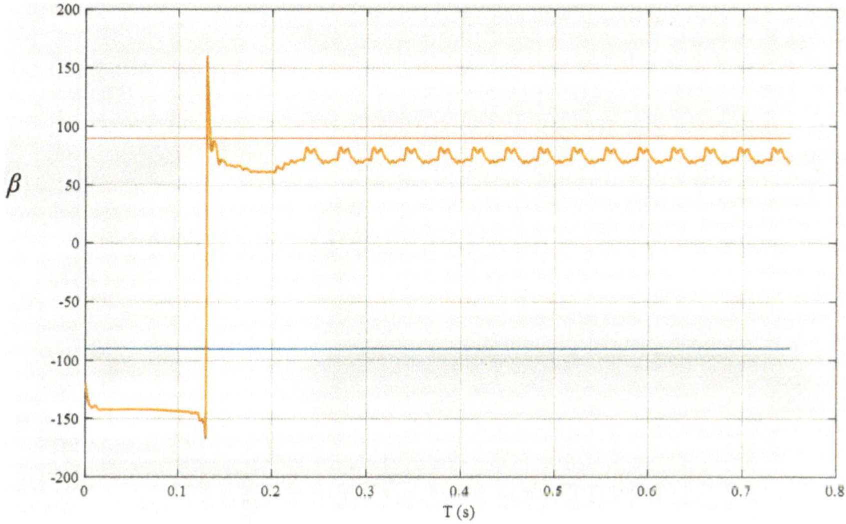

FIG. 8.18 Angular comparison for fault detection.

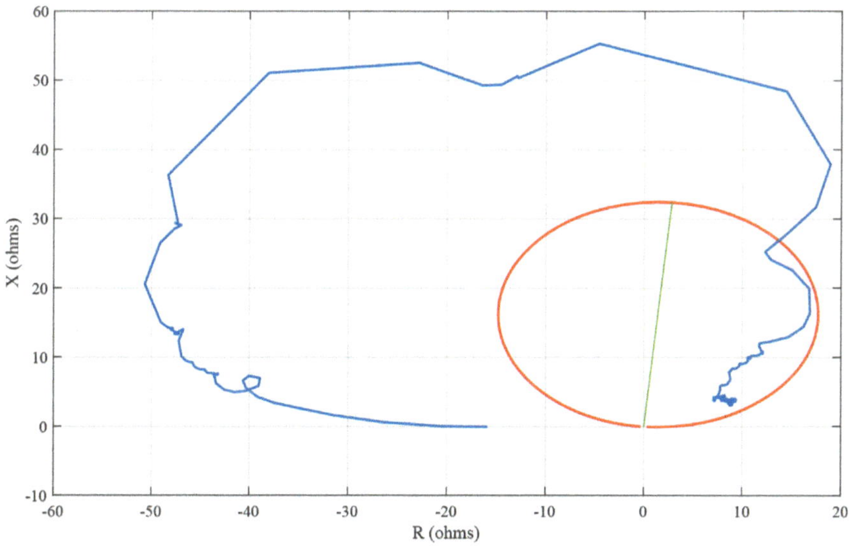

FIG. 8.19 Complex impedance plain for a fault detection visualization.

The operation process of overcurrent relay consists of an integration of the time function (Chapter 1), from this process the value of the integrator is determined. When its value of G_k is equal to 1 the relay operation is obtained (see Fig. 8.20). To determine the operating time, the product of the number of samples must be determined by the sampling period. In the case of the fault event, the DOCRs operates in 1.11 s or 66.75 cycles.

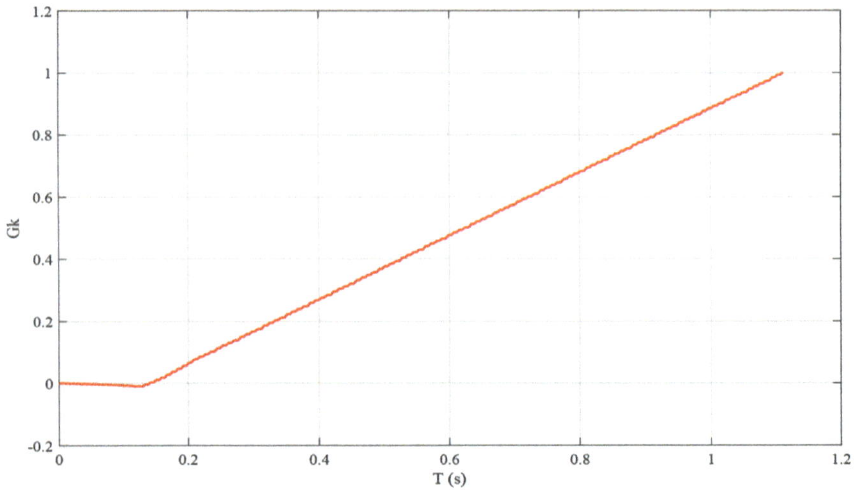

FIG. 8.20 Dynamic performance of DOCR operation.

If we compare the dynamic processes of operation of both relays, we can see that for the fault at line L34, the distance relay R23 operates before the DOCR R34, losing coordination. This can be seen in Fig. 8.21 where the operation time of both protections is shown in the same graph. To maintain coordination, the overcurrent relay should operate in 0.45 s, but since the fault current is lower than that considered in the adjustment process, which is very common, the relay

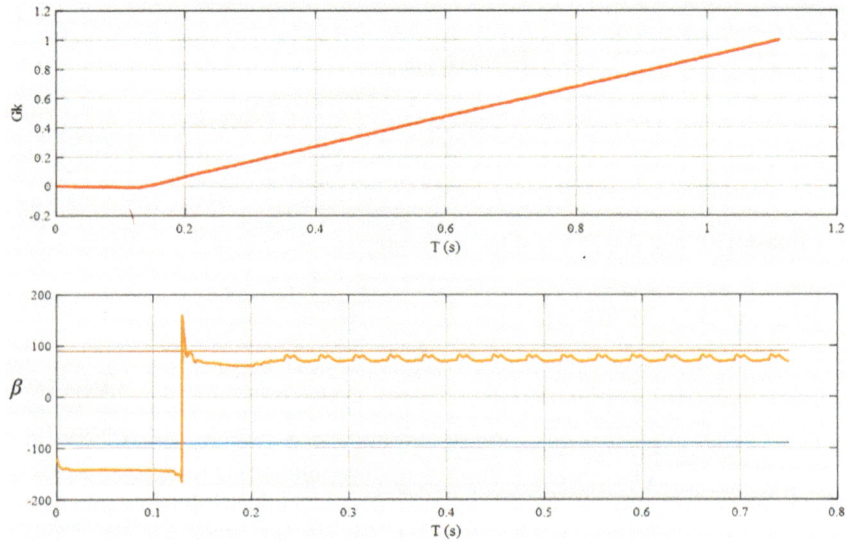

FIG. 8.21 Dynamic behavior of operation of distance relays and DOCRs.

operates slower. While the distance relay has a timer, its operation time is set at 0.65 and does not change due to the evolution of the fault.

The flowchart of the dynamic evaluation process of both relays presented in this section is described in Fig. 8.22. After reading the voltage and current signals and phasor estimation, the dynamic processes of both relays are independent. In the distance relay, the comparison signals are formed, and the evaluation of the operation criterion is carried out by means of the angular comparison. On the other hand, in the overcurrent relay, the integration of the inverse of the function of the time curve is performed to determine the operation threshold. Depending on the dynamics of the fault current will be the time of operation of the relay.

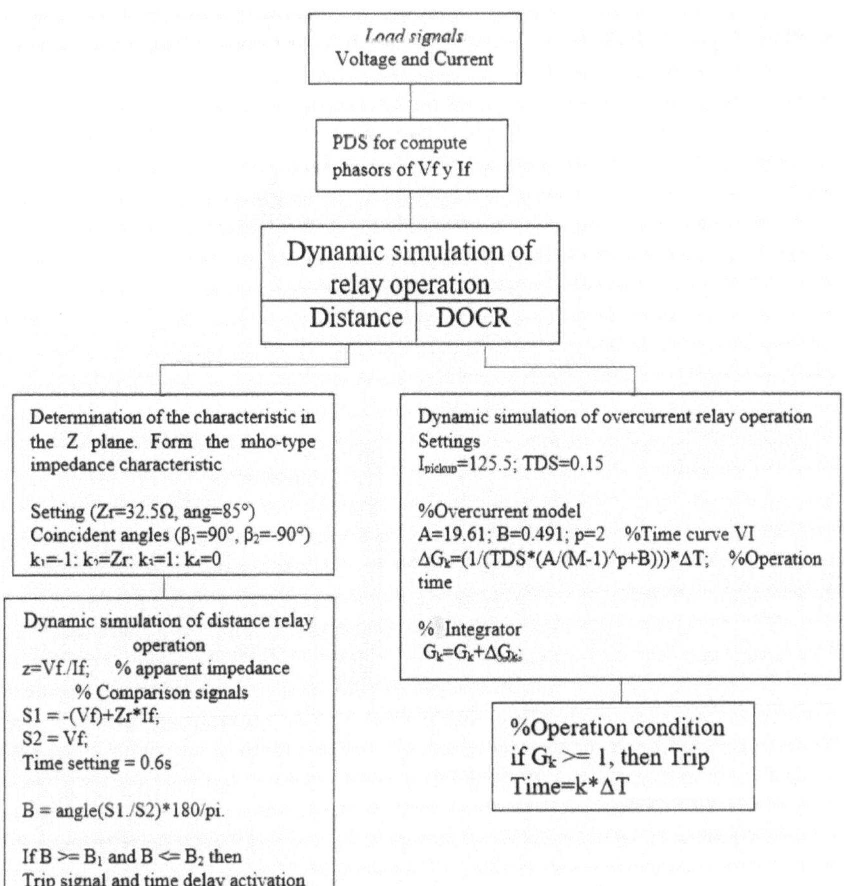

FIG. 8.22 Flowchart of dynamic relay evaluation.

8.6 Summary

In this chapter we propose the assignment of distance relays to replace the DOCR in places where sensitivity has been lost. To demonstrate the robustness of the solution and its application as backup protection schemes in subtransmission networks, the evaluation of the proposed method was carried out in large and highly interconnected networks. For purposes of improving the operating times and the sensitivity of the protections, the model of nonstandard time curves and underreach for infeed source on distance relays that replace the insensitive relays are included in the formulation. With these considerations, the results obtained were positive, both the operation times and the sensitivity of the relays of the protection scheme were improved.

The IWO method with the proposed formulation offered a very adequate behavior in the solution of large problems subject to a large number of restrictions. The incorporation of SQP offered an improvement in the results. The distance relays were setting with constant constraints in the optimization algorithms. In adaptive protection applications, it is possible to use dynamic time intervals to update the setting of the second zone.

References

[1] R.B. Gastineau, R.H. Harris, W.L. Woodside, W.V. Scribner, Using the computer to set transmission line phase distance and ground back-up relays, IEEE Trans. Power Appar. Syst. 96 (2) (March 1977) 478–484.

[2] M.J. Damborg, R. Ramaswami, S.S. Venkata, Computer aided transmission protection system design part I: algorithms, IEEE Trans. Power Appar. Syst. 103 (1) (January 1984) 51–59.

[3] R. Ramaswami, S.S. Venkata, M.J. Damborg, J.M. Postforoosh, Computer aided transmission protection system design part II: implementations and results, IEEE Trans. Power Appar. Syst. 103 (1) (January 1984) 60–65.

[4] D.E. Schultz, S.S. Waters, Computer-aided protective device coordination, a case study, IEEE Trans. Power Appar. Syst. 103 (11) (November 1984) 3295–3301.

[5] R. Ramaswami, M.J. Damborg, S.S. Venkata, A.K. Jamp, J. Postforoosh, Enhanced algorithms for transmission protective relay coordination, IEEE Trans. Power Deliv. 6 (1) (January 1986) 280–287.

[6] L.G. Pérez, A.J. Urdaneta, Optimal computation of distance relays second zone timing in a mixed protection scheme with directional overcurrent relays, IEEE Trans. Power Deliv. 16 (3) (July 2001) 385–388.

[7] A.J. Urdaneta, R. Nadira, L.G. Pérez, Optimal coordination of directional overcurrent relays in interconected power systems, IEEE Trans. Power Deliv. 3 (3) (July 1988) 903–910.

[8] M. Khederzadeh, Back-up protection of distance relay second zone by directional overcurrent relays with combined curves, in: IEEE Power Engineering Society General Meeting, Pages 1–6, June 2006, 2006.

[9] L.A. Kojovic, J.F. Witte, A new method in reducing the overcurrent protection response times at high fault currents to protect equipment from extended stress, in: Proceedings of the IEEE/PES Transmission and Distribution Conference and Exposition, vol. 1, October 2001, pp. 65–70.

[10] J. Sadeh, V. Aminotojari, M. Bashir, Optimal coordination of overcurrent and distance relays with hybrid genetic algorithm, in: Proceedings of the 10th International Conference on Environment and Electrical Engineering, pages 1–5, May 2011.

[11] J. Sadeh, V. Amintojjar, M. Bashir, Coordination of overcurrent and distance relays using hybrid particle swarm optimization, in: Proceedings of the International Conference on Advanced Power System Automation and Protection, vol. 2, October 2011, pp. 1130–1134.

[12] M. Singh, B.K. Panigrahi, A.R. Abhyankar, Combined optimal distance to overcurrent relay coordination, in: Proceedings of the IEEE International Conference on Power Electronics, Drives and Energy Systems, December 2012, pp. 1–6.

[13] D.S. Nair, S. Reshma, Optimal coordination of protective relays, in: Proceedings of the International Conference on Power, Energy and Control, February 2013, pp. 239–244, https://doi.org/10.1109/ICPEC.2013.6527658.

[14] M. Farzinfar, M. Jazaeri, F. Razavi, A new approach for optimal coordination of distance and directional over-current relays using multiple embedded crossover PSO, Int. J. Electr. Power Energy Syst. 61 (May 2014) 620–628, https://doi.org/10.1016/j.ijepes.2014.04.001.

[15] IEEE Guide for Protective Relay Applications to Transmission Lines, IEEE Standard C37.113-1999, 2002, pp. 1–112.

[16] V. Cook, Analysis of Distance Protection, John Wiley & Sons, 1985.

Chapter 9

Online centralized control for adaptive protection system

9.1 Introduction

Subtransmission and distribution systems are the most numerous and dynamic parts of the entire power system, hence economic and robust protection scheme must be implemented; the overcurrent relay (OCR) is the ideal solution to this problem. Despite of its simple operation scheme, it is very complicated when the coordination of OCRs involves short-circuit currents and load demands of huge difference in the interconnected system. But the nature of the subtransmission and distribution systems cannot be avoided. When there are bilateral fault current contributions, it is necessary to incorporate directional functions to the OCR; therefore, the application of directional overcurrent relays (DOCR) is required. Therefore this time-consuming task which is traditionally done as manual labor should be replaced by computational resources. The computational resources here refer to optimization algorithms.

As it was mentioned before, the distribution system is numerous and with dynamic performance. In other words, it is complicated and changes unpredictably as time goes by, so an efficient algorithm that encounter quality relay settings in a reasonable computational time is indeed needed to face this challenge. The objective is online coordination of the protections, which is very different from all reported literatures of coordination because such literatures coordinate offline and on fixed networks.

Conventionally, relay settings are not reset very often, it depends on the complexity of the system. In radial or industrial systems there may be more resetting frequency, but in a meshed system, recoordination of DOCRs is a very rare practice. Therefore this led to big advantage of the online coordination compared to conventional practice, as load flow changes drastically from winter to summer and vice versa, relays can be coordinated with the latest system information which enhances sensitivity and operation time of DOCRs. There would also be a significant benefit if DOCRs are recoordinated every preestablished time. It is necessary to evaluate the coordination algorithms to determine their application in the face of topological changes and changes in the operation of the electrical network. The criteria must be established based on the robustness of the solution, obtaining the results under different operating conditions with

Overcurrent Relay Advances for Modern Electricity Networks. https://doi.org/10.1016/B978-0-323-91777-3.00009-X

the minimum of violations and reduced operating times. Furthermore, being metaheuristic algorithms, the execution time must be weighed as well as the deviation of the results obtained.

To accomplish the objective, three algorithms are studied and compared. The Genetic Algorithm (GA) [1–4] because of its wide implementation and well-known fame in the protection area was taken as comparison reference. On the contrary, Ant Colony Optimization (ACO) [5–7] and Differential Evolution (DE) [8] are little used algorithms in the protection area. But the ACO has been used lately in different studies such as reactive power flow planning, economic dispatch, power generation scheduling, and has been reported to be a powerful tool in solving complex problems in different areas. The DE is reported to be very efficient in different areas. Their advantages and disadvantages are reported and documented, and the outstanding algorithm was selected [9].

As the optimization algorithm coordinates the DOCRs, the real-time algorithm updates input data (network topology, contingency load flow currents, and fault currents) for the optimization algorithm [10,11]. This stage monitors the network operation; it is assumed that the hardware has already been manufactured and just needs the installation of an appropriate online algorithm. The different operating conditions of the electrical network where it is feasible to justify the implementation of online protection systems are shown in the following section.

9.2 Dynamic operating conditions of electrical systems

The subtransmission and microgrid systems suffer constant changes, the consequence of the changes affects power flows and fault currents. The different topological conditions and the connection or disconnection of important loads modify the power flows of the electrical network. It is important that the relays do not operate due to power transfer conditions, in the event of changes in load flows. The adaptive setting of the relays can be implemented; these dynamic settings manage to increase the sensitivity of the relays and, in turn, reduce operating times. Adjustment changes can be established at different time intervals or upon detection of some topological change.

In the case of connection or disconnection of generators, the fault current (I_{sc}) values are modified, but in the coordination process only the maximum values are considered; then, when the current contributions are reduced, the DOCRs maintain coordination but their operation time increases. The updating of the coordination in the event of variations of the magnitudes of fault currents due to Thevenin impedance allows the relays to be adjusted to the current operating conditions of the electrical network. In the event of a fault, the relays will be able to have better, more selective coordination and faster operation. The protections directly suffer the consequences of the changes of electrical system (Table 9.1).

TABLE 9.1 Changes of electrical system affecting DOCRs.

Dynamic changes	Consequences
Topological	I_{load}, I_{sc}
Input/output of generators	I_{load}, I_{sc}
Change of demand	I_{load}

9.2.1 Topological changes

Fig. 9.1 illustrates the effect of system's I_{pickup} variation when L_3 is frequently out of service. If L_2 and L_3 have the same impedance, then they will have the same I_{load} flow. Now, if both L_2 and L_3 are in service most of the time then the established $I_{pickupB}$ and $I_{pickupC}$ have no problem. But if one of them is out of service frequently, then the I_{load} will transfer other relay, thus the increase of current flow with the other line out of service must be considered. So, this will make $I_{pickupB}$ equal to $I_{pickupA}$, the disadvantage will be when both L_2 and L_3 are in service and when a fault F occurs on L_2, relay B will respond to the fault with a greater time (become less sensitive due to a bigger $I_{pickupB}$).

With real-time coordination, the above can be avoided. And all relays are going to be more sensitive because they will get the latest settings every time according to the latest change of the network topology.

Another possible scenario is illustrated in Fig. 9.2, where the effect of trip line L_3 in relay coordination is illustrated. Suppose that all lines are in service and all relays are coordinated. If line L_3 is suddenly out of service, and a fault occurred at point F, then the relay D will be sensing less short-circuit current than it normally sees, because the path impedance is incremented with the output of L_3.

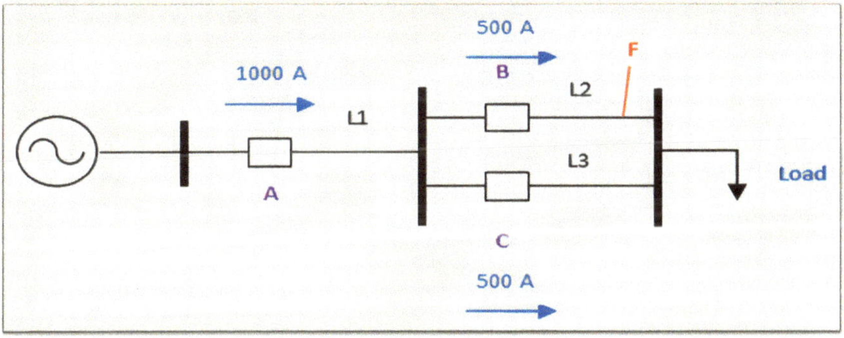

FIG. 9.1 Effect of topological change of distribution line.

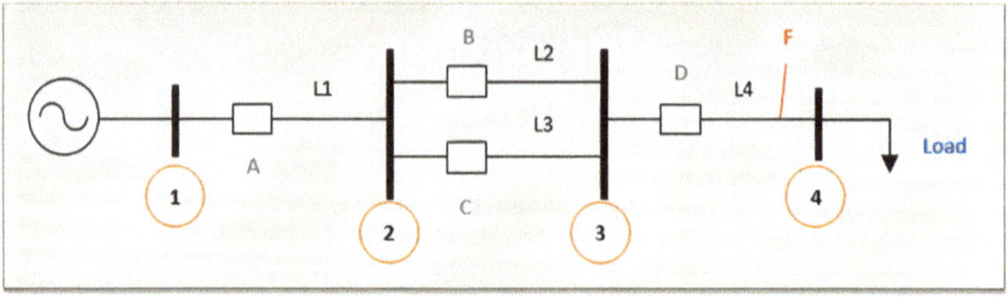

FIG. 9.2 Effect of input/output of distribution line.

Under this circumstance the relay *D* will become slower to operate (less sensitive) or even loss of coordination (loss of relays' selectivity) because relay *B* might operate before relay *D*. In this case, the output of a line can compromise the coordination between the relays D and B. That is why the coordination of relays must be carried out with the analysis of contingencies. However, this example shows the benefit that can be had with an online protection system. Since before a topological change, as the exit of the line L_3, the adjustment of the relays will be modified, improving the sensitivity taking less I_{pickup} and reducing the operating times.

9.2.2 Input/output of generators

The connection or disconnection of generators affects fault current profiles. Fig. 9.3 illustrates the effect of system's I_{sc} variation when sources input/output to/from the system. When we coordinate two relays based on a coordination

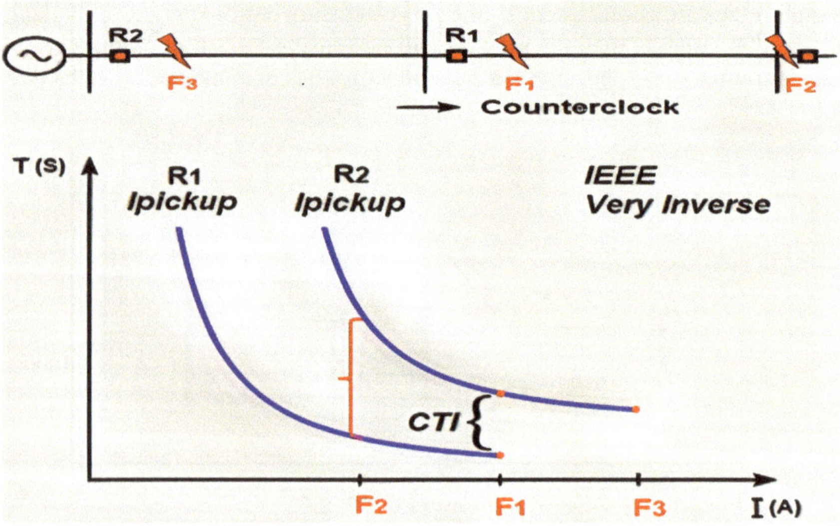

FIG. 9.3 Effect of the variation of the fault current by the output/input of generators.

point F_1 ($I_{sc\ max}$), a *CTI* must be considered for warranty in the time sequence of operation between relays. The reduction of the fault current results in a displacement of the coordination point F_2, the divergence of the time curves guarantees the coordination is maintained, but longer operating times result. This condition is very feasible with the contribution of DG sources, for coordination they must be considered, but it is very likely that they will only be connected for specific operating conditions.

On the other hand, the additional contributions not contemplated in the coordination due to the incorporation of new generation sources, without due updating of the adjustments can result in a loss of coordination due to the convergence of the time curves. It is possible that the *CTI* is not fulfilled and there is simultaneous operation of both DOCRs. It is necessary to confirm the loss of coordination, since it depends on the magnitude of the contribution of the fault current and the location on the time curve, since it can be in the asymptotic zone of the curve. Thus the additional fault contribution can modify the operation time very little, and the coordination can be accomplished.

The update of the DOCRs coordination during the connection and disconnection of power sources allows to reduce the operating times in case of the reduction of fault currents and to avoid the loss of coordination in case of unforeseen contributions. In active networks, this condition can be very common, that is, the fault current magnitudes can be highly variable due to the connection and disconnection of sources can be very frequent. Compromises are regularly presented in the adjustment and coordination process of DOCRs, since in highly dynamic conditions of the electrical network, conventional adjustments that consider load and fault current boundary conditions can result in insensitive and very slow protection. Online relay coordination avoids adjustment compromises, adjusting and coordinating relays in the different operating conditions for connection or disconnection of generators of the electrical network.

9.2.3 Change of demand

The application of DOCRs in microgrids presents serious limitations in terms of sensitivity. High dynamic load conditions combined with minimal faults due to DERs compromise the correct fault detection, thus the lack of sensitivity can be very common. The maximum I_{load} is present for only a few minutes per day for the remaining time, the I_{load} has lower values, but the relay setting is done considering maximum demand. The I_{load} has different contributions throughout the day, meanwhile the I_{pickup} setting is constant as shown in Fig. 9.4. During low demand condition, the I_{pickup} results in poor sensitivity. In case of electrical systems with high connection of CIGR sources, the low-fault contribution combined with high I_{pickup} can result in lack of sensitivity too.

The I_{pickup} can be modified by the I_{load}, so that in minimum demand, greater sensitivity is obtained and less operating time of the relay (see Fig. 9.4).

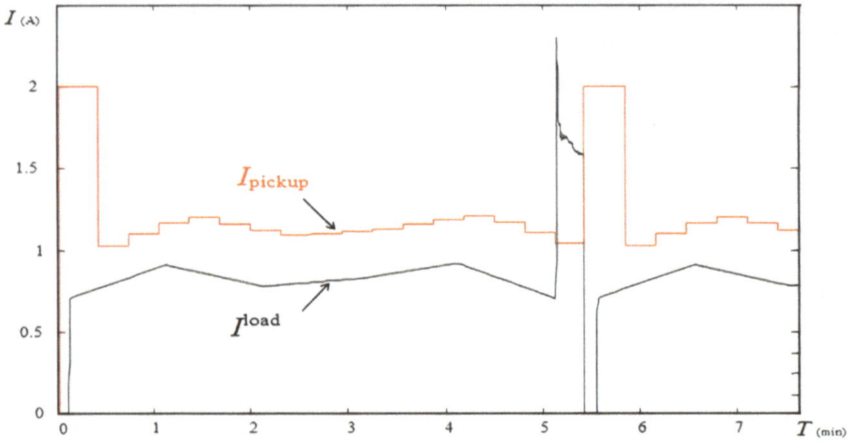

FIG. 9.4 Pickup current setting.

The operating conditions of active networks can result in load and fault conditions that affect the quality of protection, both for fault detection at low values of fault current and in the increase in operating times for large values of I_{pickup}. In many operating scenarios it is possible that coordination is lost affecting the selectivity of the protection. The incorporation of online protection systems is a viable alternative to improve the protection of electrical networks; having dynamic adjustments and robust coordination algorithms, it is possible to obtain selectivity, security, and adequate operating times of the protection system.

9.3 Online protection system

Relay setting is commonly done off-line, and it is constant under the dynamic operation of the electrical grid. However, in applications for MG protection and active networks, due to dynamic operating conditions such as topological changes and the intermittent contributions from renewable sources, applications are needed that give solutions to enable online coordination [10,11].

To carry out the proposed online coordination and the correct operation of the protections in the preestablished times, the adjustment update system must be implemented in centralized mode with measurement processing and monitoring of the system topology. The time updating before any topological change must be almost in real time, and the update by variation of the I_{load} can be carried out in intervals of no more than 15 min, like the times used for the calculation of the demand or in longer intervals. The I_{pickup} is updated with the measured I_{load} within the time interval considered, and the I_{sc} will be modified in case of a topological change or the connection of generation sources. The latency of communication available in the electricity grid must be considered.

It is possible to carry out the application completely online or independently with the two alternatives. The first alternative is to update only the topology and fault current of the system. And the second alternative is to modify the I_{pickup} at the desired time, in periods of every 15 min, in hourly periods or monthly periods.

The update of the relay settings must be carried out in a stable state, so no change or update is recommended during a fault condition to maintain the safety of the protection scheme. The use of units of measurement, for example, μPMU, is recommended and can greatly reduce the amount of calculation.

A flowchart for online coordination applications is presented in Fig. 9.5. Updating of the relay settings must be carried out in a stable state, so topological changes and the demand currents must be considered when updating the I_{pickup} and the coordination of relays. It is important to constantly calculate I_{sc} since the

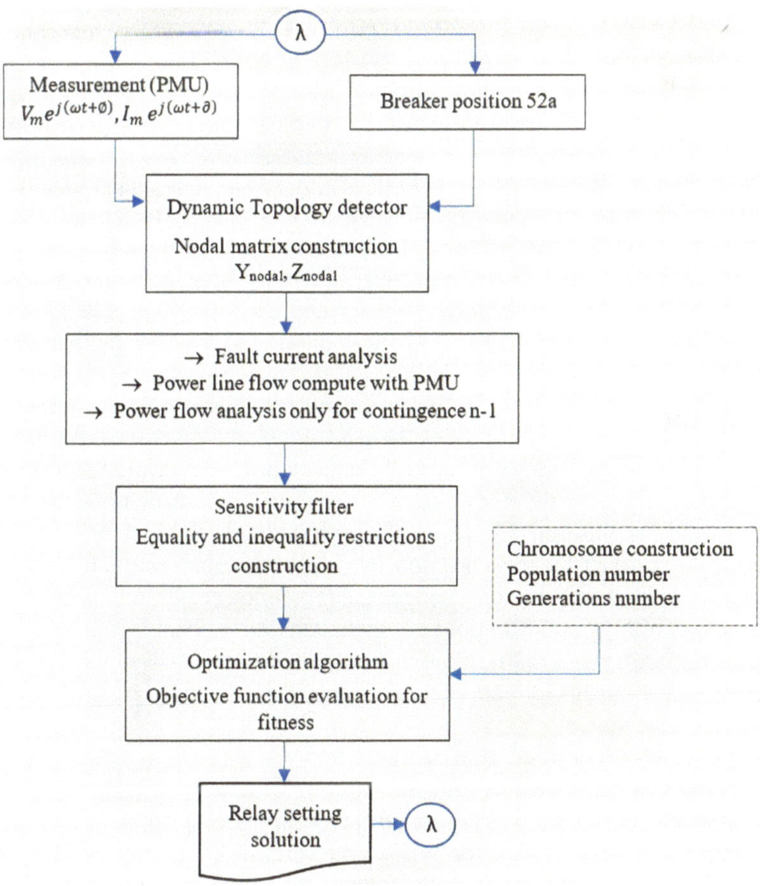

FIG. 9.5 Implementation of the proposed method of coordination.

contributions from renewable sources and topological changes have a great impact on the criteria for setting relays. The detailed description of the algorithm is as follows: first, the system's data is updated according to elements and network changes. Then, the Y_{bus} is constructed or modified from the obtained data using the incident method and the inverse of inspection method. Next, both lists of "Relay Names" and "Coordination Pairs" are generated automatically. After that, the load flow analysis is run using the Newton-Raphson or another method. Then, the Z_{bus} is constructed or modified by the block construction method and partial inversion. Fault analysis is run using Thevenin's method or symmetrical components [12]. Finally, the coordination algorithm will have defined using the coordination pairs and computed the maximum load currents and fault currents of each relay.

The optimization algorithm that will be selected for online execution must show the robustness necessary for protection systems. To carry out the proposed coordination and the correct operation of the protections in the preestablished times, the adjustment update system must be implemented in centralized mode with measurement and monitoring of the system topology.

9.3.1 Coordination algorithm

In this approach, an objective function is established and used to perform a search, to minimize the operation time of the relays and thus ensure that the coordination requirements are met. In this way, equality and inequality restrictions are established to comply with the relay time calculation and the sequence of operation between the primary and backup relays. The automatic formation of these functions has a low computational load for their execution online.

Heuristic optimization methods can search for solutions over a larger area than deterministic ones, which gives them an advantage in terms of finding better solutions, although they are typically expected to be slower. The operation times of the metaheuristic algorithms are slow since they depend on the search space to find the solution. However, the update times of the settings in stable state are determined by the variation in current demand and can be several minutes, and before topological changes the protection requirements must adapt to the new operating condition but not instantaneously. That is, the relay readjustment must always be carried out in a nonfailed state. It is a defensive action that must offer sensitivity and adequate protection times against the possible occurrence of a fault.

In the search for new solution methods, approaches such as differential evolution (DE) and gray wolf optimization (GWO) have been proposed, which offer adequate calculation times for online coordination applications. In the weighting of the optimization method used for online coordination, the dispersion of the results obtained must be considered, since heuristic methods will give different results from each execution. Some approaches such as genetic algorithms (GAs) have a high level of dispersion in the results, i.e., with substantial differences in each run, while DE gives a small standard deviation, with very similar results in each run.

9.4 Special considerations for online coordination system

Before solving the coordination using an optimization method, it is of critical note to discuss the steps carried out to obtain the data used as input. The initial steps include the input of the data for the tested networks considering the conversion of these to a common system base. Typical data contains the number of buses, bus type, bus voltages, generator, shunt and line impedances, generation, and load powers. The use of nodal matrices allows to represent the topology of the electrical network in steady state. The expressions are algebraic for the calculation of power flows and faults; the formulation is computationally efficient for topological and operational changes that must be considered during real-time operation. The output and input of elements such as generation sources, lines, and loads must be considered; the intermittency of the RES can be considered as power injection in power flows and the reactance of the source for faults analysis.

Special algorithm considerations are necessary to consider for online protection systems. Topological detector, parallel protection lines, and transformer protection application must be evaluated. Automatic formation of these algorithms should be implemented; the computational load is generally low, but additional constraints should be incorporated into the optimization algorithms. The execution time must be evaluated, that is why the choice of the metaheuristic algorithm and its specifications such as the number of individuals and the total number of generations are not trivial.

9.4.1 Network topology

Obtaining nodal matrices is the basis for studies of the electrical network, thus the nodal admittance matrix can be obtained by means of incidence matrices to obtain a computationally efficient method, in the same way the nodal impedance matrix can be obtained by construction algorithms [12]. These methods can consider mutual coupling scenarios and parallel lines very easily (Fig. 9.6).

The structure of the matrices allows updating the topological or operation variations by direct methods in the case of the Y_{nodal} and by inversion lemma in the case of the Z_{nodal}. The topological detection systems must have the position of the auxiliary contacts of the switch (52a) and the current measurement in that branch (Fig. 9.6). This information verified at both ends will confirm the current topology and will allow to represent the topological modifications in the nodal arrays. The voltage phasor and the sample time will allow to obtain the directionality of the measured current.

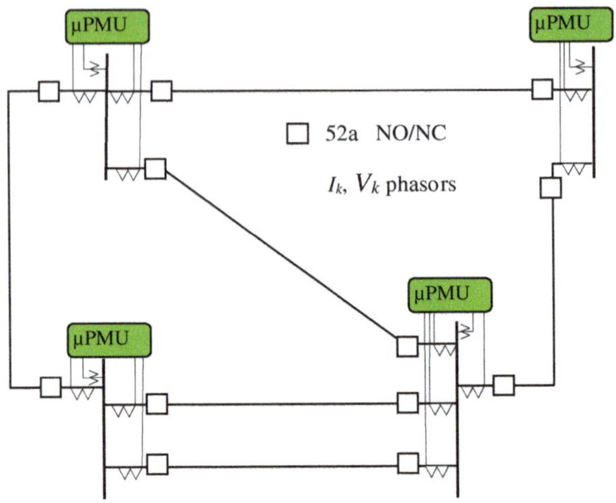

FIG. 9.6 Electrical network information for topology record.

9.4.2 Load flow and fault current analysis

Load flow analysis is used to solve the nonlinear power flow equations to obtain node voltages and currents flowing in each direction of the network lines. The obtained voltages at each network bus and at each branch current are used to get the line loads each relay sees during the steady-state condition of the interconnected system. DOCR relays require that the I_{pickup} be set above the protected line maximum I_{load} to prevent any unnecessary trip due to overload or transformer inrush currents. In smart grids, the measurement obtained by μ-PMU is available, avoiding the execution of iterative flow algorithms; however, for contingency analysis it is necessary to run load flows algorithms. For identification purposes, all relays are assigned a name based on the network topology. If the relay is located between buses 1 and 4, with bus 1 being the closest to the given relay, the assigned relay name is $R14$.

The different current levels seen by each relay on interconnected systems increase the difficulty in conventional coordination methods, especially if there is more than one coordination pairs formed by the same relay. Fault analysis routine is used to determine the maximum and minimum fault currents for each location of interest. The maximum fault current is calculated assuming that a bolted three-phase fault (Zf=0) occurs at bus nearest to the relay, which is assumed to be of the same magnitude as a fault occurring directly in front of the relay. Minimum fault currents are used for a sensitivity analysis and are obtained for the far end of the adjacent line. The faults are simulated using the Z_{th} of the diagonal of impedance nodal matrix. The remote line-end fault can be simulated like distribution systems (forward and reverse method), add the impedance line to Z_{th} because the remote end of the line is considered open.

9.4.3 Special considerations

The operation of DOCRs is very simple which makes this protection scheme to be widely used especially on interconnected systems. However, special considerations must be taken into account in coordination formulation, equipment protection like transformer, reactors, or capacitor bank can be included in the formulation. The algorithm will activate the procedure for setting relays for transformer protection to backup differential protection or like primary protection. Also, parallel lines must be included in the formation of nodal matrices, losing the information of the relays is avoided.

9.4.3.1 Setting of relays for transformer protection

The application of DOCRs as transformer protection can be located depending on the capacity (IEEE C37.91-2000). The limit considered is 10 MVA, when the capacity of the transformer is lower only the overcurrent protection is used, when it is higher the differential protection is added, and the backup protection is the overcurrent protection [13].

Two winding power transformers <10 MVA. Protection for transformers less than 10 MVA is basically overcurrent; these schemes must ensure the relay operation for internal transformer faults and guarantee backup for faults in the feeders (Fig. 9.7).

Protection on primary side, 51H. Primary transformer protection must operate for transformer faults, have coordination with low-side protection equipment. Relay operation should be avoided to hot load pickup, inrush current on instantaneous reclosing of source side circuit breaker and cold load pickup. The setting of the I_{pickup} of the primary protection must be:

For Fuse: 150% and 175% of low-speed *E* type high voltage side primary current.

For Relay: 125%–150% of maximum kVA nameplate rating of a transformer is common.

Both protections must coordinate with the damage curve of the transformer and allow the magnetization current, being greater between 8 and 12 times the nominal current in a time greater than 0.1 s. Only the application of relays for the generation of restriction functions is considered.

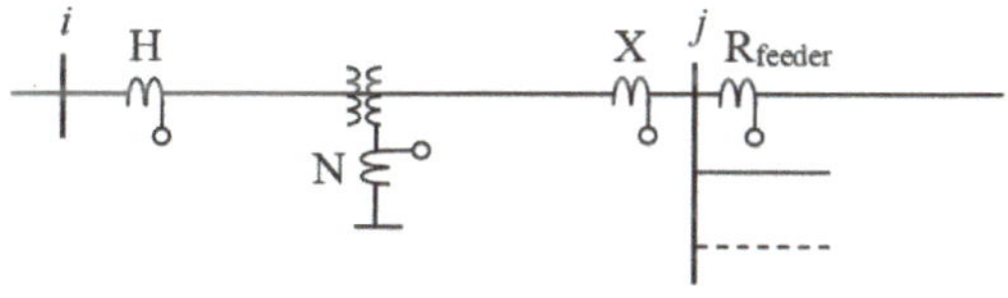

FIG. 9.7 Protection scheme for power transformer less of 10 MVA.

$$I^{ij}_{pickup\ 51H} = I_{load\ rating}*PSM_{Trans} \tag{9.1}$$

$$1.25 \leq PSM_{Trans} \leq 1.5 \tag{9.2}$$

$$T^{ij}_{relay} \geq 0.1 \text{ for } I_{inrush} = 8*I_{nom} \tag{9.3}$$

The relay must protect the transformer for internal faults; to obtain this, its time curve must be less than the damage curve and greater than the inrush point. To guarantee the protection function of the transformer, it is only necessary to verify points *1*, *2*, and *4* for Cat II-IV and points *1* and *4* for Cat I (Fig. 9.8). This verification is carried out by determining the times of the damage curves for the transformer and its category as presented in Table 9.2.

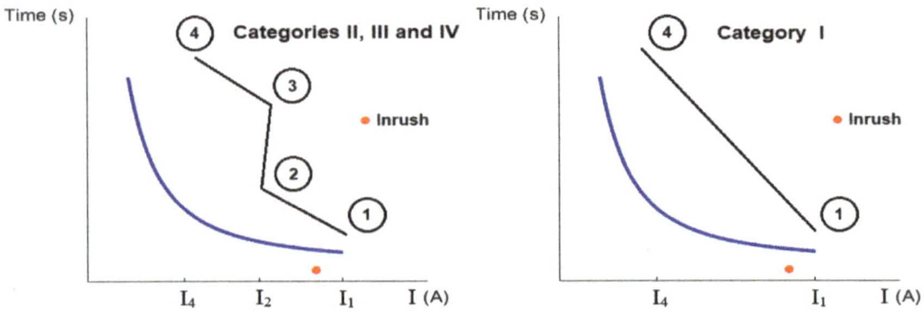

FIG. 9.8 Transformer damage curve category.

TABLE 9.2 ANSI curve points.

Point	Category of the transformer	Time (s)	Current (A)
1	I	$t=1250*(Z_t)^2$	$I_f=\frac{I_n}{Z_t}$
	II	$t=2$	$I_f=\frac{I_n}{Z_t}$
	III, IV	$t=2$	$I_f=\frac{I_n}{Z_t+Z_s}$
2	II	$t=4.08$	$I_f=0.7*\frac{I_n}{Z_t}$
	III, IV	$t=8$	$I_f=0.5*\frac{I_n}{Z_t+Z_s}$
3	II	$t=2251*(Z_t)^2$	$I_f=0.7*\frac{I_n}{Z_t}$
	III, IV	$t=5000*(Z_t+Z_s)^2$	$I_f=0.5*\frac{I_n}{Z_t+Z_s}$
4	I, II, III, IV	$t=50$	$I=5*I_n$

where Z_t is the transformer impedance, I_n is the nominal current, Z_s is the Thevenin impedance, and I_f is the fault current for damage curve for plotted purposes.

For Cat II, III, and IV:

$$T^{ij}_{relay}(I1) > T1 \text{ and } T^{ij}_{relay}(I2) > T2 \text{ and } T^{ij}_{relay}(I4) > T4 \tag{9.4}$$

For Cat I:

$$T^{ij}_{relay}(I1) > T1 \text{ and } T^{ij}_{relay}(I4) > T4 \tag{9.5}$$

Phase overcurrent protection on secondary side, 51X. The relays must coordinate with the relays installed in the R_{Feeder} feeders and must operate with time delay for three-phase or phase-to-phase faults in the feeders ($I^{3\varnothing}_{sc\ x}$), maintaining the *CTI* of the slowest protection of the feeders.

$$I^{ij}_{pickup\ 51X} = I_{load\ rating}{*}2.0 \tag{9.6}$$

$$0.5 \le T^{ij}_{relay\ 51X}\left(I^{3\varnothing}_{sc\ x}\right) \le 0.8 \tag{9.7}$$

Two winding power transformers ≥10MVA. The primary protection function of transformer is differential protection. Adjustment considerations do not affect DOCRs coordination and they are not described. Transformer backup and external fault backup functions are performed by DOCRs.

Phase overcurrent protection on primary side, 50/51H. The function of the instantaneous element (50H) is to protect the transformer without overreaching the feeders on the secondary side (Fig. 9.7); two conditions must be met:

$$I_{load\ rating}{*}10 < I^{ij}_{pickup\ 50H} \text{ and } 2{*}I^{3\varnothing}_{sc\ x} < I^{ij}_{pickup\ 50H} \tag{9.8}$$

Protection on primary side, 51H. It must allow the primary protections of the bank and the feeders to operate first. It must operate with time delay *CTI* for faults on the low side of the bank or on feeders and must operate as transformer overload protection.

$$I^{ij}_{pickup\ 51X} = I_{load\ rating}{*}PSM_{51H} \tag{9.9}$$

$$2.0 \le PSM_{51H} \le 2.2 \tag{9.10}$$

Its time curve must be below the damage curve (for frequent fault) of the transformer. The operating time must be between 0.8 and 1 s for a three-phase fault on the low voltage bus.

$$0.8 \le T^{ij}_{relay\ 51H}\left(I^{3\varnothing}_{sc\ x}\right) \le 1.0 \tag{9.11}$$

Transformer secondary side backup protection, 51X. It must allow the bank's primary protections and feeder protections to operate first. It must operate with a time delay for three-phase or phase-to-phase faults in the feeders (see Fig. 9.7).

$$I^{ij}_{pickup\ 51X} = I_{load\ rating}{*}2.0 \tag{9.12}$$

$$0.5 \le T^{ij}_{relay\ 51x}\left(I^{3\varnothing}_{sc\ x}\right) \le 0.9 \tag{9.13}$$

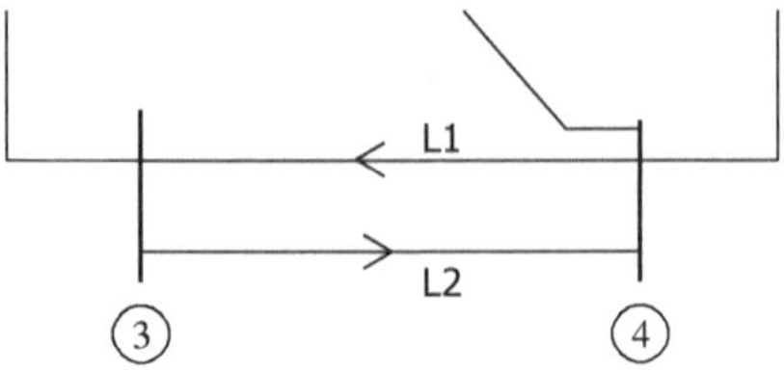

FIG. 9.9 Parallel lines.

9.4.3.2 Parallel lines

In the case of parallel lines (Fig. 9.9), the corresponding pair of relays will be assigned to the equivalent branch. The adjustment will be made considering the fault current and I_{load} in each line in proportion to the impedance ratio of each line, the most common case is that the lines have the same impedance. With these values, the I_{pickup} in each relay is determined and the corresponding coordination pairs are generated.

Coordination between relays $R43$ and $R34$ on different lines will be established by coordinating both relays. With the conventional coordination method:

$$T_{R43}^{L1} = T_{R34}^{L2} + CTI \text{ for Fault } L34 \text{ end line open} \tag{9.14}$$

$$T_{R34}^{L1} = T_{R43}^{L2} + CTI \text{ for Fault } L43 \text{ end line open} \tag{9.15}$$

With the proposed method, only compliance with the intervals will be necessary.

The fault analysis for relay setting is done considering the line-end open, the fault contribution from only one end. This criterion is done under the consideration that the probability that relays located at each end fail to operate is very low. In the case of faults on parallel lines, the contribution from the remote end is given by the nonfault line. That is why the fault calculation for the coordination of relays must be carried out considering the remote end connected with the impedance of a single line, the Y_{nodal} must consider this increase in the impedance of the branch or use the inversion lemma for modifying the Z_{nodal} (Fig. 9.10).

9.4.3.3 Radial lines detection

Radial line protection is located at the load size or downstream protection; therefore, this condition must be recognized by the algorithm and reflected in the formulation. The connection of a single branch to a node is an indicator

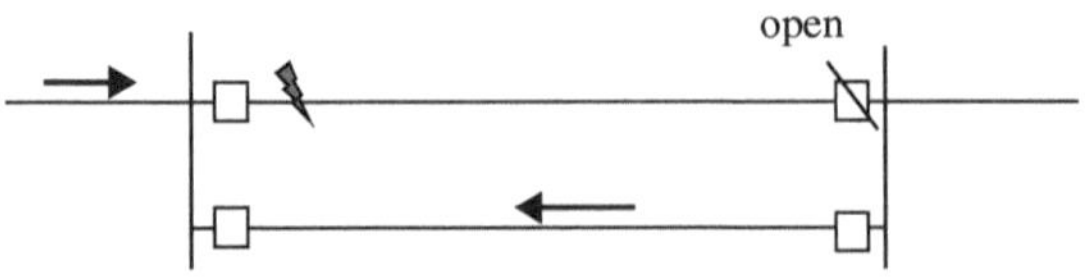

FIG. 9.10 Fault analysis for parallel lines.

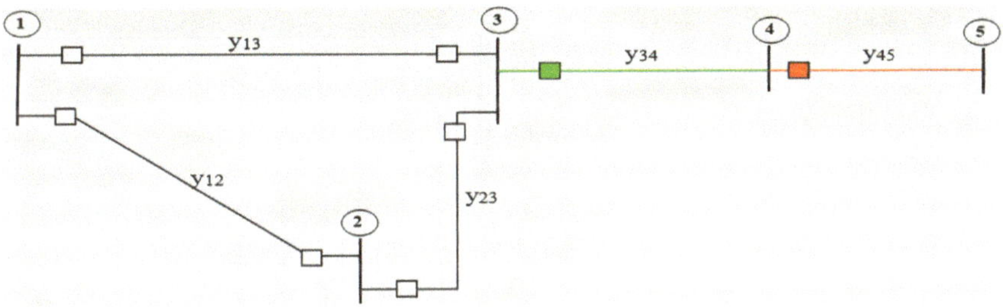

FIG. 9.11 Radial line detection example.

of being the last node of the load end. The Y_{nodal} construction algorithm must indicate when a single branch is connected to a node on the diagonal of the matrix or when there is only one mutual element (y_{45}) in the column vector or row vector (Fig. 9.11). From this node the algorithm performs a search toward the next value on the diagonal or in the next column vector that has the previous branch and an additional branch (y_{34}). This process detects lines with a radial nature, ending when there is an element of the diagonal with three or more connected branches (Eq. 9.16). The algorithm's relay assignment will be done from the source node, omitting the remote end relay. This procedure is performed with the Y_{nodal} for power flow studies because the impedance of the sources is not considered.

$$Y_{nodal} = \begin{bmatrix} (y_{13}+y_{12}) & y_{12} & y_{13} & 0 & 0 \\ y_{12} & (y_{23}+y_{12}) & y_{23} & 0 & 0 \\ y_{13} & y_{23} & (y_{13}+y_{23}+y_{34}) & -y_{34} & 0 \\ 0 & 0 & -y_{34} & (y_{34}+y_{45}) & -y_{45} \\ 0 & 0 & 0 & -y_{45} & (y_{45}) \end{bmatrix} \tag{9.16}$$

9.4.4 Sensitivity analysis

The sensitivity analysis is performed to detect all the coordination pairs that due to system restrictions will not be able to coordinate for remote line-end faults. Backup relays should have the ability to detect minimum fault currents that occur at the far end of its protected line. Eq. (9.17) is used to analyze each coordination pair to determine whether coordination can be achieved. The coordination pairs that do not overcome the sensitivity analysis are relays with either high or unacceptable operating times. The lack of sensitivity is a limitation of the overcurrent principle used, the use of distance relays for backup functions can be an alternative. In case of having numerous relays that do not pass the filter, a lower percentage of coverage of the backup function can be considered (Fig. 9.12). For example, 80%—this percentage will depend on the impedance

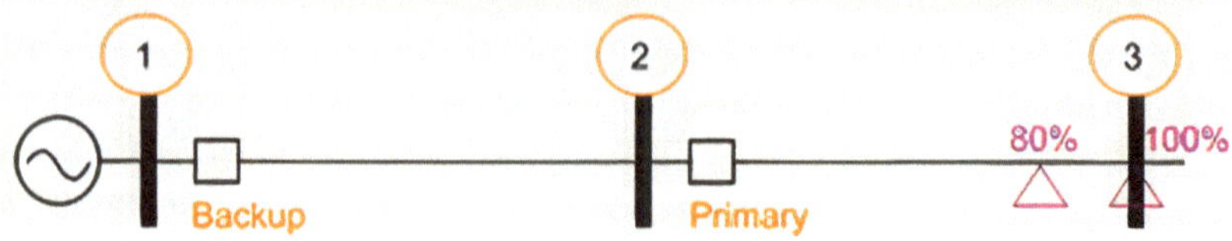

FIG. 9.12 Sensitivity analysis for minimum fault currents at the remote line-end.

of the backup line. This scheme should be supplemented with transferred trip schemes to ensure that 100% of the line is backed up.

$$S_b^{NF} \geq \frac{V_{nom} / \left(Z_{line}^{backup} * 0.8 + Z_{th} \right)}{I_{pickup}} \tag{9.17}$$

The pseudocode shown in Fig. 9.13 is used to evaluate the relay sensitivity and to active o disable relays. This step is convenient for coordination since when a relay lacks sensitivity, it will be an algorithmic violation that cannot be avoided, increasing the execution time of the optimization algorithm.

9.5 Code of optimization algorithms

DOCRs coordination in large interconnected systems is a highly constrained optimization problem that has been solved using various techniques. The restrictions limit the decision variables on its cost function making it even more

```
Step 1. Run initial routines
        Topology detection for nodal matrix construction
        Calculation of Fault current and power flow
Step 2. Three-phase and two-phase current seen by each relay
        Three-Phase Current through Relay – Irel_3P
        Two-phasic Current through Relay – Irel_2P
Step 3. Primary and Backup Relays Line Currents
        Primary Relays r= Start Node; s= End node;
        Backup Relays m= Start Node; n= End node;
        Current Line Primary Relay ILp(r,s)=abs(-Ybus(r,s)*(Vbus(r)-Vbus(s)))*IBase(r)
        Current Line Backup Relay ILb(m,n)=abs(-Ybus(m,n)*(Vbus(m)-Vbus(n)))*IBase(m)
Step 4. Sensitivity Calculation
        Sb=Irel_2P/(1.5*ILb(m,n))
Step 5. Determine which relay pairs can be coordinated
        Sensitivity Condition Scond=find (Sb > 1.5)
Step 6. Reduced Coordination Matrix
        Relays that they comply with the Sensitivity Irel_3Pc=Irel_3P (Scond)
        Relays that they comply with the Sensitivity Irel_2Pc=Irel_2P(Scond)
Step 7. Coordination Pairs (CPs) after Sensitivity analysis
        CPsn=CPs(Scond)
End
```

FIG. 9.13 Pseudocode of sensitivity.

complex to solve using conventional methods. The coordination viewed as an optimization problem requires the satisfaction of both hard and soft constraints on the formulation of its population and the objective function. The hard constraints are given for the ranges on which its decision variables should operate while the soft constraints aid with assisting in the final decision to choose the best individual of a population.

Equality constraints. The equality constraints are hard functions that delimit the solution space of the optimization region and guarantee that the results obtained are adequate to the conditions of the coordination problem. They are the expressions of the time curves of each relay, transformer protection, and lower and upper limits of variables.

In Fig. 9.14, the pseudocode for the equality restrictions is shown, the parameters that shape the nonconventional time curves are evaluated, as well as the formulation for the calculation of the operation times. There will be an equality equation for N active relays; each variable parameter will be bounded generating corresponding inequality equations. This could represent a computational increase but is offset by the increase in the feasibility of finding the solution. The correct operation of the relay is ensured by normally setting its I_{pickup} greater than the maximum possible I_{load} and below the minimum fault current that can be seen on the protected line by using a plug setting multiplier (PSM) as shown in Eq. (9.18).

$$\left[I_{pickup} = I_{load} * PSM\right]^N \tag{9.18}$$

```
Step 1. Run initial routines
        Calculation of Fault current and power flow
Step 2. Adjustable settings for the Algorithm
        Relay Adjustable Settings AS=5
        Agents or individuals (Pack Size) PS=80
        Total Iteration Niter=1000
        Total times code repeated Crpt=1
        Proposed CTI=0.3
        Dimension of Decision Variables DIM = Nrelays*NS
Step 3. Upper and Lower Settings for Adjustable settings
        dial [min max]       d_lim= [0.1      3]
        k factor [min max]   fk_lim=[1.4    1.6]
        A [min max]          A_lim= [1       28]
        B [min max]          B_lim= [0.1   0.49]
        p [min max]          p_lim= [0.01     2]
Step 4. Primary and Backup Relays Information
        Primary Relays    Prdata=[NRelP   IscP   IloadP]
        Backup Relays     Brdata=[NRelB   IscB   IloadP]
Step 5. Primary Relay operating time
        tp= (A/ (((Prdata (2)/ (fk*Prdata (3))) ^p)-1) +B) *dial
Step 6. Backup Relays operating time
        tb= (A/ (((Brdata (2) / (fk*Brdata (3))) ^p)-1) +B) *dial
End
```

FIG. 9.14 Pseudocode of equality constraints for time curve and I_{pickup} computation.

Inequality constraints. The *PSM* range should ensure that the relay discriminates between normal and abnormal conditions, ranges between 1.4 and 1.6 are used commonly. The *TDS* is the multiplier setting used to increase or decrease the relay operation time; the range must be selected for obtaining the operation time relay adequate for protection proposes. The upper range is defined to be 3 or 2 commonly, while the lower range has had different values, going from 0.5 in the first jobs, to 0.1 and even 0.01. It is important to assess that very small *TDS* may result in very short operating times that can affect safety, triggering for maneuvering operations or transients of the electrical network. For nonconventional time curves, the coefficients A, B, and p must be limited within ranges of the curve standards to maintain the shape of the time inverse overcurrent relays. The time constraints for each interval result in three inequality equations for each relay (Fig. 9.15). This represents an algorithmic advantage since the evaluation of the restriction is for each relay.

Conventional coordination methods result in ϑ inequality restrictions for each coordinated pair.

$$\left(t_b - t_p \geq CTI\right)_{\vartheta} \tag{9.19}$$

To ensure that relay settings obtained from the coordination algorithm are suitable for at least one element output without coordination loss, the maximum load and fault currents must be computed according to the different $n-1$

```
Step 1. Run initial routines
Step 2. Adjustable settings for the Algorithm
        lim = [d_lim (2) fk_lim (2) A_lim (2) B_lim (2) p_lim (2);
               d_lim (1) fk_lim (1) A_lim (1) B_lim (1) p_lim (1)]
Step 3. Matrix with lower and upper Boundaries for the decision variables
       lub = repmat(lim,Nrelays)
Step 4. Upper and Lower Boundary Vectors
       lb=lub (2)   ub=lub (1)
Step 5. Initial Position
        Position =rand (PS, Number relays*AS);
Step 6. Aleatory dial, fk, A, B, P for initial generation
      dial=d_lim (1,1) +(d_lim (1,2)-d_lim (1,1)) *Position
      fk=fk_lim (1,1) +(fk_lim (1,2)-fk_lim (1,1)) * Position
      A=A_lim (1,1) +(A_lim (1,2)-A_lim (1,1)) * Position
      B=B_lim (1,1) +(B_lim (1,2)-B_lim (1,1)) * Position
      p=p_lim (1,1) +(p_lim (1,2)-p_lim (1,1)) * Position
Step 7. Coordination time Interval calculation
        CTI Calculation     CTI_c=tb - tp
       Error calculation    CTI_error=CTI_c - CTI
Step 8. Penalization for CTI errors
        Zero Penalization    CTI_e >= 0 & CTI_e <= 0.3
        Slightly Penalized   CTI_e > 0.3
        Highly Penalized     CTI_e < 0
Step 9. Number of Violations
        NV=NV + 1
End
```

FIG. 9.15 Pseudocode of inequality constraints.

contingency topologies. All elements are taken out one at a time and the simulation is carried out repeatedly. Only the maximum load and fault currents are stored as data for coordination use.

9.6 Evaluation and comparison among the algorithms in IEEE test systems

Different algorithms are implemented to coordinate the overcurrent relays, but there will always be some performance advantages that one algorithm has over another. Due to this purpose, three algorithms are programmed and compared in this chapter. The GA is used as benchmark reference (because it is the most widely known in the protection area), then both ACO and DE are programmed and compared with the GA. The ACO and DE were chosen because they are reported to have outstanding performance in different literatures and had never been implemented in the protection field.

The performance of each algorithm is tested in this section using the IEEE 14-bus test system, and advantages and disadvantages of each algorithm are highlighted. Their performance is based on the execution time, result quality, robustness, and convergence ability of each algorithm.

9.6.1 Test system data

The IEEE 14-bus test system was chosen to test and compare the performance of the three algorithms. The system consists of 32 relays, and they are located as shown in Fig. 9.16. The voltages were selected to be 34.5 kV for buses at high voltage side of transformers and 22 kV for buses at low voltage side of transformers. All relays are considered to have very inverse time curve.

The relay names are not assigned as a number as was done conventionally but generated automatically by the algorithm as a string of numbers. These relay names (string of numbers) consist of three digits. The first digit is the name of the nearby bus. The second digit is the name of the remote bus, and the third digit represents the number of the lines (parallel lines) between two buses. For example, the relays between buses 1 and 2 that are nearby bus 1 are assigned as [1 2 1], [1 2 2] while the relays that are nearby bus 2 are assigned as [2 1 1] and [2 1 2], respectively.

The two lines between buses 1 and 2 have the same impedance value. Therefore the relays [1 2 1], [1 2 2], [2 1 1], and [2 1 2] sense the same amount of maximum load currents of 815 A, but due to the $n-1$ contingency analysis the maximum load currents of these relays are 1849 A. The current values are based on minimum load operation.

The fault currents are calculated with the remote line-end opened. This was done due to two considerations, to obtain the maximum fault current that the relay senses and the very small probability for the remote end relay to malоperate. Bear in mind that as elements' operation or network topology changes,

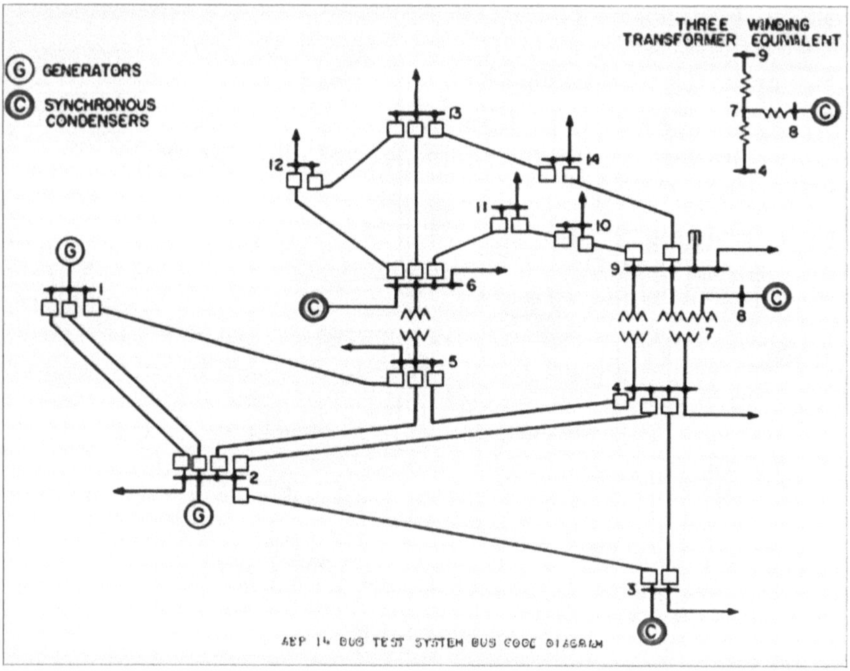

FIG. 9.16 IEEE 14-bus test system.

load flow and fault analysis must be computed again through real-time algorithm. The data X'_d for the generators one and two are 0.01 and 0.3, respectively. The maximum load is the same as the original 14 bus load data while the minimum load is the 70% of the 14 bus load data.

9.6.2 Parameter settings of coordination algorithms

The range of *TDS* was selected by running the test system several times and by observing the results, one concludes the range where the results are located, and hence narrow the search space for obtaining better results and speeding up the algorithm convergence.

Both GA and DE have continuous *TDS* and *PSM* settings, while ACO has discrete settings. Bear in mind that the AS graph is constructed by the discrete step settings; as the step size becomes more continuous (smaller), the size of AS graph increases and therefore the execution time increases. So, for real-time coordination, the step size was chosen as presented in Table 9.3, not too big (will cause bad quality result) and not too small (will cause long execution time).

The only stopping criterion is to stop when the algorithms have reached the maximum iteration of 1000. Although each algorithm has its own stopping criteria, they were disabled to be comparable between them. This maximum iteration number was selected due to the minimum improvement that the

TABLE 9.3 Parameter settings of GA, ACO, and DE.

Parameters	GA	ACO	DE
CTI	0.3	0.3	0.3
TDS	[0.5:2.0]	[0.5:2.0]	[0.5:2.0]
PSM	[1.4:1.6]	[1.4:1.6]	[1.4:1.6]
TDS step	Continuous	0.05	Continuous
PSM step	Continuous	0.01	Continuous
Q	–	100	–
R	–	5	–
r	–	–	0.5
F	–	–	0.8
Cr	–	–	0.5
Individuals	500	500	500
Iterations	1000	1000	1000

algorithms contribute after 1000 iterations when 3000 and 5000 were established as maximum simulation iteration. Both GA and DE are simulated with 500 individuals and ACO is simulated with 500 ant agents. The number of individuals and ant agents were selected due to the need of speed.

9.6.3 Results and comparison

The 14-bus test system consists of 62 relay coordination pairs before sensitivity filter. But this varies accordingly to the variation of load demand.

Case 1 Maximum load

The 14-bus test system is simulated using GA, ACO, and DE with the corresponding parameters presented in previous section at maximum load. All three algorithms were simulated 10 times. The convergence of each algorithm was averaged using the best fitness of each iteration of the 10 simulations. This is presented in Fig. 9.17 for comparison.

The average number of violations of coordination constraints; averaged fitness convergence; and averaged time of GA, ACO, and DE are presented in Table 9.4.

From these results, it is concluded that both ACO and DE have less violation of coordination constraint, better convergence, and faster performance than GA. Note that in ACO and DE all coordination pairs are coordinated for the 10

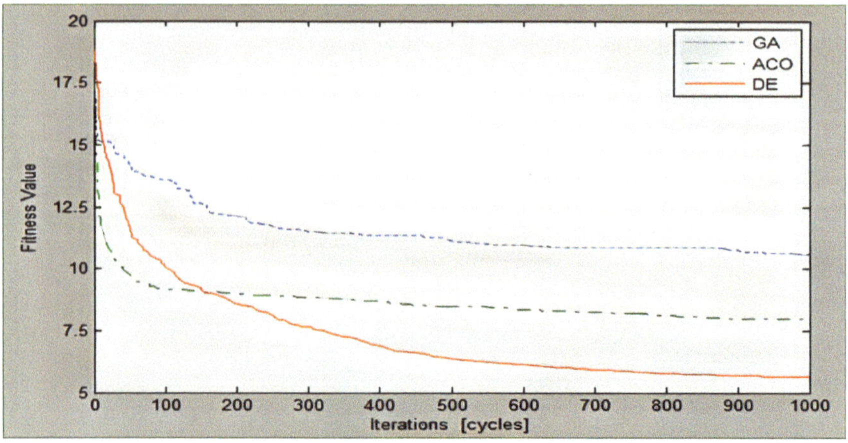

FIG. 9.17 Averaged fitness convergence of GA, ACO, and DE at maximum load.

TABLE 9.4 Comparison of averaged results of GA, ACO, and DE at maximum load.

Parameters	GA	ACO	DE
NV	0.1	0	0
Fitness	10.611	7.994	5.598
Time (s)	5507	441	89

simulations. On the contrary, some of the 10 GA simulations do not have all coordination pairs coordinated. The DE shows an outstanding performance at maximum load (see Table 9.5).

Although it is observed from Table 9.5 that the averaged *TDS* is near 1, there are many relays that use the minimum *TDS* value which leads to less operation

TABLE 9.5 Averaged relay settings, operation time, CTI, and sensitivity for 14-bus test system at maximum load.

Algorithm	TDS	PSM	Backup time	Primary time	CTI	Sensitivity
GA	1.1324	1.4952	3.1425	1.1817	1.9650	3.12005
ACO	0.9875	1.5004	2.3968	1.0383	1.3584	3.12624
DE	0.7374	1.4289	1.7015	0.7346	0.9669	3.29381

time, some use a *TDS* near minimum, and the rest use bigger *TDS*s due to the necessity for coordination purpose. All relays use small *PSM* which leads to greater sensitivity. This observation goes for GA, ACO, and DE.

Case 2 Minimum load

The 14-bus test system is simulated again using GA, ACO, and DE with the corresponding parameters presented in previous *sec*tion at minimum load. There are a total of 48 relay coordination pairs after the sensitivity filter. All three algorithms were simulated 10 times. The convergence of each algorithm was averaged using the best fitness of each iteration of the 10 simulations. This is presented in Fig. 9.18 for comparison and Table 9.6 comparison of averaged results at minimum load is shown.

Note that in ACO and DE all coordination pairs are coordinated for the 10 simulations. Some of the 10 GA simulations do not have all coordination pairs

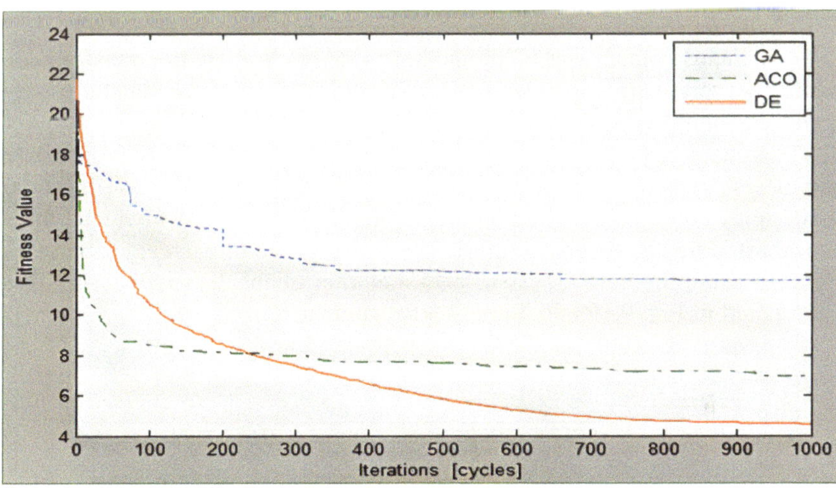

FIG. 9.18 Averaged fitness convergence of the 14-bus test system of GA, ACO, and DE at minimum load.

TABLE 9.6 Comparison of averaged results of GA, ACO, and DE at minimum load.

Parameters	GA	ACO	DE
NV	1.4	0	0
Fitness	11.748	6.968	4.612
Time (s)	5825	438	94

TABLE 9.7 Averaged relay settings, operation time, CTI, and sensitivity of GA, ACO, and DE for 14-bus test system at minimum load.

Algorithm	TDS	PSM	Backup time	Primary time	CTI	Sensitivity
GA	1.1336	1.4903	2.7546	0.8511	1.9266	4.03250
ACO	1.1165	1.4434	2.1331	0.7974	1.3357	4.17205
DE	0.7863	1.4324	1.4938	0.5297	0.9641	4.18241

coordinated. The DE shows an outstanding performance, Table 9.7 presents the averaged relay settings and operation time at minimum load.

Case 3 Considering DOCRs as protections for transformers

Even though other type of protection principle is used for transformers, overcurrent relays are implemented here for the purpose of coordination study in this section. This increased the number of coordination pairs.

The system is simulated using GA, ACO, and DE with the corresponding parameters presented in previous *sec*tion at minimum load. There are a total of 102 relay coordination pairs before sensitivity filter and 73 after. All three algorithms were simulated 10 times. The convergence of each algorithm was averaged using the best fitness of each iteration of the 10 simulations. This is presented in Fig. 9.19 and Table 9.8 for comparison.

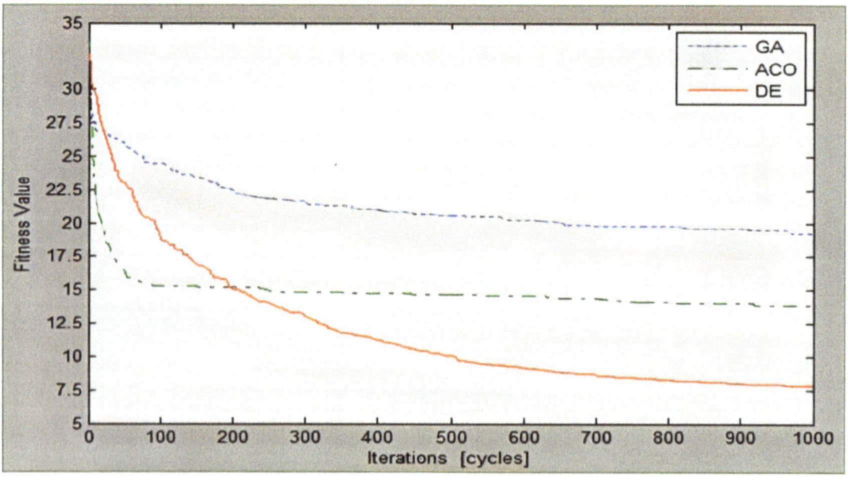

FIG. 9.19 Averaged fitness convergence of the 14-bus test system at minimum load, considering DOCRs as protections for transformers.

TABLE 9.8 Comparison of averaged results of GA, ACO, and DE at minimum load, considering DOCRs as protections for transformers.

Parameters	GA	ACO	DE
NV	3.2	1.3	0.8
Fitness	19.225	13.813	7.849
Time (s)	10,259	467	120

From these results, it is concluded that both ACO and DE have less violation of coordination constraint, better convergence, and faster performance than GA. The DE shows an outstanding performance. The averaged relay settings and operation time, as well as *CTI* and sensitivity of the 10 averaged GA, ACO, and DE simulations are presented in Table 9.9.

Case 4 Open range of TDS parameter at minimum load

In this case, the 14-bus test system is simulated at minimum load. Overcurrent relays are not considered as protection for transformers in this case. The new idea in this section is the open range of *TDS* parameter in the coordination process. So, limits of *TDS* are modified for studying the algorithms behaviors as presented in Table 9.10.

There are a total of 48 relay coordination pairs after the sensitivity filter, all three algorithms were simulated five times. The convergence of each algorithm was averaged using the best fitness of each iteration of the five simulations. This is presented in Fig. 9.20 and Table 9.11 for comparison.

It is observed that there is zero violation of coordination constraints in all three algorithms. But the fitness values of GA and ACO are very far from

TABLE 9.9 Averaged relay settings, operation time, CTI, and sensitivity of GA, ACO, and DE at minimum load, considering DOCRs as protections for transformers.

Algorithm	TDS	PSM	Backup time	Primary time	CTI	Sensitivity
GA	1.1832	1.4970	2.7396	0.9894	1.8516	3.91388
ACO	1.1707	1.5172	2.3805	0.9477	1.4661	3.88585
DE	0.9347	1.4600	1.6853	0.7581	0.9543	4.03123

TABLE 9.10 Parameter settings of GA, ACO, and DE. Open range search.

Parameters	GA	ACO	DE
CTI	0.3	0.3	0.3
TDS	[0.5:10.0]	[0.5:10.0]	[0.5:10.0]
PSM	[1.4:1.6]	[1.4:1.6]	[1.4:1.6]
TDS step	Continuous	0.05	Continuous
PSM step	Continuous	0.01	Continuous
Q	–	100	–
R	–	5	–
r	–	–	0.5
F	–	–	0.8
Cr	–	–	0.5
Individuals	500	500	500
Iterations	5000	5000	5000

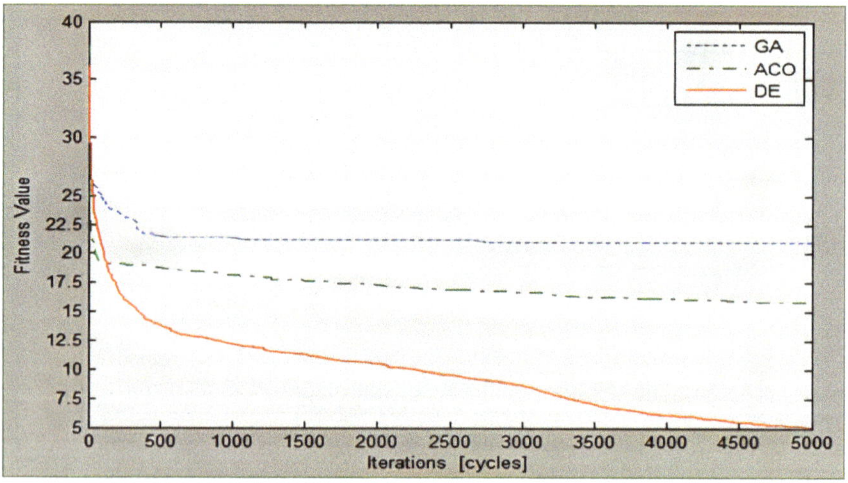

FIG. 9.20 Averaged fitness convergence of the 14-bus test system at minimum load. Open range TDS simulation.

the fitness values of Case 2. The fitness value of DE is very close to the value in Case 2. The DE shows an outstanding performance. The averaged relay settings and operation time, as well as CTI and sensitivity of the 10 averaged GA, ACO, and DE simulations are presented in Table 9.12.

TABLE 9.11 **Comparison of averaged results of GA, ACO, and DE at minimum load. Open range *TDS* simulation.**

Parameters	GA	ACO	DE
NV	0	0	0
Fitness	20.980	15.808	5.054
Time (s)	20,687	4075	515

TABLE 9.12 **Averaged relay settings, operation time, CTI, and sensitivity at minimum load. Open range dial simulation.**

Algorithm	TDS	PSM	Backup time	Primary time	CTI	Sensitivity
GA	3.8181	1.4977	7.3595	2.6492	4.7102	4.00712
ACO	2.7053	1.5815	4.9068	1.8311	3.0757	3.79766
DE	0.9475	1.4330	1.6429	0.6255	1.0174	4.18542

9.7 Discussion

The comparison purpose of Cases 1 and 2 is to observe the speed and result quality of the algorithms. The DE outperformed GA and ACO. Case 3 evaluated the robustness of the three algorithms by increasing the number of coordination pairs and maintaining the same initial settings as presented in Table 9.3. The DE outperformed GA and ACO.

Case 4 evaluated the convergence ability of the three algorithms by using an open range of dial as search space with no prior elimination. The settings are as presented in Table 9.10. The DE outperformed GA and ACO. DE found good quality results in reasonable amount of computational time. These can be observed from the averaged results of the three cases in Fig. 9.21–Fig. 9.23.

The overall performances (result quality, robustness, and convergence) of the three algorithms of all four cases are averaged and presented in Fig. 9.24.

From Fig. 9.21 it is observed that the execution time of GA increased almost double in Case 3 due to the increment of coordination pairs and almost 5 times in Case 4 due to the increment of iterations compared to Cases 1 and 2. On the other hand, both ACO and DE did not increase significantly in Case 3 (are not affected significantly) due to the increment of coordination pairs. But in Case 4, ACO increased more than 5 times compared to Cases 1 and 2, this is because the

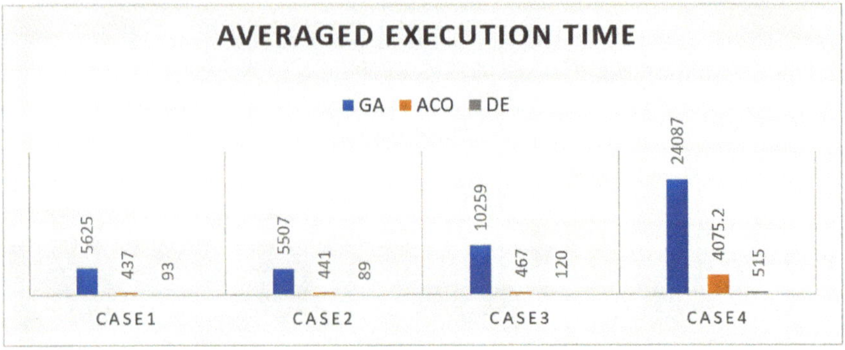

FIG. 9.21 Averaged execution time of GA, ACO, and DE for the four cases.

open dial search increased the size of AS graph, therefore decreased the speed of the algorithm dramatically. The DE increased reasonably in Case 4 compared to Cases 1 and 2.

From Fig. 9.22, it is observed that as there are more coordination pairs in the system, the fitness values increase: Case 3 > Case 1 > Case 2. This is reasonable because as there are more coordination pairs, there are more coordination constraints and therefore more difficult to coordinate. GA and ACO seem not to be as robust as DE in Case 3 due to the increment of coordination pairs. In Case 4, it is observed that DE tends to have much better convergence ability than GA and ACO because its fitness value approaches the fitness of Case 2.

From Fig. 9.23, it is observed that as coordination pairs increase the number of violation increases as well. This can be easily observed from Case 3 compared to Cases 1 and 2. The number of violations increased for

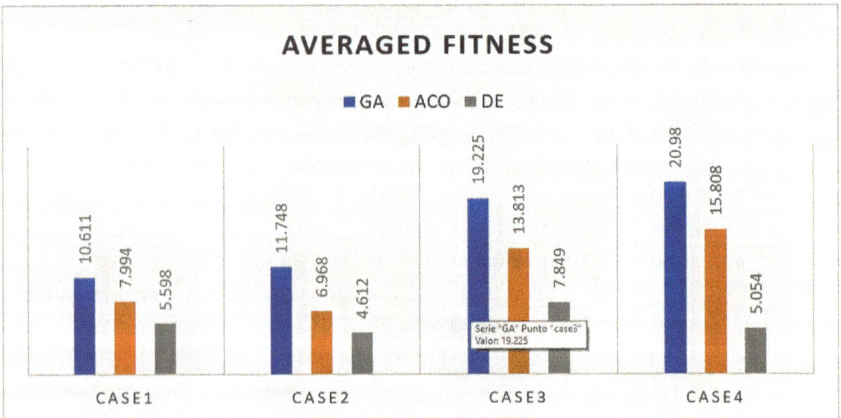

FIG. 9.22 Averaged fitness of GA, ACO, and DE for the four cases.

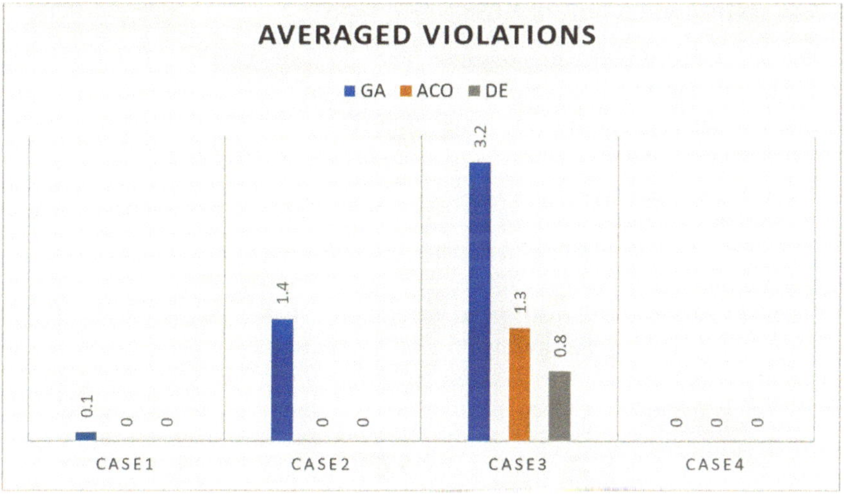

FIG. 9.23 Averaged number of violations of GA, ACO, and DE for the four cases.

all three algorithms because the number of coordination pairs was increased but the number of individuals/ant agents or number of iterations was not increased.

From Fig. 9.24, it is observed that the overall performance of DE outstand GA and ACO. DE tends to converge to the minimum during the 5000 iterations, on the contrary, GA and ACO tend to converge to a fixed result and maintained throughout the 5000 iterations.

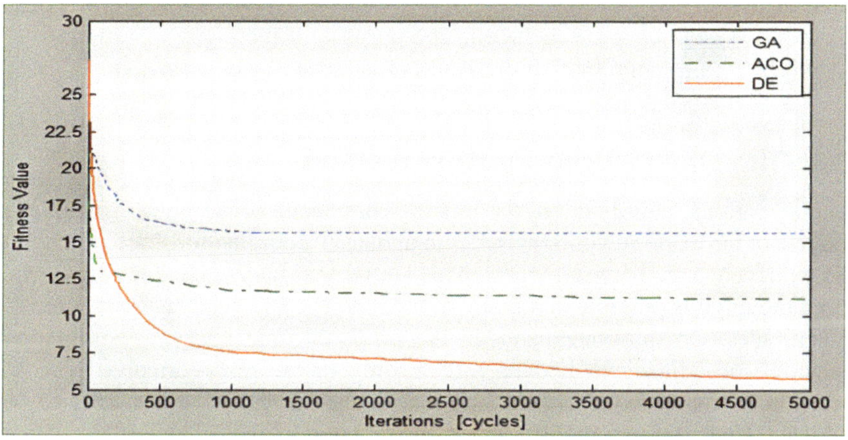

FIG. 9.24 Overall performance of GA, ACO, and DE.

9.7.1 Interconnection of RES

As the different sceneries are presented, one can easily understand the effect of intermittent source. For example, wind, solar, and hydraulic powers are very dynamic in the aspect of energy generation. They connect to the system when they have energy and disconnect from the system when they don't. Therefore, in order to avoid inappropriate relay operations, their generations are considered in the coordination. But this causes the DOCRs to become less sensitive when these intermittent sources are not connected on to the system. So as a conclusion, an online coordination will become very effective for this case.

Actually, the effect of distributed generation (wind, solar, hydraulic, etc.) is seemed like a normal energy injection. The intermittent characteristic is the major factor of these sources. Therefore the benefits of real-time coordination can be easily understood, since the intermittent source can suddenly reach maximum generation because there is a lot of wind and also reach minimum generation because there is no wind, the load flow changes drastically. Hence, the criteria of including the distributed generation are critical.

The distributed generations of small capacity (solar) can simply be omitted from the analysis, since their fault contribution and generation are very small compared to the fault contribution and generation of the entire system. They are insignificant.

The distributed generations of big capacity (wind, hydraulic, etc.) will be considered to have maximum fault (as if they are generating at maximum to have maximum fault), and at the same time considered to have minimum generation (as if there is almost no wind or almost no water). These considerations are to avoid the loss of coordination for DOCRs that are located on the distributed generator bus which offer backup for relays on adjacent lines and DOCRs of the system.

The considerations of distributed generators of big capacity cannot follow or base on prognostic data, since the nature is unpredictable. Therefore worst conditions that can occur are included. The purpose of these considerations is to avoid the loss of coordination of DOCRs that are to be coordinated.

9.8 Time interval coordination method

In conventional coordination, the topological change of the power grid or the change of the load/fault current forces to update all coordination due to the functional dependence between relays. This adjustment update includes all relays, even though it is possible that many of them do not appreciate any variation in magnitudes. On the other hand, in the proposed time interval coordination method, the adjustment update will be only in the relays involved. Those who do not have an appreciable change in the currents will not have to be adjusted, reducing computing and communication times. The update of the relay settings must be carried out in a stable state, so no change or update is recommended during a fault condition to maintain the safety of the protection scheme. A sensitivity analysis is performed to eliminate relays that do not have the ability to detect minimal faults that occur at the remote line-end.

The GWO algorithm is shown only to show the interaction of the optimization algorithm with the other functions of the implementation, the position vectors define a circular area that estimates the location of the prey, then the wolves search agents update its position. When the best position of the prey is identified by elitism, the wolves $\overrightarrow{X_\alpha}$, $\overrightarrow{X_\beta}$, and $\overrightarrow{X_\delta}$ surround the prey. Results and comparison section with DE are shown.

The proposed coordination method presents advantages in its implementation in an online protection system because the modification of the settings of one or more relays is independent, and it is not necessary to carry out the coordination of all the relays of the protection scheme. The optimization algorithm to be selected for online execution should show the robustness necessary for protection systems. To carry out the proposed coordination and correct operation of the protections at preestablished times, the adjustment update system must be implemented in a centralized mode with measurement and monitoring of the system topology. When a topology change is detected, the system updates the settings and coordination of the relays for the new configuration.

The latency of the communication available in the electrical network must be considered; however, in smart grid applications, the latency is of a few cycles [14], allowing the execution of the proposed protection functions. The update of the relay settings must be carried out in a stable state; thus, no change or update is recommended during a fault condition to maintain the safety of the protection scheme. The I_{pickup} is updated with the maximum current resulting at the location of each relay in the new configuration, and I_{sc} is modified in the event of a topological change or the connection of the generation sources. Therefore it is necessary to conduct power flow studies and fault analysis. The use of units of measurement, such as μPMU, is recommended as they can verify and validate the topology of the electrical network. Fig. 9.25 presents the pseudocode for implementation of the proposed method.

Step 1. Topological detection *Validation of topology by:* *Real-Time measurement* *Breaker position (52a)* *Nodal matrices construction* *Step 2. Calculation of fault current and power flows* *Power flows consider (n – 1) contingency analysis.* *Fault currents (Close-in, Bus-end, Remote Line -end)* *Detection of relays* *Sensitivity filter*	*Step 3. DE algorithm formulation* *Initial population and setting* *(NP – F – CR – Strategy – Refresh)* *Coordination constraints* *TDS and PSM* *Non-conventional curves (A, B and p)* *Establish Operation Time Intervals for all relays* *Step 4. DE Algorithm* *Evaluate the fitness function values of all individuals (Eq. 6)* *Step 5. Solution* *Relay setting actualization through communication*

FIG. 9.25 Pseudocode of the proposed coordination method using a DE algorithm.

9.9 Summary

All three algorithms showed the ability to encounter reasonable results for the tested system. But due to the need of speed in real-time coordination, the number of individuals and ant agents was reduced to only 500 for the complicated problem. This leads to bad results for GA, but at the same time also showed the outstanding performance of DE. DE encountered the best results with 500 individuals out of the three algorithms and is also the fastest. DE is the most suitable candidate for real-time coordination.

The online coordination surely increased the sensitivity of DOCRs and relay operation time is reduced. These are all due to the operation of the network. The increase of sensitivity or reduction of relay operation time depends on the network operation condition. Therefore the system is prepared for another unknown contingency. This made the system to always stay strong.

In this way, the adjustment of the protections has an adaptable behavior to the operating conditions of the electrical network. Topological or operation changes of the electrical network will determine the action of the online protection system. These schemes represent a paradigm shift in protection schemes. However, with the communication and programming facilities of digital relays, the implementation of these schemes is feasible. The overall result is an increase in protection performance.

References

[1] R. Mohammadi, H.A. Abyaneh, F. Razavi, M. Al-Dabbagh, S.H.H. Sadeghi, Optimal relays coordination efficient method in interconnected power systems, J. Electr. Eng. 61 (2) (2010) 75–83.

[2] M. Singh, B.K. Panigrahi, Optimal Overcurrent Relay Coordination in Distribution System, IEEE, 2011.

[3] D.K. Singh, S. Gupta, Optimal coordination of directional overcurrent relays: a genetic algorithm approach, in: IEEE Students' Conference on Electrical, Electronics and Computer Science, 2012.

[4] D. Uthitsunthorn, T. Kulworawanichpong, Optimal overcurrent relay coordination using genetic algorithms, in: International Conference on Advances in Energy Engineering, 2010.

[5] K.Y. Lee, J.G. Vlachogiannis, Optimization of power systems based on ant colony system algorithms: an overview, in: ISAP, 2005.

[6] Y.-H. Hou, Y.-W. Wu, L.-J. Lu, X.-Y. Xiong, Generalized Ant Colony Optimization for Economic Dispatch of Power Systems, IEEE, 2002.

[7] S.-J. Huang, Enhancement of hydroelectric generation scheduling using ant colony system based optimization approaches, IEEE Trans. Energy Convers. 16 (3) (2001).

[8] S. Das, P.N. Suganthan, Differential evolution: a survey of the state of the art, IEEE Trans. Evol. Comput. 15 (1) (2011).

[9] M.N. Alam, B. Das, V. Pant, A comparative study of metaheuristic optimization approaches for directional overcurrent relays coordination, Electr. Pow. Syst. Res. 128 (November) (2015) 39–52.

[10] M.Y. Shih, A.C. Enríquez, L.M.T. Trevino, On-line coordination of directional overcurrent relays: performance evaluation among optimization algorithms, Electr. Pow. Syst. Res. 110 (2014) 122–132.

[11] R. Corrêa, G. Cardoso Jr., O.C. de Araújo, L. Mariotto, Online coordination of directional overcurrent relays using binary integer programming, Electr. Pow. Syst. Res. 127 (2015) 118–125.

[12] G.W. Stagg, A.H. El-Abiad, Computer Methods in Power System Analysis, McGraw-Hill, 1968.

[13] IEEE C37.91-2000, IEEE Guide for Protective Relay Applications to Power Transformers, PE/PSRCC - Power System Relaying and Control, 2000. 2000-06-14.

[14] Nguyen, C. P., et al., Modeling of communication latency in smart grid., IEEE Power and Energy Society General Meeting (2011).

Chapter 10

Virtual relay design and lab tests of an adaptive overcurrent relay

10.1 Introduction

The performance of protective relays is evaluated by determining their ability to detect faults within a zone of protection (selectivity and sensitivity) at the correct time. The level of complexity of a postmortem analysis depends on the variables used by the relay to perform the function of protection. On the principles of protection as the differential and pilot, the protection zone is constant and insensitive to the operating state of the grid being an absolute selectivity protection. It does not provide backup functions, simplifying the analysis of its operation that focuses on the certainty of the decision of operation. In these protection principles, the uncertainty in the relay's operating time is restricted to a small range and in many cases is not the focus of analysis. Another type of protection, the distance relay, has a limited dynamic zone of protection due to its polarization. The real and imaginary parts of fault impedance represent a source of error that results in uncertainty in the location of the fault. The determination of the apparent impedance and its relationship with the electrical distance to the point of fault is a central theme in an analysis. In most studies on differential relays, pilot, or distance, the dynamic modeling of the relay is not required, and only the dynamic conditions of the network are needed.

For the specific case of an overcurrent relay, the current signal is the only measurement variable, and it is sensitive to the load effect, as well as the location and dynamics of the fault. Thus determining the location of the fault is difficult because its protection zone is dynamic and sensitive to the operating conditions of the electrical system. The postmortem analysis of the operation of an overcurrent relay is complicated because it has a greater dynamic reach than other protection principles, and because the time at which the protection should operate is unknown, as this depends on the dynamics of the fault current. It is common practice to perform a reconstruction of the operation of the relay by injecting event records in the laboratory to validate the operating time of protection. Other alternative methods are to reproduce the dynamics of operation by developing hardware prototypes or with the use of relay modeling software.

The relay prototype development is a difficult task that involves a considerable investment of time, mainly due to the signal conditioning challenges and

Overcurrent Relay Advances for Modern Electricity Networks. https://doi.org/10.1016/B978-0-323-91777-3.00010-6

the issue of hardware stability. Refs. [1,2] describe the designs of an analog electronic overcurrent relay and a microprocessor overcurrent relay; the relays proposed there are intended for laboratory applications. While prototyping might be a valuable experience, it is not recommended for validation of relay operation given the development time and the dependence on physical simulators for playing events.

To evaluate and model relay operation using software, simulators in both academic and commercial platforms have been developed that allow an evaluation of relay response in various test scenarios. The feasibility of relay performance analysis through digital simulation has been demonstrated in Ref. [3], and the requirements for this are specified in Ref. [4]. The academic simulation platforms must model the system components and perform calculations to obtain values for variables such as flows and faults [5,6]. The analysis of overcurrent relay coordination and test modules for distance relays developed in PSCAD and Matlab are presented in Refs. [7,8], respectively. However, with the academic software it is difficult to achieve the level of modeling reached in recognized commercial packages such as EMTP. Moreover, the analysis of relay protection performance should be done for a variety of operational conditions, which can hardly be covered by a single simulation platform.

The analysis of relay operation is facilitated by the development of software tools consisting essentially of the dynamic model of the relay and its associated protection functions, which allow the inclusion of signals from real events or records of simulations of electrical systems. Another benefit resulting from the development of relay modeling applications is found in education and training activities. The authors in Ref. [9] point out the importance of engineering design courses for power systems and refer to an urgent need to identify and develop the materials needed for the education and training of future engineers of protection.

Frequently, graphical resources are used to aid in education and in the study of electrical power systems [10–12]. The performance evaluation of protection is well served by visually comparing the operating conditions of an electrical system with the characteristics of relays.

Many of the tests to evaluate the functionality of OCR are possible to be carried out in modest laboratories, real-time simulator tools are desirable, but many of the functional evaluations do not require platforms with real-time models because of the high OCR operation times. The interest of the tests is focused on the algorithmic validation of new proposals and on the functional validation having results within acceptable time ranges.

10.2 Relay programming

The relay programming was done in Labview; this application development program uses a graphical programming language (G language) to create programs in the form of block diagrams. Labview programs are called VI (Virtual Instruments), because by their appearance and operation they can mimic real instruments. Labview includes libraries of data acquisition functions, digital signal analysis, filters, curve approximation, probability and statistics, linear algebra, array operation, numerical methods, and communication functions among others. These qualities are very attractive from the point of view of analysis, design, and testing of algorithms in real time.

The characteristics mentioned before make it advisable to carry out the programming of the adaptive overcurrent relay in this computational environment. The main advantage of using this program is the possibility of managing the acquisition of input signals in real time and carrying out control actions.

The development of the adaptive overcurrent relay was carried out considering the algorithms described in Chapters 2 and 7 for the relay adaptation of the operating time and the I_{pickup}, respectively. Algorithms for instantaneous- and inverse time-relay, resetting operation, and cold load tolerance were also included. The block diagram of the virtual relay is shown in Fig. 10.1. It consists of subroutines for the acquisition or generation of input signals, digital signal conditioning and filtering, as well as protection and output.

The input signal acquisition subroutine controls the acquisition of signals in real time; it performs the A/D conversion at 960 Hz (16 samples per cycle) with a buffer of 40,000 samples. A test module with two alternatives is also included: (a) acquisition by reading file signals: this option allows extract data from external files in ASCII format generated in simulation programs (such as EMTP) or from real fault log data files; (b) generation of internal signals: this variant offers great versatility for the simulation of different operating states,

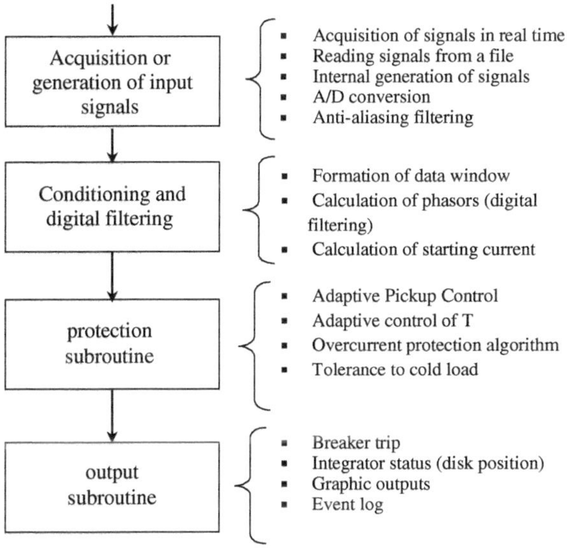

FIG. 10.1 Program block diagram.

parameter variation, digital processing, and noise contamination of the signal. A Butterworth filter with adjustable order and cutoff frequency was programmed to optionally emulate the effect of analog prefiltering of the signal.

In the digital conditioning and filtering subroutine, the formation of the rectangular data window of one cycle length is carried out, with 16 samples per cycle. A Butterwort low-pass filtering of the signal is performed. Also, a digital filtering of the signal is performed to obtain the corresponding phasor of the fundamental component; this phasor is supplied to the digital relay. The nonfundamental components like harmonics are eliminated by digital filter of relays. However, the effect on the operation of the fuses in the presence of harmonics of the signal commonly is not evaluated, and it is considered that its operation is due with greater influence on the fundamental component. The behavior of reclosers is very similar to that of relays. Finally, in this subroutine, the calculation of the I_{pickup} of the relay is performed.

The protection subroutine oversees controlling the I_{pickup} and the operating time of the protection. The setting update process for adaptive relay is carried out according to the calculated value of the I_{pickup} and the integration process of the $H(I_k)$ function to determine the relay operation time. The cold load tolerance subroutine performs a temporary change in the setting of the relay; two steps of adjustment are used both in the magnitude of the I_{pickup} and in the duration of the setting process.

The output subroutine transmit the trip signal through an analog channel of the acquisition card, the status of the relay integrator is represented graphically. This function is useful in coordination tests with reclosers. Two graphical outputs are available, the input signal to the relay and the relay setting along with the system load current (I_{load}). It is also possible to store it in a data file (Excel).

Dynamic algorithms of overcurrent relays take as input signal the phasor estimate of a current. Even when a digital filter can reject harmonic components, interharmonic and subharmonic signals are merely modulated and are not rejected totally, representing a source of error in the phasor estimate. This too can be evaluated in the proposed application.

The protection functions consist of the dynamic model of an overcurrent relay, the instantaneous relay (50), the cold load function (CL), and a module for editing a time curve (Fig. 10.2). The subroutine of time curve design allows the generation of curves by introducing an analytical expression, and in this way, it is possible to reproduce and evaluate new operating characteristics of relays, fuses, and restorers. The routine generates the current vector bounded by the I_{pickup} and the short-circuit current. The time vector is calculated from the analytical expression that has been entered. Where $T(I_k)$ is the inverse time curve, Δt is the sample period, k_{op} is the operating sample, and ε is the current detector.

10.3 Virtual OCR

The proposed application has been made to contain a graphical user interface, and its functional structure consists basically of three modules: signal

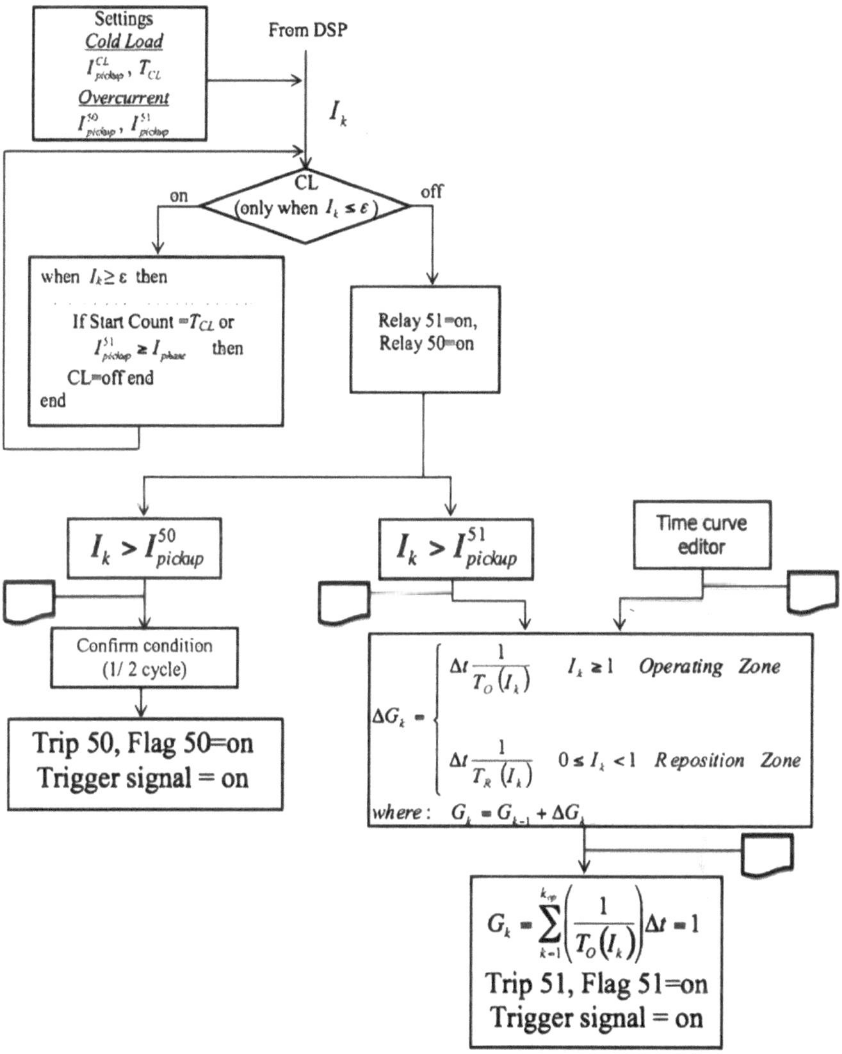

FIG. 10.2 Structure of the protection functions.

generation, phasor estimation, and protection functions. The relay evaluation algorithm takes as input either internally generated signals or it reads data from external files in ASCII format. Such files may have been generated either in simulation programs (EMTP, PSCAD, Matlab, etc.) or they might contain data consisting of real fault records (Fig. 10.3). The superposition principle is used for insertion of the fault signal and additional sources of different frequency components.

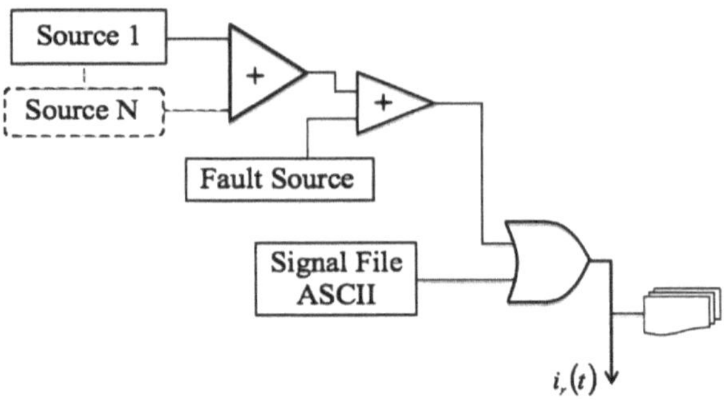

FIG. 10.3 Signal generation module.

The electrical input signal to a relay from a current transformer is composed of signals that are representative of electrical phenomena. They should not be used to determine the relay operating criterion, because they exhibit a random character that precludes the correct determination of the location of a fault. Digital processing must remove unwanted signal quantities and retain the quantities of interest. The overcurrent relay operation should be established only by using the fundamental component of the signal, because it is that component which is proportionally affected by the location of the fault. The structure of data development permits the evaluation of relay operation under nonfundamental components.

10.3.1 Virtual relay

The computer application is divided into sections according to the thematic content of the functions found within it (Fig. 10.4). The order of sections in the application is consistent with the parameter setting process as well as the evaluation sequence of the relay. Thus, in the signal format section, the test type is defined, there are two options: internal simulation using *signal control*, or importation of a signal file in *simulation mode*. In the *current* and *time curve setting* sections, the setting of I_{pickup} of instantaneous element (Ipickup 50), inverse time (Ipickup 51), and the current coordination (Icc coordination) are set. The time curve is defined by selecting a time curve and specifying some standard coefficients, and curve editing is performed in the *formula editor* section. The load current and the fault current are determined by the manual controls in the *signal control* and *fault current* sections. Definitions of magnitude and time steps are made in the function of *cold load* section. Time breaker (52), induction disc inverse type emulation, and trip signal are in the *trip* section. The computer application is independent of the programming platform.

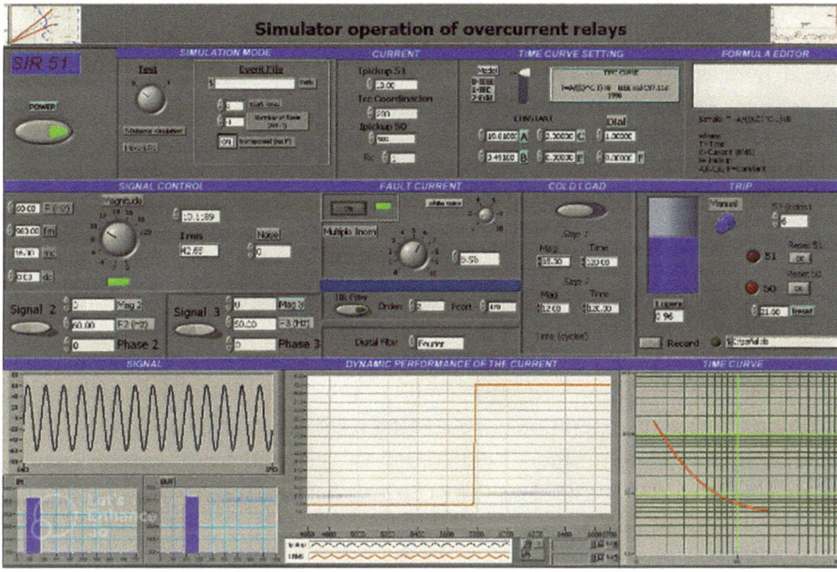

FIG. 10.4 Front panel of the computer tool.

10.3.2 Modules of virtual relay

Fig. 10.4 shows the real-time acquisition front panel of the virtual relay; it is made up of the following subpanels: Signal Control, Simulation mode, Current, Time curve setting, Fault current, Signal filtering, Cold load, Trip, and Record and Graphic Outputs.

The **Signal Control and Simulation mode** subpanels define the control and supervision elements of the acquisition parameters. The device and the acquisition channel used are defined; the sampling frequency sets the number of samples acquired in 1 s. The file event writes the record of event file. In the test type, the selection is made between the internal simulation of the signal and signal file. An option required for the configuration of the acquisition modules is the buffer of samples. Two types of noise can be added to the signal, white noise and periodic noise. Finally, the type of signal coupling and the type of measurement used are selected.

In the **Current** subpanel, there are the indicators of the relay setting (I_{pickup}) for time and instantaneous, the coordination current (Icc coordination) is defined. In the **Time curve setting** subpanel, the time curve is defined. Three alternatives are presented: (1) curve according to the IEEE standard [13]; (2) curve according to the IEC standard [14]; and (3) editing any relay, fuse, or recloser operating characteristics. Six controls are included to define the coefficients of the selected fit model.

In the **Fault Current** subpanel, the magnitude of the fault current is controlled as a multiple of the nominal value of the current in stable state. In the **Signal processing** subpanel, the Butterworth filter and the cutoff frequency are defined. This section emulates the effect of an analog filter on the acquired signal. The digital filter is defined too, the base case is Fourier but other types of digital filter can be evaluated. The **Cold Load** subpanel temporarily activates two levels of adjustment for load reset tolerance, both in the magnitude of the adjustment current and in the time that this transient adjustment is in effect. The logic only allows activation of these settings after a deenergized state of the line; during a normal load state or fault state, the cold load logic is automatically disabled.

In the level indicator of the **Trip** subpanel, the induction disc of an electromechanical relay is simulated. The integration value is normalized to the unit and includes both the operation zone and the reset zone. There are indicators of operation time and reset time. The recorded values are the load current and the setting values of the relay I_{pickup} are recorded in EXCEL format.

The **graphic outputs** are included in the virtual relay; the adaptive relay I_{pickup} and the fundamental phasor of the input current are displayed. The waveform of the signal and the frequency spectrum at the prefiltering (IN) and postfiltering (OUT) stage are also deployed.

10.4 Stand-alone examples

10.4.1 Operating time validation

The operating time of an overcurrent relay is determined by the dynamics of the fault current, often there is uncertainty about whether the operating time will be right. This section presents the validation of the operation of a ground directional overcurrent relay 67 N based on the record of the real fault as captured by an event recorder. Fig. 10.5 depicts the signal record of a real event involving a single-phase fault on a 230-kV line; the voltage, ground current, and the relay trip contact status are deployed. The validation of the relay operating time can be accomplished in two ways, first by injecting the event signal into a similar relay setting using relay test equipment or alternatively by emulating the relay operation in software. The second option is presented in this section, using the computational tool that we are proposing.

The 67N relay settings were as follows:

$$I_{pickup} = 92\,\mathrm{A}$$

Moderately inverse time curve [13], $TDS = 0.2$

After defining these relay settings in the software interface and performing the relay operation simulation, the records of the resulting operation time of the

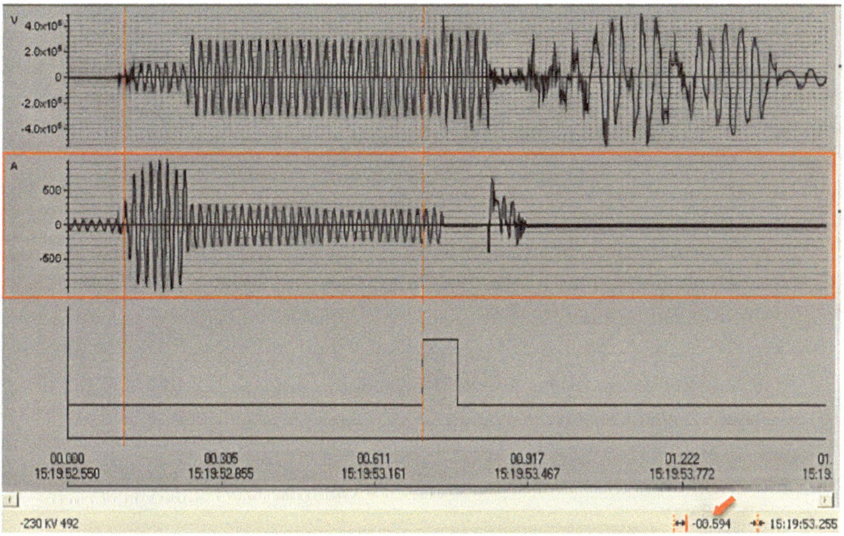

FIG. 10.5 Register of a real event from a 230-kV network in a commercial application.

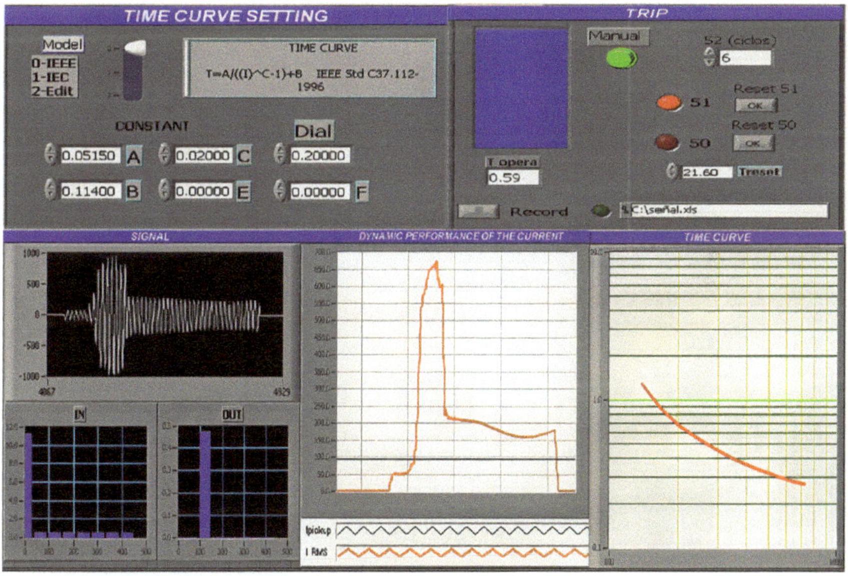

FIG. 10.6 The output obtained by the proposed application.

relay are produced. The application shows (Fig. 10.6) that the calculated operation time is very close (within one cycle) to the registered time:

$$\text{Time} = 0.59 \text{ s}, 35.4 \text{ cycles } (f = 60 \text{ Hz})$$

The relay operation was correct.

10.4.2 Cold load current

A cold load condition may occur during energization of an electrical network, manifesting as a temporary increase in current that can cause improper tripping of overcurrent relays [15]. The cold load function must temporarily disable protection functions, thereby allowing the evolution of the transient. However, even during the blockade of the protection functions, a continuous monitoring of the magnitude of the phase current and I_{pickup} of the relays should be maintained in order to enable protection functions before the end of the wait time. Fig. 10.7 shows a diagram of the cold load function logic as programmed in the proposed application.

In Fig. 10.8, the behavior of the cold load function is illustrated for the case of a three-phase motor (2 kW, 120/208 V) that is coupled to a generator, which was obtained from a physics laboratory. Data acquisition was performed at a sampling frequency of 960 Hz. The cold load function is defined with two settings, one for maximum allowable current, set in steps to values greater than the

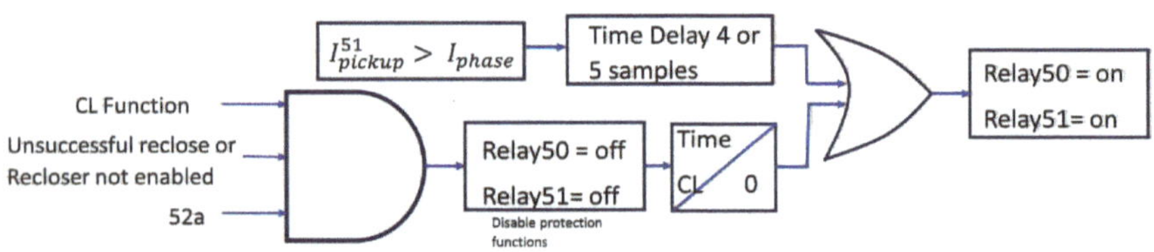

FIG. 10.7 Diagram of the cold load function logic.

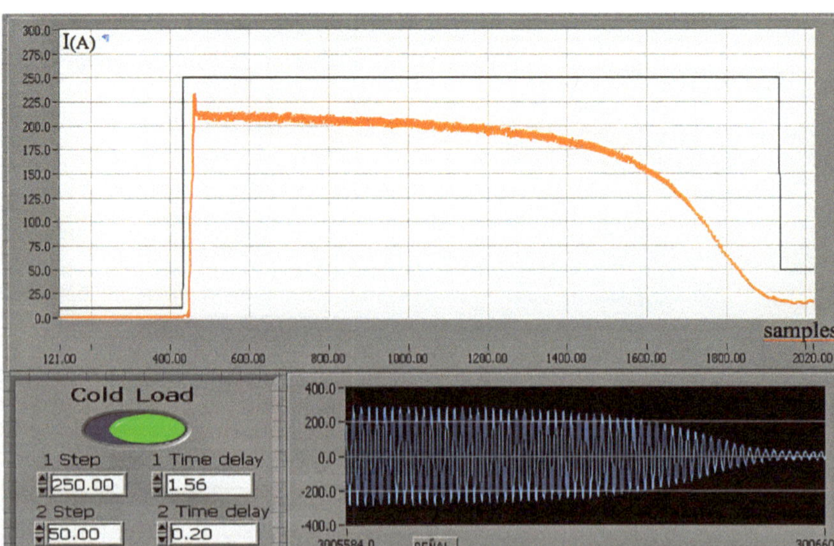

FIG. 10.8 Cold load function results.

expected maximum current, and another for a timer that is set to allow the dynamic event to unfold. The functions 50 and 51 were disabled during this simulation by cold load logic. Based on the expected maximum value of the inrush current for the motor, the maximum allowable current is assigned values in two steps, the first being 250 A out to 1.56 s (1497.6 samples), and the second at 50 A until 0.2 s.

10.4.3 Subharmonic currents

By manipulating the variables used for signal generation, several phenomena related to the effects of digital signal processing can be produced. The effect of aliasing is accomplished by manipulating the sampling frequency and the frequency of a second signal. Reproducing the effect of harmonic, interharmonic, or subharmonic signals in overcurrent relays [16,17] is also possible using features of the software application. For example, Fig. 10.9 shows the front panel parameter settings for a subharmonic signal that is injected into the overcurrent relay, the simulation run with a fault current of 350 A. Such a scenario may be the result of a coincidence between a 50-Hz frequency component and a resonant frequency, for example. The time difference is due to the 50-Hz frequency component cannot be filtered by the relay.

The results of that simulation were as follows:

$$\text{Trip time with current at } 60\,\text{Hz} : 0.50\,\text{s}$$

$$\text{Trip time with current at } 60\,\text{Hz} + 50\,\text{Hz} : 0.47\,\text{s}$$

The result of this comparative study shows that the relay time difference is due to the subharmonic signal. Also, this example illustrates why the sub- and interharmonic signals do not have a significant impact in the OCR operation. It is therefore that this effect is commonly ignored for overcurrent protection.

10.4.4 Coordination examples

The part of the computer application responsible for relay coordination is structured to use required information for both protection devices, such as I_{pickup} of

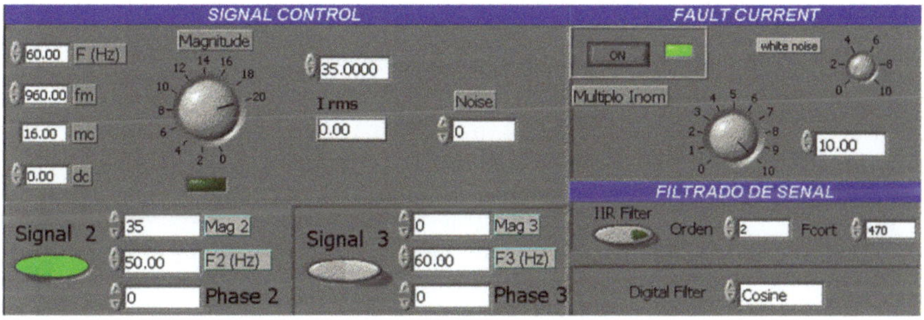

FIG. 10.9 Subharmonic effect on the input signal of a relay.

the primary device (*Ipickup P*), the backup device (*Ipickup B*), and coordination short-circuit current seen by both relays (I_{sc}). The time curve information is duplicated for both the primary and the backup protective devices. The signal format, signal control, and fault control sections and the fields for these setting for each relay were configured similar to those of the panel described in Section 10.3.

The proposed application is designed specifically for simulations involving the coordination between overcurrent protections whether they are relays, fuses, or reclosers. So, some of the functions described and presented in Section 10.3 are not included, such as the Instantaneous relay and Cold load functions. The order of sections in the application is consistent with the process of setting parameters and relay coordination.

10.4.4.1 Relay-relay coordination

For the coordination between overcurrent protective devices, the proper time sequence of maximum fault current as seen by both relays must be ensured [15]. A diagram of the electrical system for which coordination was desired and the dynamic state of simulated fault current during the coordination are shown in Fig. 10.10. More complexity in this network is not required, because the coordination involves only a pair of relays, and multiple scenarios can be achieved by utilizing the Thevenin equivalent of the source. In the case where both ends of the line are fed, the relays were assumed to be directional.

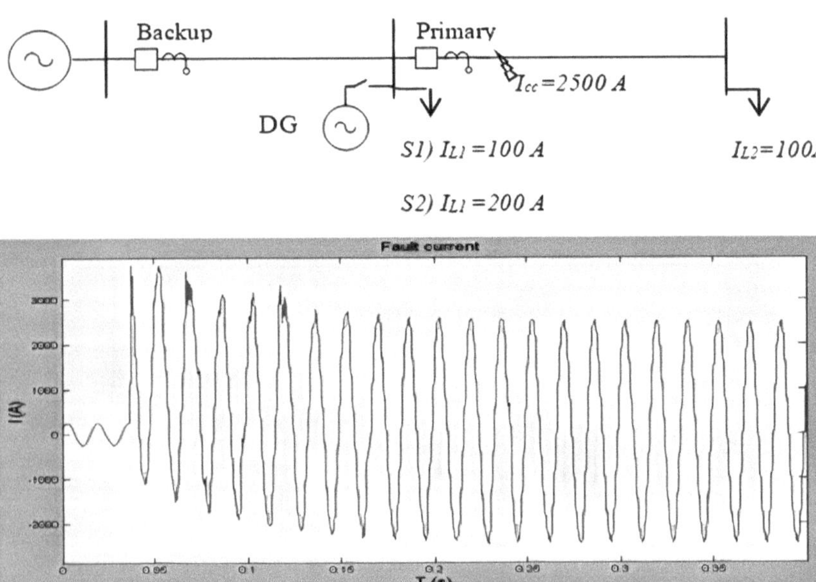

FIG. 10.10 Relay coordination, radial system, and simulated fault record.

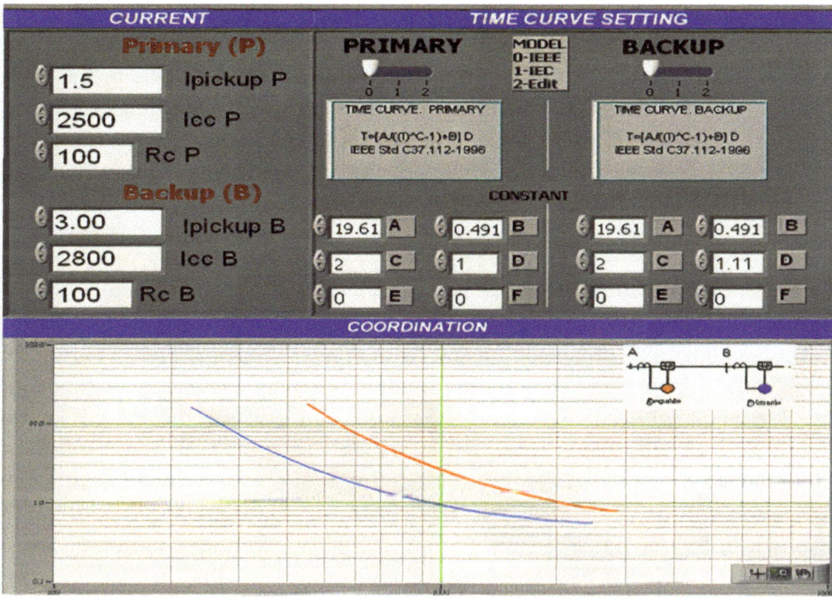

FIG. 10.11 Relay coordination, setting of relays, and coordinated time curves.

Fig. 10.11 shows the settings for the relays in the simulation, as well as the resulting coordination. The time curves of the relays were defined as "very inverse" [13], and the settings were calculated as recommended in Ref. [15] with a coordination time interval (CTI) of 0.3 s. The interval between operating times of the relays as obtained in the coordination process is met only if the fault current is a constant signal; for dynamic fault current, the relay operation times are different, as presented in Table 10.1. Additionally, the load current in the network affected the time of operation, due to the fact that the time curve is asymptotic to the I_{pickup}, so that larger load current results in longer operating times. Two different load conditions, S1 and S2, were evaluated in Fig. 10.10 (see results in Table 10.1).

TABLE 10.1 Relay operating times for two load scenarios.

	S1		S2	
	Primary (TDS = 1)	*Backup (TDS = 1.11)*	*Primary (TDS = 1)*	*Backup (TDS = 0.749)*
Constant fault current	0.57 s	0.87 s	0.57 s	0.87 s
Dynamic fault current	0.67 s	1.20 s	0.67 s	1.40 s

The fault signal utilized was generated in Matlab. It should be noted that the operating times of relays are much higher than those in the coordination, because coordination is established in a pessimistic scenario where the fault current is maintained at its maximum value. Such a consideration is suitable for the setting of relays; however, it is important to evaluate the actual operation times since it impacts directly on the mechanical and thermal stress of components and on the quality of voltage.

10.4.4.2 Infeed in coordination

The introduction of dispersed generation sources to the electrical system do not establish an unknown condition for setting of relays. The short-circuit current of the network is increased [18], and in principle the relay coordination should be reviewed and modified. However, a particular case where the infeed source effect may be negligible is when the increments of short-circuit current lie in the linear region of the time curve. Modeling this operating condition in the application will be illustrated. The settings of the relays are again configured as shown in Fig. 10.11. The fault current is 2240 A, with an additional infeed fault current that is expressed as a percentage of the primary source (Fig. 10.10). The infeed source current is seen only by the primary relay, and this accelerates a trip and decreases its operating time. The results for the simulation are presented in Table 10.2. The difference in operating times for the two relays is increased marginally as the infeed source is more representative. Since the current coordination should be the maximum current to be seen by both relays, the example shows that coordination is not compromised by additional infeed signals.

10.4.4.3 Recloser-relay coordination

The proposed application also includes an equation editor that can be used to plot curves describing relays, fuses, and reclosers; this feature is useful for evaluating new curves or curves fitted to theoretical models. The time curve

TABLE 10.2 Operating times of relays with infeed.

Infeed %	Primary relay	Backup relay
0	0.55 s	0.85 s
25	0.54 s	0.85 s
50	0.52 s	0.85 s
75	0.51 s	0.85 s
100	0.51 s	0.85 s

of a recloser was obtained from the information provided by the manufacturer. An analytical expression for this time curve was constructed by using similar criteria to those of the relays [19,20]. The coordination between the time curve of the recloser and that of the relay is shown in Fig. 10.12. The coordination is simplified by coordinating with a single curve of the recloser in order to show the usefulness of the proposed application. The analytical expression of the recloser was evaluated using nonlinear regression and polynomial algorithms. The relay was defined with a "very inverse" curve, with a time *dial* of 0.5. The electrical system of Fig. 10.10 was again used to determine the settings of all protective devices, while the I_{sc} was chosen as 200 A, and I_{pickup} currents for recloser as well as relay were 12 A and 16 A, respectively.

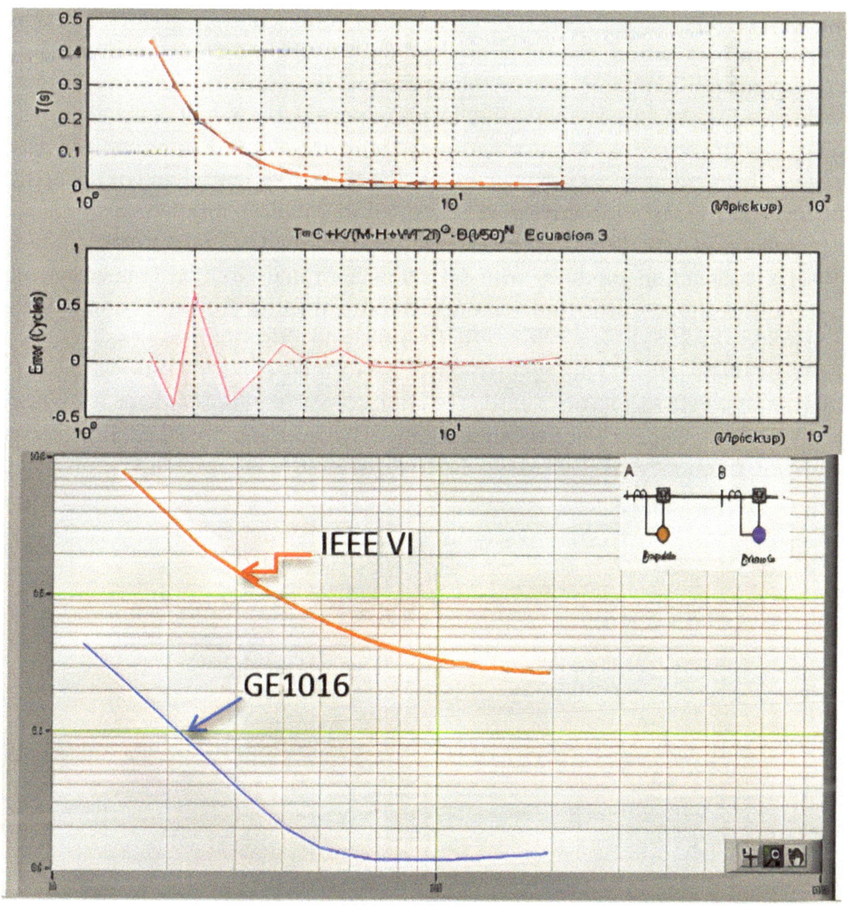

FIG. 10.12 Error of the analytical expression and coordination between recloser and relay.

The analytical expression obtained using a model proposed by Heiber [20] was as follows:

$$T = -17.326 + \frac{1413.664}{\left(I + 2.667 - 0.253I^{-2I}\right)^{5.756}} + 17.341\left(\frac{I}{50}\right)^{0.00014} \qquad (10.1)$$

The evaluation of protection performance under dynamic fault currents conditions is made possible by considering that the dynamic behavior of the recloser is similar to that of a relay. Experiments have shown that analysis of signals simulated or real events is feasible using the proposed application. The accuracy of the fitted models obtained is independent of the application.

10.5 Laboratory test structure

The general application structure of a virtual relay consists of a connection module, to concentrate the input and output signals, a data acquisition card, and a personal computer, where the operating programs of the virtual relay reside (see Fig. 10.13). The virtual relay is connected to a model of the electrical power system, receives the information from the current transformers, and acts to cause the tripping of a system switch in the event of a fault. Each of the components used in the virtual relay is described in detail as follows.

Connection module. The virtual relay has a National Instruments model TBX-68 connection module, with 68 screw terminals for the connection of input/output signals. It includes a male SCSI connector for direct connection to SH6868, R6868, SH1006868, PSHR68-68, and PR68-68F cables.

Acquisition card. Data acquisition is performed using a National Instruments model PCI-MIO-16E-1 card. The characteristics of the card are 16 analog inputs without reference or 8 differential inputs, resolution of 12 bits, maximum sampling frequency of 1.25 MHz, interval of ±10 V for bipolar inputs, and

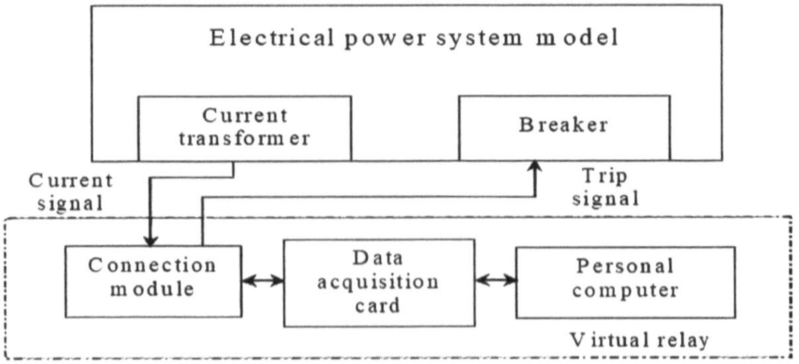

FIG. 10.13 General structure of the implementation of a virtual relay.

bandwidth of 1.6 MHz. It has two analog output channels with a voltage of ±10 V and with eight digital inputs/outputs; the timer resolution is 24 bits.

Model of the electric power system. The representation of the electrical power system was carried out in three variants depending on the origin of the input signals to the virtual relay: (a) signal acquired in real time from a physical model of the electrical power system (Fig. 10.14), (b) reading of a data file with the registry of generated or real events, and (c) internal generation of the signal with virtual controls.

In Fig. 10.15, the digital structure of the virtual relay is presented. The current signal of the electrical system model is obtained through current transformers in the case of actual acquisition of the signal, by reading file data or internal generation by manipulating internal controls. The acquisition of the signal is carried out by the acquisition card through a multiplexing process to take the reading of the defined acquisition channels, later the A/D conversion of the signal is carried out. To avoid the effect of aliasing, the current signal can be filtered analogically or the high sampling frequency available on the data acquisition card can be used to oversample the signal.

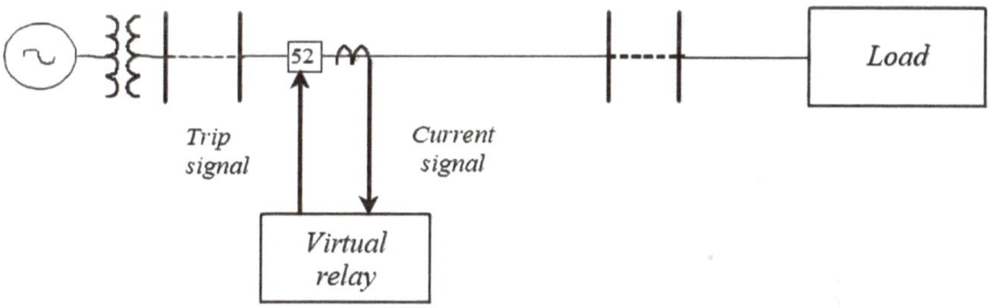

FIG. 10.14 Connection of the virtual relay to a radial system.

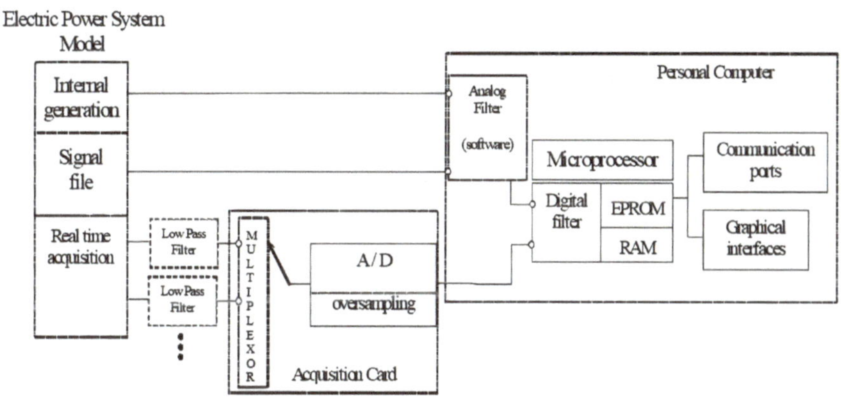

FIG. 10.15 Digital structure of the virtual relay.

For signal file reading and internal generation, an optional module is included that emulates the analog filtering process. Within the programming environment of the virtual relay, a digital filtering process of the sampled signal is carried out to obtain its fundamental component. The algorithms of the virtual relay (EPROM memory) and the input information processing capacity (microprocessor, RAM memory) are established. The graphical interfaces and the communication ports of the computer are available as output elements of the virtual relay. The switch trigger signal is made through the analog or digital output channels of the acquisition card.

The objective of the laboratory tests was the evaluation of the protection algorithms for the adaptive control of the I_{pickup} (Chapter 7) and for the adaptation of the operating time (Chapter 2) of the overcurrent relay. The tests carried out focused solely on the evaluation and validation of the proposed adaptive protection algorithms. Given the available platform, it is not intended to evaluate the virtual relay to execute tripping actions on the electrical power system. Due to limitations of the laboratory equipment, the tests were carried out only with overcurrent relays. Electromechanical and digital overcurrent relays were used for coordination tests (current and time) with the adaptive relay.

The adaptive relay algorithm was evaluated against three types of input signals, the first with signals acquired in real time from the physical module of the electrical power system, the second with signal data in file generated in Matlab® and EMTP, and finally tests were performed with internally generated signals.

10.5.1 Acquisition of signals in real time

Signal acquisition was performed on a physical model of an electrical power system. The objective of the tests was to evaluate the behavior of the relay algorithms against real signals acquired. For the physical representation of the electrical power system, an equivalent Labvolt single-phase voltage source with 120/240 V and 30 A maximum was used. Two segments of 10 AWG cables were used to simulate the distribution lines and interconnect the equivalent source and load (Fig. 10.16). The system load was represented by a GE variable single-phase resistor rated at 0–20 A and 100-V supply. The system parameters are indicated on the one-line diagram.

Steady-state and fault current were obtained by varying the load resistance. All faults are considered at bus C. In all cases, a flat voltage profile was

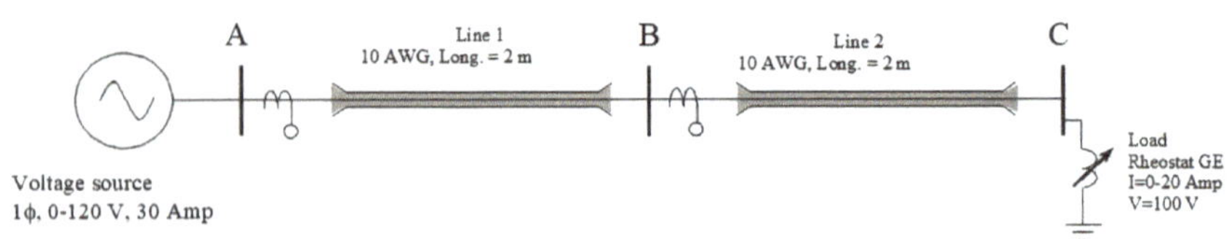

FIG. 10.16 Physical model of the electrical power system.

considered. The tests were carried out considering the adaptive overcurrent relay located at bus A in Fig. 10.16 performing backup functions to an electromechanical relay and a digital relay located at bus B.

In these tests, the entire analog-digital acquisition and conversion process of the input signal to the relay is evaluated, as well as the I_{pickup} control logic and the adaptation of the relay's operating time. To validate the adaptation of the adaptive relay operating time, constant fault currents are considered, and the coordination interval between the adaptive relay and conventional relays is verified. The results are presented graphically on a log-log time-current scale. Due to the delay in processing the signals acquired in real time, tests with variable fault current are not presented. Validation of the I_{pickup} control logic is verified by simulating a load current sequence of events. The quantitative differences of the I_{pickup} of the three types of relays analyzed are established.

10.5.1.1 Operating time verification

The tests for constant fault current consist of the comparison of the operation time of each overcurrent relay (adaptive, electromechanical, and digital) for a constant value of fault current. The adaptive relay performs backup functions according to Fig. 10.16, when it has an operation time CTI (0.3 s) greater than the other relays for any fault current value. This guarantees an automatic coordination between these relays for each constant value of fault current.

The first test was performed between an electromechanical overcurrent relay supported by the adaptive relay. Fig. 10.17 shows the electromechanical relay curve given by the manufacturer [21]. Ten pairs of time-current values were taken from the curve with time lever five; current values were defined with uniform spacing as proposed in Ref. [22]. An interval of 2.7 was established from the multiple 2 to 20.

The curve fitting program proposed in Chapter 2 was used to obtain the analytical expression of the relay's operating characteristic. The pairs of values entered in the program were taken directly from Fig. 10.17. Table 10.3 presents the selected time-current values.

The analytical expression obtained from the program with the least error was the polynomial equation of the form:

$$T = A_0 + \frac{A_1}{(I-1)} + \frac{A_2}{(I-1)^2} + \frac{A_3}{(I-1)^3} + \frac{A_4}{(I-1)^4} + \cdots$$

And the coefficients were as follows:

$A_0 = 0.39825200918131, A_2 = 11.04277679324150, A_4 = -52.60081458091736$
$A_1 = 3.60116781108081, A_3 = 30.11577248573303, A_5 = 29.55377423763275$

The resulting times obtained by the analytical and physical expression of the relay were validated. The operation time of the electromechanical relay was taken directly with a stopwatch in hand; four samples were taken for each fault

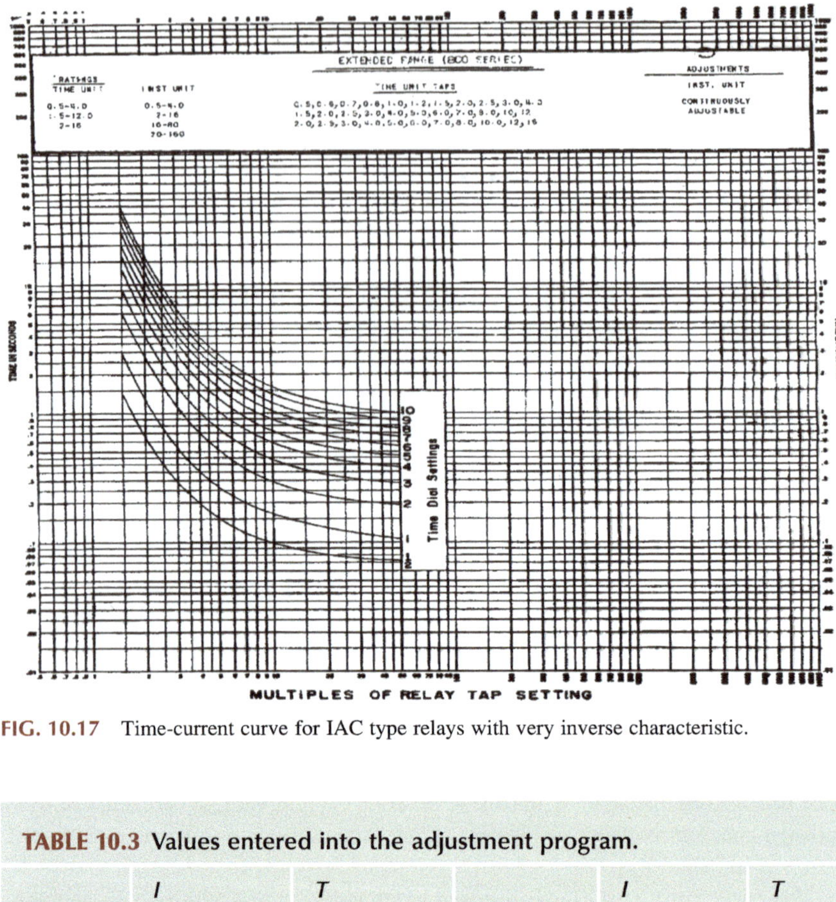

FIG. 10.17 Time-current curve for IAC type relays with very inverse characteristic.

TABLE 10.3 Values entered into the adjustment program.

	I	*T*		*I*	*T*
(1)	2.15	16.3	(7)	16.5	0.68
(2)	3	6.36	(8)	19.2	0.64
(3)	5.7	1.86	(9)	21.9	0.6
(4)	8.4	1.14	(10)	24.6	0.57
(5)	11.1	0.9	(11)	27.3	0.55
(6)	13.8	0.75	(12)	30	0.54

current value. The operating times were not similar (Table 10.4) probably due to a calibration problem of the electromechanical relay since it was taken from the warehouse with a long time without service.

The operating time values were taken directly from the relay to perform the curve fitting. The average value of the sample obtained is taken as valid,

TABLE 10.4 **Comparison of operating times for the electromechanical relay.**

I_{sc}	Operation time		
	Measure		*Analytical expression*
3.18	6.26	6.25	5.55
	6.28	6.28	
4.12	3.69	3.69	3.22
	3.73	3.70	
5.10	2.47	2.58	2.20
	2.55	2.60	
6.44	1.71	1.80	1.56
	1.74	1.78	
0.23	1.32	1.34	1.16
	1.28	1.30	
10	1.05	1.08	0.96
	1.08	1.09	
11.9	0.99	1.00	0.84
	0.97	1.01	
13.9	0.88	0.91	0.75
	0.94	0.89	
14.7	0.90	0.89	0.72
	0.88	0.85	
15.7	0.79	0.89	0.695
	0.77	0.83	

the highest and lowest value is discarded, and the two remaining values are averaged. With these values, curve fitting is performed to obtain the analytical expression that corresponds to these values (Table 10.5).

The analytical expression obtained from the program with the least error was the polynomial equation of the form:

$$T = A_0 + \frac{A_1}{(\log I)} + \frac{A_2}{(\log I)^2} + \frac{A_3}{(\log I)^3} + \frac{A_4}{(\log I)^4} + \cdots$$

TABLE 10.5 Corrected values entered into the adjustment program.

	I	*T*		*I*	*T*
(1)	3.18	6.26	(6)	10	1.08
(2)	4.12	3.69	(7)	11.9	1.0
(3)	5.10	2.56	(8)	13.9	0.9
(4)	6.44	1.75	(9)	14.7	0.88
(5)	8.23	1.31	(10)	16.4	0.81

And the coefficients were as follows:

$$A_0 = -37.60884046555, A_2 = -1342.03836321831, A_4 = -2098.37408494949$$
$$A_1 = 362.41678184271, A_3 = 2418.57128095627, A_5 = 710.27802491188$$

The graphical result of the adjustment is presented in Fig. 10.18.

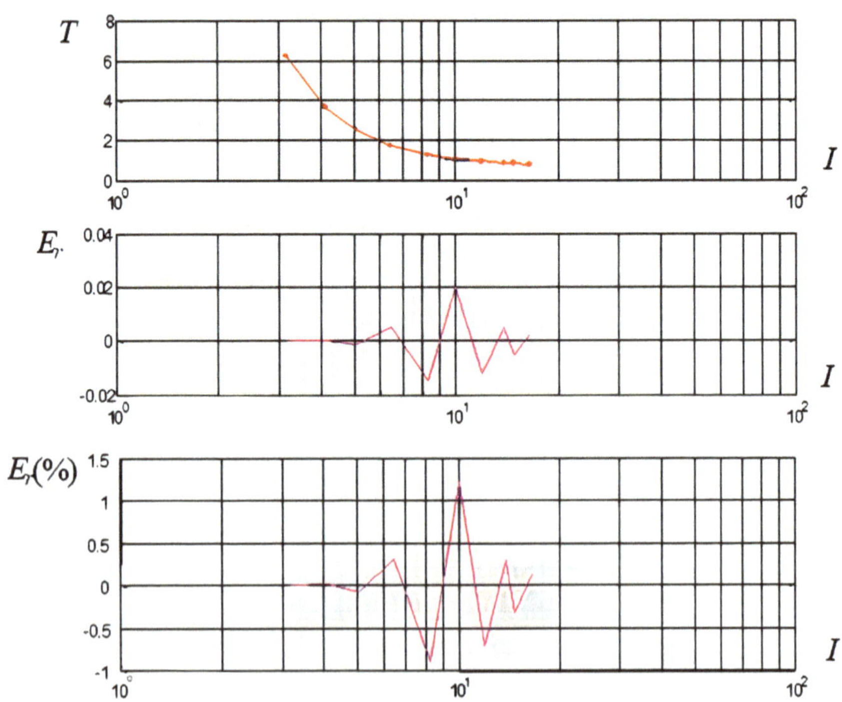

FIG. 10.18 Graphical result of the best fit.

TABLE 10.6 Analytical expression validation with rounding on coefficients.

Input data		Validation of the analytical expression		
I	*T*	*No rounding*	*Round to three decimal places*	*Round to a decimal*
3.18	6.26	6.259	6.260	6.297
4.12	3.69	3.690	3.690	3.713
5.10	2.56	2.558	2.559	2.576
6.44	1.75	1.755	1.755	1.769
8.23	1.31	1.295	1.295	1.306
10.00	1.08	1.100	1.100	1.110
11.9	1.00	0.988	0.988	0.997
13.9	0.90	0.904	0.905	0.914
14.7	0.88	0.874	0.875	0.883
16.4	0.81	0.812	0.813	0.821

Table 10.6 presents the validation of the expression obtained for the input values; the deviation is obtained for the exact values of the coefficients, as well as the values rounded to three and one decimal place of accuracy. It is observed that the error due to rounding the values of the coefficients to three decimal places is minimal, but not so for rounding to one decimal place. Therefore coefficients with three decimal places of accuracy are used for these tests.

Considering the rounding to three decimal places, the analytical expression is of the form:

$$T = -37.608 + \frac{362.416}{(\log I)} - \frac{1342.038}{(\log I)^2} + \frac{2418.571}{(\log I)^3} - \frac{2098.374}{(\log I)^4} + \frac{710.278}{(\log I)^5}$$

A sample of the signal acquired in the laboratory is illustrated in Fig. 10.19. The noise in the signal is random and present at the power outlet located in the test lab. In the frequency spectrum shown in Fig. 7.5, the IN graph shows the frequency content present in the acquired signal; in addition to the harmonic content of the signal, its amplitude is variable and random.

Due to the above, four samples were taken for each fault current value, and the same procedure as in the electromechanical relay was carried out to determine the average value. Table 10.7 presents the values registered in the virtual relay. For comparison purposes, the calculated times of the electromechanical relay are shown (Table 10.4) and the times calculated by the resulting analytical expression. It is observed that despite the noise present in the test signal, the

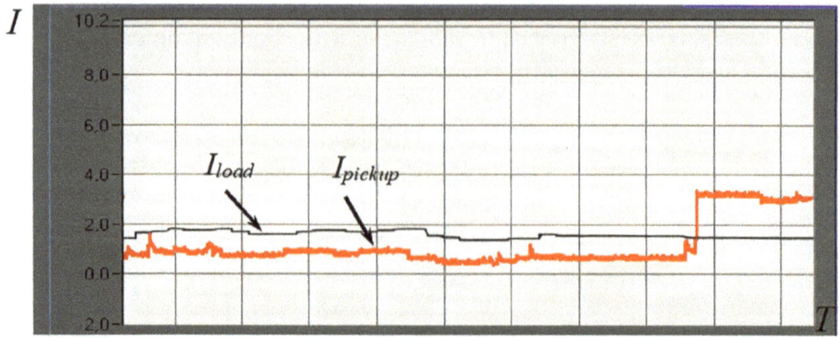

FIG. 10.19 Test signal shape.

TABLE 10.7 Measured operation times of the electromechanical relay, virtual relay, and adjustment analytical expression.

	Operation time				
I_{sc}	*Virtual relay*		*Electromechanical relay*		*Analytical expression*
3.18	6.45	6.60	6.26	6.25	6.259
	6.68	6.67	6.28	6.28	
4.12	4.0	3.91	3.69	3.69	3.690
	4.10	3.98	3.73	3.70	
5.10	2.93	2.93	2.47	2.58	2.558
	2.95	2.88	2.55	2.60	
6.44	2.08	2.15	1.71	1.80	1.755
	2.15	2.12	1.74	1.78	
8.23	1.62	1.62	1.32	1.34	1.295
	1.60	1.58	1.28	1.30	
10	1.42	1.40	1.05	1.08	1.100
	1.40	1.43	1.08	1.09	
11.9	1.32	1.33	0.99	1.00	0.988
	1.30	1.30	0.97	1.01	
13.9	1.22	1.22	0.88	0.91	0.904
	1.22	1.22	0.94	0.89	
14.7	1.18	1.18	0.90	0.89	0.874
	1.20	1.18	0.88	0.85	
15.7	1.15	1.15	0.79	0.89	0.838
	1.13	1.13	0.77	0.83	

time resulting from the solution of the analytical expression is very similar to the time measured in the electromechanical relay. This result validates the analytical expression obtained from the adjustment program proposed; also the integration of this expression emulates the operation of the electromechanical relay for constant and variable fault current. Due to the random nature of the acquired signal, the electromechanical relay operating time and analytical expression have a small margin of variation.

The average operation times of the virtual relay and the electromechanical relay, as well as the time interval are presented in Table 10.8. The time interval is very close to the time defined as the coordination interval (0.3 s). The observed variations are due to the noise present in the test signal. Although the test is considered as constant current, it actually has limited dynamics due to the noise present in the signal; the relays perform the integration of this variable signal to determine the actual operating time. It is observed that for low current values, the random character affects to a greater extent than for higher current values where the interval is very close to 0.30 s.

In Fig. 10.20, the results presented in Table 10.8 are represented graphically. The time-current values obtained in a bi-logarithmic plane of the electromechanical relay and the adaptive relay are presented. It is observed that the coordination interval is preserved, thus guaranteeing an automatic coordination between these two relays.

TABLE 10.8 Average operating times and coordination interval.

	Operation time		
I_{sc}	*Virtual relay*	*Electromechanical relay*	*CTI*
3.18	6.63	6.26	0.37
4.12	3.99	3.69	0.30
5.10	2.93	2.56	0.37
6.44	2.13	1.75	0.38
8.23	1.61	1.31	0.30
10.0	1.41	1.08	0.33
11.9	1.31	1.00	0.31
13.9	1.22	0.9	0.32
14.7	1.18	0.88	0.30
15.7	1.14	0.81	0.33

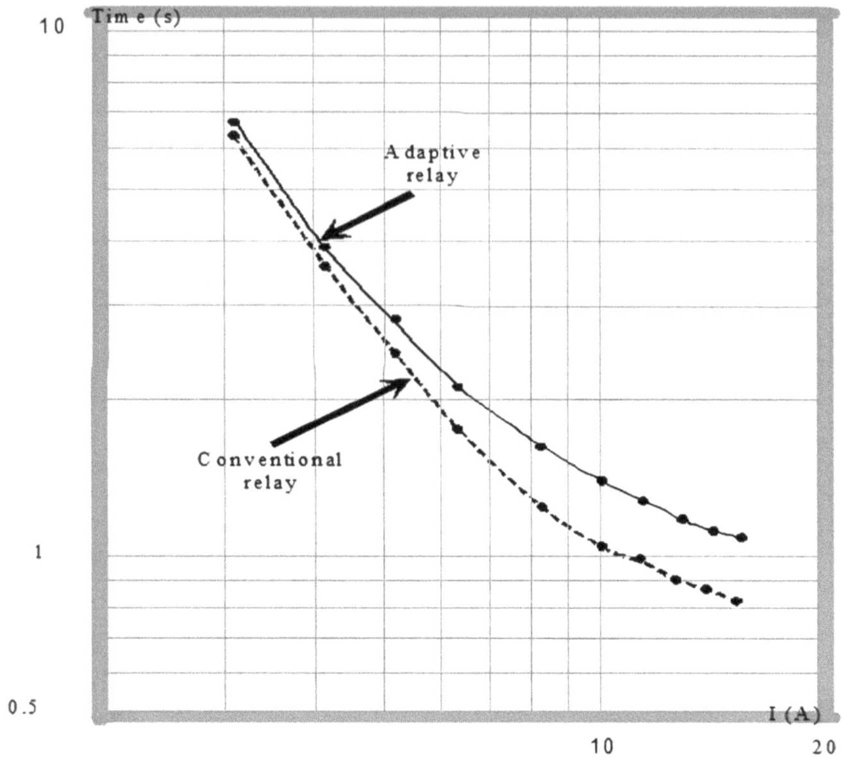

FIG. 10.20 Operating characteristics of the electromechanical and virtual relays obtained in laboratory tests (results of Table 7.6).

The earlier mentioned test was also performed between a digital overcurrent relay and the adaptive relay. The curve defined in the digital relay was as follows [23]:

$$T = \frac{0.052}{I^{0.02} - 1} + 0.113$$

Fig. 10.21 shows the family of curves defined by the manufacturer, time lever five was used. For this test it was not necessary to fit the curves since the analytical expression of the digital relay is entered as data in the adaptive relay. Fig. 10.22 shows the settings programming defined in the digital relay.

Table 10.9 presents the results obtained from the operation times of the virtual and digital relay. The test signal presented the same characteristics of the signal shown in Fig. 10.20, so the time given by the solution of the analytical expression does not coincide with the operation time obtained. Similar to the results of the tests with the electromechanical relay, the random character of the signal affects it to a greater extent for low magnitudes of current. In Fig. 10.23, the results are presented graphically.

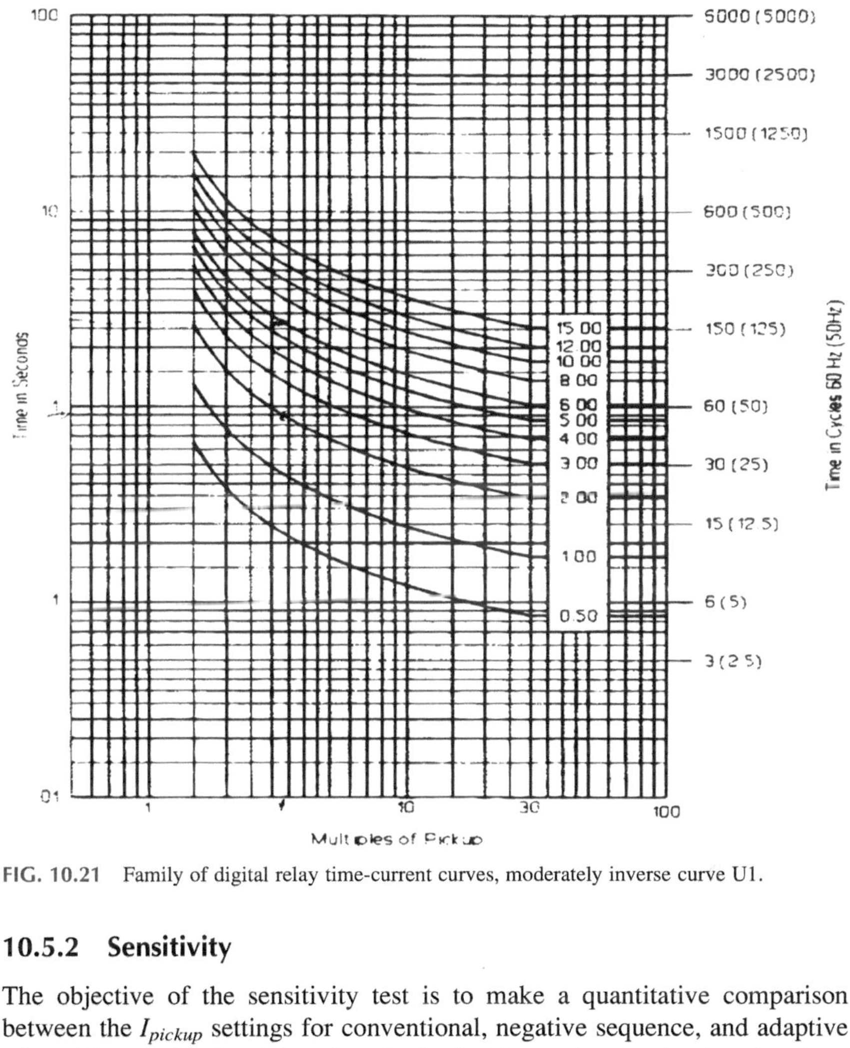

FIG. 10.21 Family of digital relay time-current curves, moderately inverse curve U1.

10.5.2 Sensitivity

The objective of the sensitivity test is to make a quantitative comparison between the I_{pickup} settings for conventional, negative sequence, and adaptive relays. The sensitivity of the relays is checked by performing backup functions to another relay. The physical model of the electrical system used for the sensitivity tests is the same as described in Fig. 10.16. The load was added to the bus B to emulate the system's contribution to other feeders. The operating conditions of the system are shown in Fig. 10.24. Supported relay $n-1$ settings and maximum line and load current magnitudes are displayed.

The settings of the conventional and negative sequence relays located on bus A (relay n) are as follows:

Conventional Relay

$$I^{n}_{pickup_c} = 1.5\, I_{load\_m Æx} = (1.5)(8.33) = 12.5\text{ A}$$

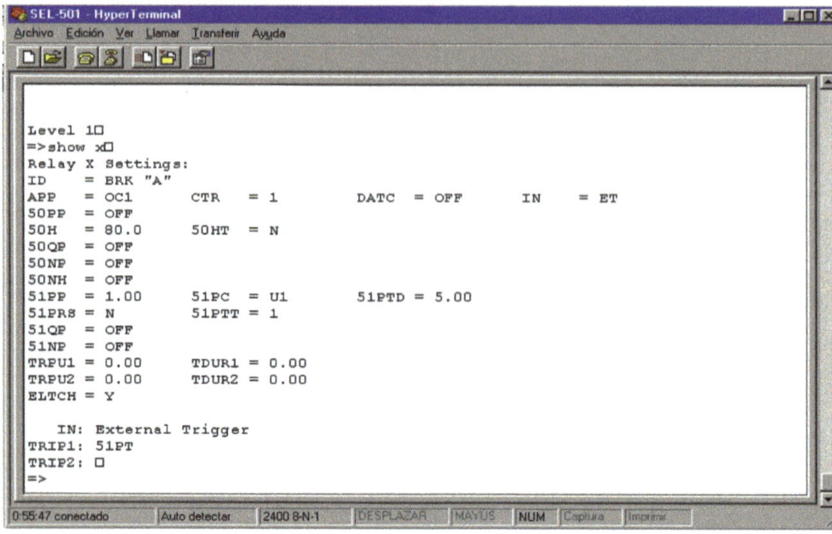

FIG. 10.22 Digital relay settings programming.

TABLE 10.9 Average operating times and coordination interval.

	Operation time			
		Digital relay		
I_{sc}	*Virtual relay*	*Measured*	*Analytical*	Δt
3.1	2.63	2.25	2.38	0.38
4.1	2.20	1.87	1.92	0.33
5.1	2.0	1.65	1.68	0.35
6.4	1.80	1.45	1.48	0.35
8.2	1.65	1.29	1.32	0.36
10.0	1.52	1.19	1.21	0.33
11.9	1.45	1.11	1.13	0.34
13.6	1.40	1.05	1.08	0.35
14.5	1.37	1.04	1.05	0.33
15.5	1.35	1.02	1.03	0.33

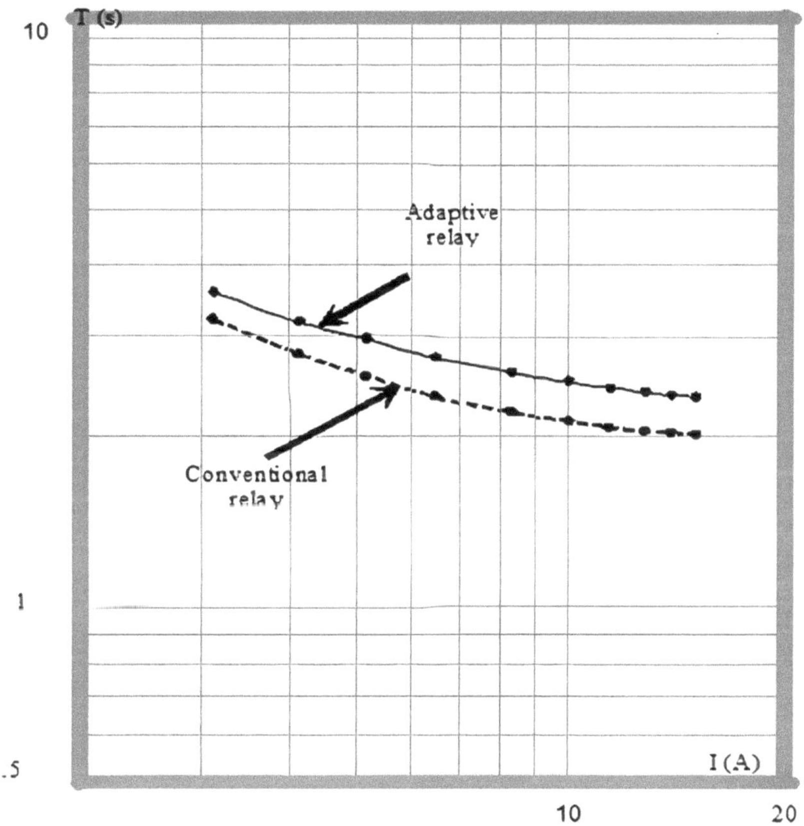

FIG. 10.23 Operating characteristics of the digital and virtual relays obtained in laboratory tests (results of Table 10.9).

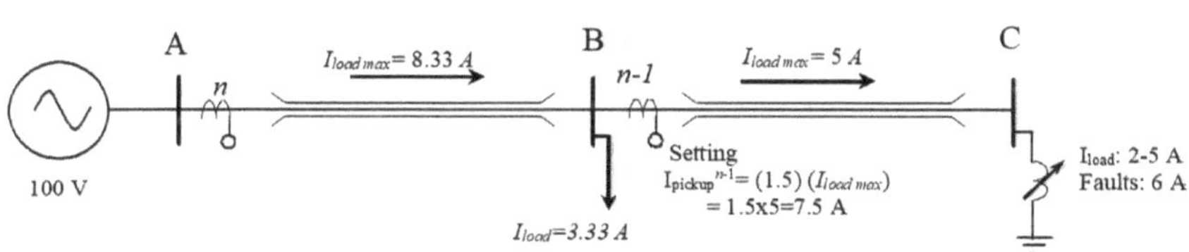

FIG. 10.24 Physical model of the electrical power system.

Negative sequence relay

$$I^{n}_{pickup_{(-)}} = 1.2 I^{n-1}_{pickup} = (1.2)(7.5) = 9\ \text{A}$$

The minimum I_{pickup} setting of the conventional overcurrent relay, located on bus A is calculated for the load current of 12.5 A. During the entire sequence of the event, the conventional relays located on bus A and bus B will not have enough sensitivity to detect fault currents lower than this value: relay B for primary protection and relay A as backup protection. The adjustment of a negative sequence relay will have the same sensitivity problems for the detection of faults with a magnitude of current less than 9 A. The relay sensitivity problems are commonly avoided by preventing the relay from negative sequence is more sensitive than the conventional relay, causing time curve crossing and loss of coordination.

Acquisition tests were carried out against signals with energizing, fault and trip operation sequences. In Fig. 10.25, initially the line is open ($I_{load} = 0$), and the I_{pickup} is fixed at its maximum value, which in this case is 12.5 A. When the breaker closes, a variable value of $\mathrm{I}_{\mathrm{load}}$ appears, and the floating of I_{pickup} on I_{load} begins. Then a 6-A phase-to-phase short circuit occurs, through fault impedance between phases. This short circuit, according to the calculated load current setting values, is not detected by the primary protection (relay $n-1$), nor by the backup relays in n (conventional relay and negative sequence relay). This short circuit is detected and eliminated by adaptive relay n. The value of I_{pickup} is that corresponding to the previous demand calculation period, remaining constant; subsequently, when the short-circuit current is reduced, the value of I_{pickup} is adjusted to its maximum value of 12.5 A.

The increase in I_{pickup} is limited to the value of I_{pickup_max} by the logic of the relay. This value is equal to the I_{pickup} of a conventional relay. This implies that the most the relay will desensitize during the demand interval following the fault is at the sensitivity level of a conventional relay. The adaptive relay for this event will have a more sensitive I_{pickup} than the conventional relay $n-1$; according to what is described in Chapter 7, it is convenient to allow the loss of selectivity so that the adaptive relay operates for the fault in the bus C. The adaptive relay will operate with the operating characteristic equal to that

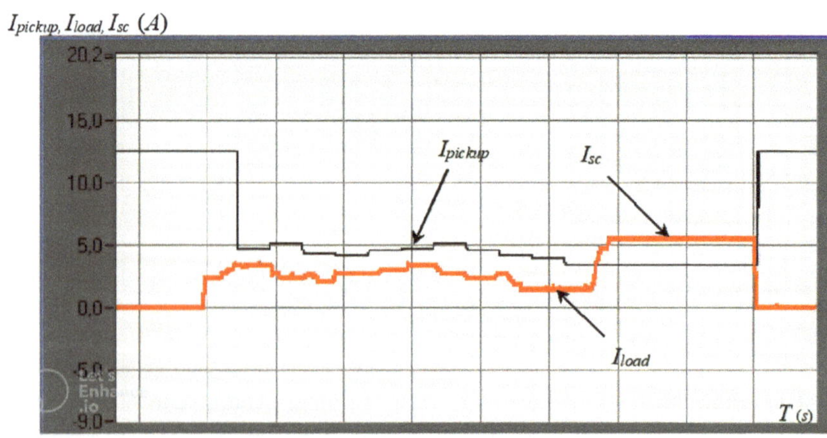

FIG. 10.25 Test with signal acquisition in real time.

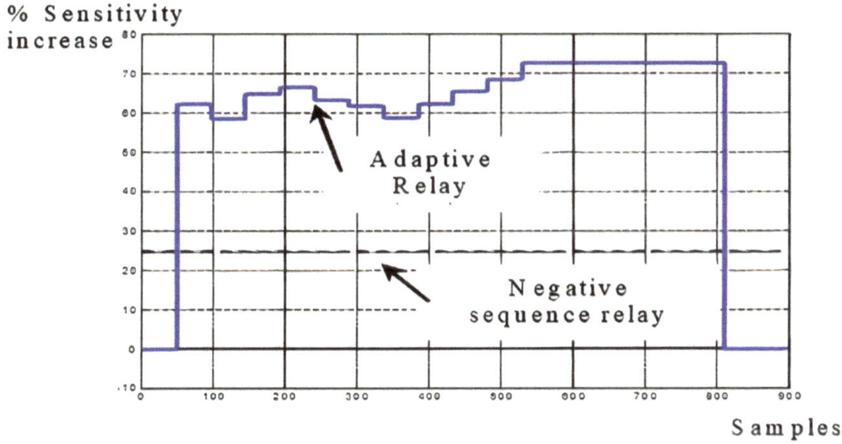

FIG. 10.26 Sensitivity percentage between the types of relays.

assigned to relay $n-1$, but with the I_{pickup} calculated by the adaptive relay itself, thus allowing the operation of the relay for fault in bus C.

The graph shown in Fig. 10.26 shows the percentage of the increase in sensitivity of the adaptive and negative sequence relay with respect to the conventional relay. For the deenergized state, the conventional and adaptive relays have the same sensitivity; when the line is energized the sensitivity of the adaptive relay is clearly superior to the conventional relay and the negative sequence relay, presenting the maximum sensitivity during the fault for the analyzed example. The negative sequence relay has 25% sensitivity compared to the conventional relay.

10.6 Signal file

The files were generated in stable state in Matlab® and in fault state by means of a simulation using the Electromagnetic Transients Program (EMTP). The objective of these tests is to evaluate the operating logic of the adaptive relay in response to signals generated in the file. Since there is no other relay, the tests were aimed at establishing sequences of events through signal files generated in professional packages.

The control panel configuration for the files inserted to the relay was the one shown in the panel of Fig. 10.27. Test type one is selected for the signal file test. In the signal file subpanel, the path of the test file is indicated in Path; if this control is blank, a Windows window opens to select the file manually. In reading start, the element within the data array in which you want to start reading is specified, in this case the tests started from the first element, therefore 0 was indicated. The input data are arranged in vector row or matrix arrays. The number of rows is specified in the row number section, with -1 it is indicated that

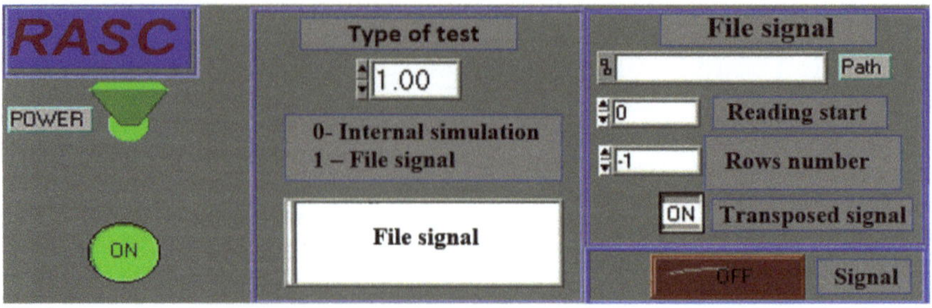

FIG. 10.27 Signal file control panel.

the reading is carried out in all the rows of the array, no distinction is made between matrix and vector. If the array is arranged in a column vector, it is possible to perform a transposition to accommodate the data, as was the case in this example.

The signal files were recorded in ASCII format with a *.txt extension. The results obtained are presented later. In Fig. 10.28, the line is initially deenergized, and the relay has the maximum value of I_{pickup} (10 A in this case). When the line is energized, the floating regime of I_{pickup} on I_{load} begins. When the fault occurs, the value of I_{pickup} remains fixed and later takes the maximum value (10 A) when the fault disappears. Finally, the line is successfully reclosed and the I_{pickup} begins to float on I_{load} again.

Fig. 10.29 shows the previous sequence but with the tripping of another protection. In Fig. 10.30, an evolutionary fault is shown with the opening of the local switch and its consequent reclosing. The evolutionary fault released by another protection is shown in Fig. 10.31.

Fig. 10.32 shows the activation of the cold load subroutine before starting the motor shown in Fig. 10.8. Two adjustment steps were assigned, the first with

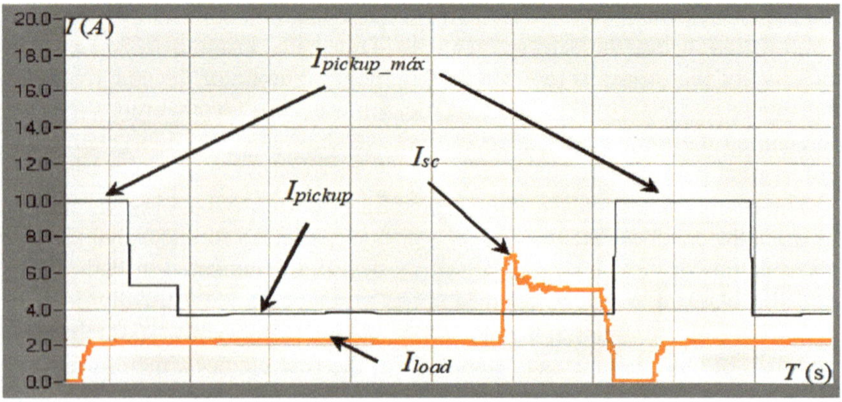

FIG. 10.28 Line opening sequence, energization, and fault insertion (demand current).

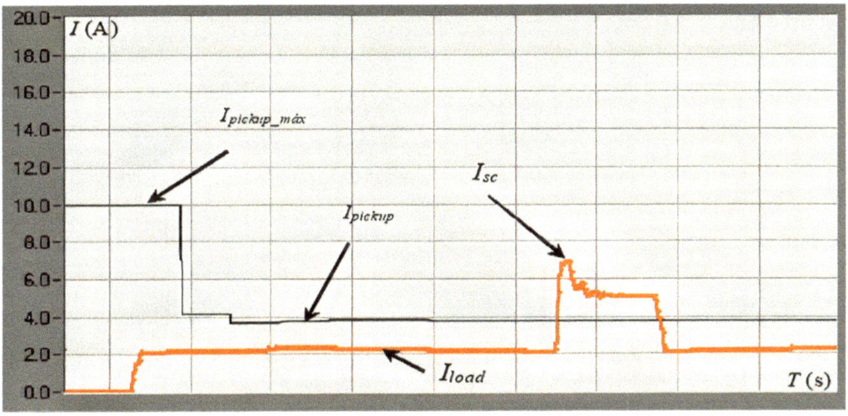

FIG. 10.29 Protection operation supported.

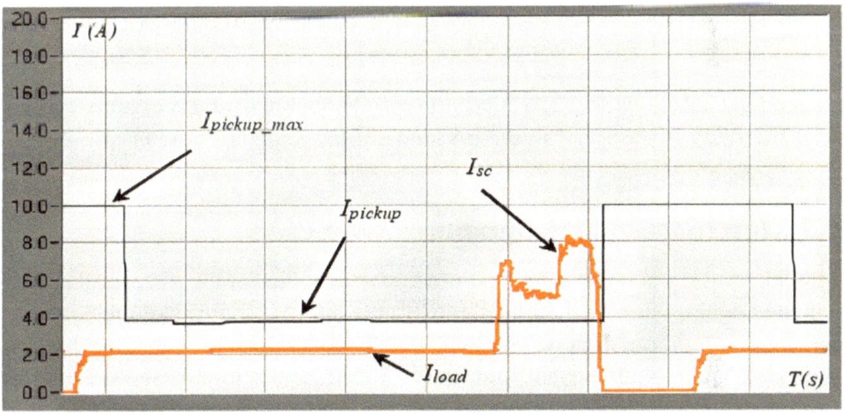

FIG. 10.30 Evolutionary fault. Self-protection operation.

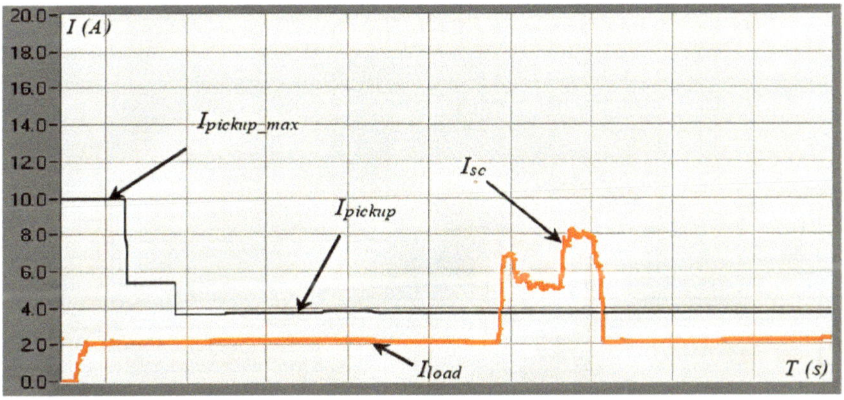

FIG. 10.31 Evolutive fault. Backup protection operation.

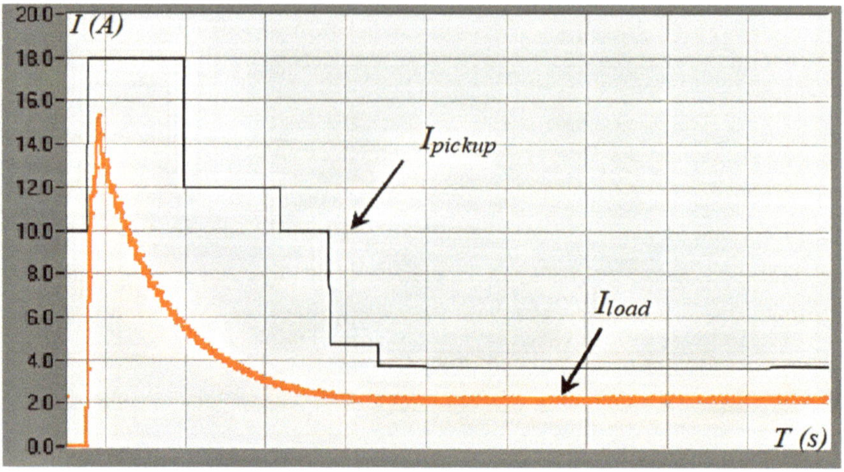

FIG. 10.32 Cold load current.

18 A and the second with 12 A. From the results obtained, an acceptable behavior of the relay is observed in the presence of the signals studied.

10.7 Internal signal generation

In the internal signal generation test mode, the behavior of the virtual relay was evaluated against signals manipulated with the internal controls of the adaptive relay, operating sequences were tested, signals contaminated with noise both in stable state and in fault current, and finally, the algorithmic stability of the relay against current pulses was evaluated.

In Fig. 10.33, the control panel used for the internal generation of signals is shown. In this panel, the values of frequency (F), samples/cycle, F sampling,

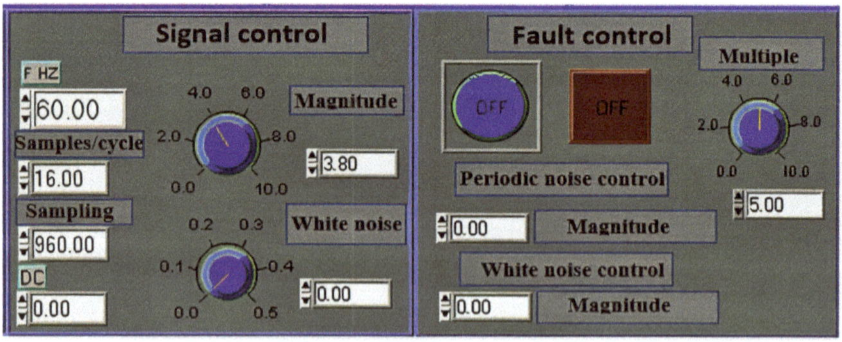

FIG. 10.33 Control panel for load and fault current generation.

and DC are defined. The variation of the steady-state current was done by varying the manual magnitude control. Noise in the signal was introduced with the white noise control. The controls for the fault state are defined by Multiple control, which is defined as times of the nominal current; signal noise is inserted either periodically or randomly, with the periodic noise and white noise controls, respectively.

Fig. 10.34 shows an operation sequence with load variation, fault, and protection tripping. The variation of the load was carried out through the direct manipulation of the current control, for the insertion of the fault the increase in current was defined as times of the prefault current and it was inserted with the corresponding control, finally the trip was simulated activating the trip function within the time-current characteristic module.

Fig. 10.35 illustrates the virtual relay characteristic when its I_{pickup} is allowed to be less than that of the supported device. When the short-circuit current value is greater than the I_{pickup} of the supported device, the coordination interval is preserved as presented in Table 10.10. For smaller values, coordination is not possible; the virtual relay will operate for its value of I_{pickup} and the defined operation time characteristic.

Fig. 10.36 shows the verification of coordination in reset zone between Relay-Recloser shown in Section 2.8.3. This example is presented in Ref. [24]; the test is performed by verifying that the relay does not operate for the entire operation sequence of the recloser. For the definition of the recloser operation sequence, the controls of Fig. 10.37 are used. The adaptive relay registers the dynamics of the relay before the operations of the recloser. For this test, an IAC relay has a very inverse characteristic, the curve fitting was performed for obtaining the analytical expression of the relay. This expression was introduced in the virtual relay. The replenishment zone was as recommended in [24].

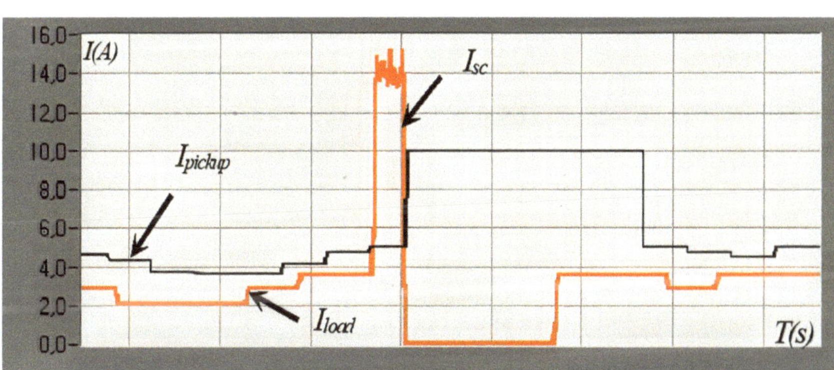

FIG. 10.34 Test for a case of internal signal generation.

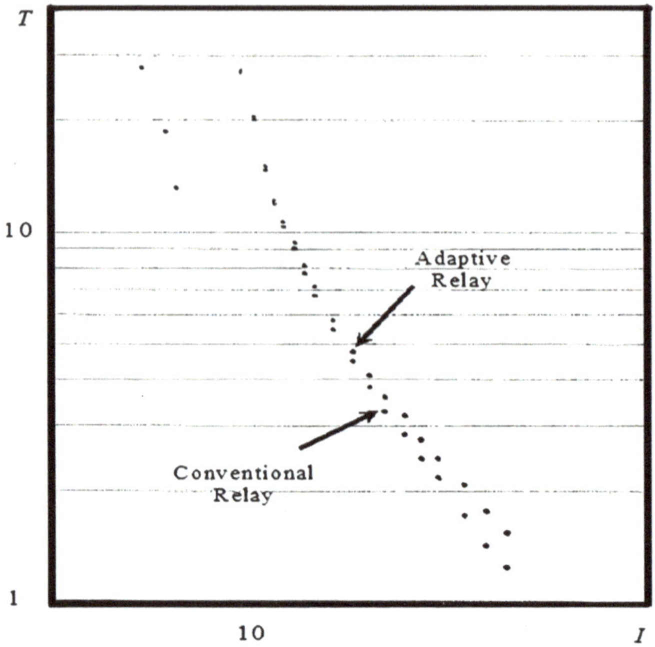

FIG. 10.35 Adaptive relay time-current characteristic.

TABLE 10.10 Setting of the virtual relay, before I_{sc} lower than the I_{pickup} of the supported device.

I_{sc}	I	Tadap	Tconv	I_{sc}	I	Tadap	Tconv
35.35	5.05	1.6	1.29	13.43	1.91	8.10	7.80
31.82	4.54	1.79	1.488	12.72	1.81	9.3	9.00
28.28	4.04	2.08	1.77	12.02	1.71	10.86	10.55
24.74	3.53	2.5	2.197	11.31	1.61	12.96	12.66
22.98	3.28	2.8	2.496	10.60	1.51	15.93	15.65
21.21	3.03	3.19	2.88	9.89	1.41	20.4	20.176
19.79	2.82	3.6	3.29	9.19	1.31	27.86	27.56
18.38	2.62	4.12	3.81	6.78	1.56	13.93	13.94
16.97	2.42	4.82	4.51	6.22	1.43	18.88	18.93
15.55	2.22	5.78	5.47	5.65	1.30	28.19	28.19
14.14	2.02	7.16	6.85				

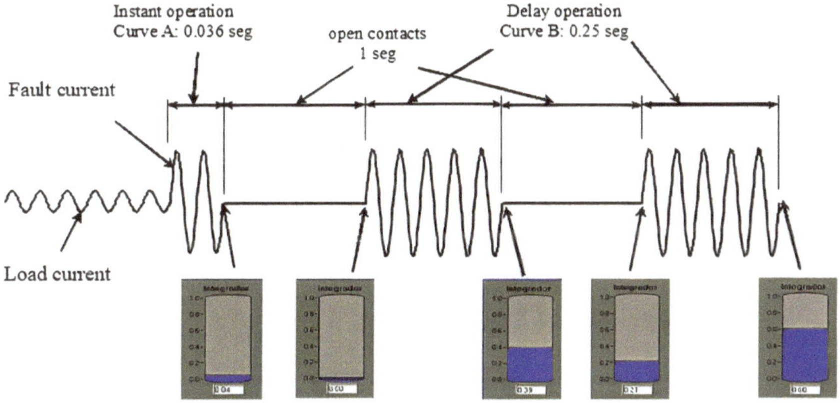

FIG. 10.36 Register of the virtual relay integrator before the sequence of operation of a recloser.

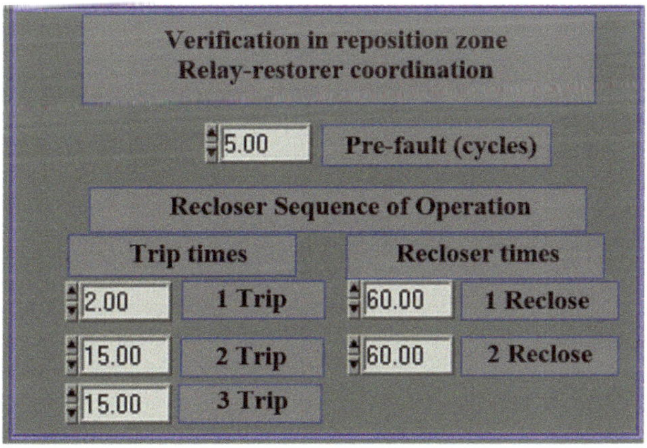

FIG. 10.37 Test module coordination in reset zone relay-restorer.

10.8 Summary

In the acquisition of signals in real time with electromechanical and digital relays, compliance with the coordination criterion is obtained for any fault current value. This demonstrates compliance with the automatic adaptation of operating time.

For the test model used for the adaptation of the starting current, it was observed that the electromechanical and negative sequence relays are insensitive to the fault current established in the test; the adaptive relay does detect the fault current and is even more sensitive than the relay supported.

For all the cases studied using signal files, the behavior of the adaptive relay was satisfactory. The response of the adaptive relay using the internal signal generation test mode was satisfactory too.

The test was performed when the I_{pickup} of the virtual relay is allowed to be less than that of the supported device. The results obtained demonstrate the feasibility of this consideration.

The verification of coordination in the replenishment zone between Relay and Restorer is also presented. The test is performed by verifying that the relay does not operate for the entire operating sequence of the recloser. This tool is useful in the coordination process with restorers.

References

[1] W.-J. Lee, G. Jyh-Cherng, R.-J. Li, P. Didsayabutra, A physical laboratory for protective relay education, IEEE Trans. Educ. 45 (2) (2002) 182–186.

[2] T.S. Sidhu, M.S. Sachdev, H.C. Wood, Design of a microprocessor based overcurrent relay, in: IEEE Western Canada Conference on Computer, Power and Communications Systems in a Rural Environment, Regina, Sask, 29–30 May, 1991, pp. 41–46.

[3] T.S. Sidhu, M.S. Sachdev, M. Hfuda, Computer simulation of protective relay designs for evaluating their performance [in power systems], in: Canadian Conference on Electrical and Computer Engineering, Meeting Date: 26–29 May, 1996, pp. 720–723. Calgary, Canada.

[4] Working Group F-8 of the Relay Input Sources, Subcommittee of the IEEE Power System Relaying Committee, Digital simulator performance requirements for relay testing, IEEE Trans. Power Deliv. 13 (1) (1998).

[5] A. Isono, K. Okuda, T. Kubo, Digital simulation of power distribances and protective relaying, IEEE Trans. Power Appar. Syst. PAS-90 (5) (1971) 2039–2048.

[6] M. Kezunovit, M. Aganagic, V. Skendzic, J. Domaszewicz, J.K. Bladow, D.M. Hamai, S.M. McKenna, Transients computation for relay testing in real-time, IEEE Trans. Power Deliv. 9 (3) (1994).

[7] A.E. Efthymiadis, O.E. Roennspiess, J.A. Guerra, Simulation of protective relay performance under short-circuit and transient swing conditions, IEEE Trans. Ind. Appl. 26 (6) (1990) 1108–1115.

[8] X. Liu, A.H. Osman, O.P. Malik, Advanced simulation tool for relay testing, in: 40th North American Power Symposium 2008, NAPS '08, Calgary, Canada, 28–30 Sept., 2008, pp. 1–6.

[9] S. Brahma, J. De La Ree, J. Gers, A.A. Girgis, S. Horowitz, R. Hunt, M. Kezunovic, V. Madani, P. McLaren, A.G. Phadke, M.S. Sachdev, T.S. Sidhu, J.S. Thorp, S.S. Venkata, T. Wiedman, The education and training of future protection engineers: challenges, opportunities, and solutions, IEEE Trans. Power Deliv. 24 (2) (2009) 538–544.

[10] R. Bayindir, E. Irmak, I. Colak, A. Bektas, Development of a real time energy monitoring platform, Int. J. Electr. Power Energy Syst. 33 (1) (2011) 137–146.

[11] E.S. Ibrahim, A comparative study of PC based software packages for power engineering education and research, Int. J. Electr. Power Energy Syst. 24 (10) (2002) 799–805.

[12] T. Nagata, H. Sasaki, Personal computer based expert system for power system operation education, Int. J. Electr. Power Energy Syst. 18 (3) (1996) 195–201.

[13] IEEE C37.112, IEEE Standard Inverse-Time Characteristic Equations for Overcurrent Relays, 2018.

[14] IEC Standard 255-4, Single Input Energizing Measuring Relays with Dependent Specified Time, IEC Publication 255-4, first ed., 1976.

[15] IEEE Recommended Practice for Protection and Coordination of Industrial and Commercial Power Systems (IEEE Buff Book), IEEE Std 242-2001 (Revision of IEEE Std 242-1986) [IEEE Buff Book], 2001, pp. 1–710, https://doi.org/10.1109/IEEESTD.2001.93369.

[16] P.M. Donohue, S. Islam, The effect of nonsinusoidal current waveforms on electromechanical and solid-state overcurrent relay operation, IEEE Trans. Ind. Appl. 46 (6) (2010).
[17] R.M. Chabanloo, H.A. Abyaneh, A. Agheli, H. Rastegar, Overcurrent relays coordination considering transient behaviour of fault current limiter and distributed generation in distribution power network, IET Gener. Transm. Distrib. 5 (9) (2011) 903–911.
[18] N. Hadjsaid, J.-F. Canard, F. Dumas, Dispersed generation impact on distribution networks, IEEE Comput. Appl. Power 12 (2) (1999) 22–28.
[19] M.S. Sachdev, J. Singh, R.J. Fleming, Mathematical Models Representing Time-Current Characteristics Of Overcurrent Relays for Computer Application, IEEE Paper A78, 1978, pp. 131–135.
[20] J.E. Hieber, Empirical equations of overcurrent relay curves for computer application, in: IEEE Winter Power Meeting, New York, NY, January/February, 1965, pp. 1–11. Paper No. 31 CP 65-91.
[21] General electric Co., Time Overcurrent Relays - Instruction Manual GEH-1788L, 2014. https://www.gegridsolutions.com/app/viewfiles.aspx?prod=iac&type=3.
[22] IEEE Committee Report, Computer Representation of Overcurrent Relay Characteristics, 1989. https://ieeexplore.ieee.org/document/32656.
[23] Schweitzer Engineering Laboratories, Inc., Instruction Manual SEL-501, Dual Universal Overcurrent Relay (1995). https://selinc.com/products/501/.
[24] J.J. Burke, Power Distribution Engineering, Marcel Dekker, 1994.

Index

Note: Page numbers followed by *f* indicate figures and *t* indicate tables.

A

Academic test system, 138–139, 219–228, 219*f*, 220–222*t*, 223*f*, 225*t*, 227–228*f*, 230*t*
Active networks, lack of sensitivity in, 235–238, 237*t*
  negative sequence relays, 236–238, 237*f*
Adaptive location, 76
Adaptive logic, current values used, 263
Adaptive operating time
  curve fitting program, 53–60, 54*f*
  definition, 38–41
  evaluation of, 41–43
  implementation, 44–46, 45–46*f*
  proposed relay, functional structure of, 43–46, 44*f*
  time curve fitting algorithms, 46–53
Adaptive pickup current, 240–244, 255–258, 256*f*, 258*f*
  relay setting groups, 241–242, 242*f*
  utility, microgrid connected to, 242–244, 242*t*, 243–244*f*
Adaptive protection system, online centralized control for
  code of optimization algorithms, 310–313, 311–312*f*
    coordination algorithms, parameter settings of, 314–315
    RES, interconnection of, 324
    results and comparison, 315–320
    test system data, 313–314
  electrical systems, dynamic operating conditions, 296–300, 297*t*
    change of demand, 299–300, 300*f*
    input/output of generators, 298–299, 298*f*
    topological changes, 297–298, 298*f*
  online coordination system, special considerations for, 303–310
    load flow and fault current analysis, 304
    network topology, 303, 304*f*
    parallel lines, 308
    radial lines detection, 308–309
    sensitivity analysis, 309–310, 310*f*
    special considerations, 305–309
    transformer protection, setting of relays for, 305–307
  online protection system, 300–303, 301*f*
    coordination algorithm, 302–303
  time interval coordination method, 324–325, 325*f*
Adaptive relay, 70–71
Ant colony optimization (ACO), 81–82

B

Battery energy storage system (BESS), 201–205, 204–205*f*
Battery management system (BMS), 201, 203*f*
Battery technologies, 200
BESS. *See* Battery energy storage system (BESS)
Biogeography-based optimization (BBO), 81–82
Breaker failure scheme (BF), 208

C

CIGRs connection, 235–238
Code of optimization algorithms, 310–313, 311–312*f*
  coordination algorithms, parameter settings of, 314–315
  RES, interconnection of, 324
  results and comparison, 315–320
  test system data, 313–314
Coincident angles, 276
Cold load current, 338–339, 338*f*
Constant fault current, 41
Continuous adaptive overcurrent, 245–250, 245*f*
  adaptive relay, control logic for, 249–250, 249*f*
  overcurrent relay, adaptive characteristic of, 246–249, 246–248*f*
  RMS, pickup current with, 250–258, 252–253*f*
Conventional-adaptive-conventional relay set, 69–70

Conventional and interval coordination methods, 178–182, 178*t*, 180–182*t*, 181*f*, 183–184*f*
Conventional relay, 70–71
Coordinating relays, general procedure for, 7–9
Coordination, 5–10, 6*f*
  algorithm, 302–303
    parameter settings of, 314–315
  IWO, 127–132, 127–128*f*, 129*t*, 130–132*f*
Coupling Capacitor Voltage Transformer (CCVT), 274
Covariance matrix evolution strategy (CMA-ES), 81–82
Current transformer (CT), 274
Curve fitting program, 53–60, 54*f*
  functions H(I) and T(I), 63–66
  fuse fitting, 55–59, 56*f*, 56–57*t*
  H(I) function, 60–66, 63*f*, 64*t*
  recloser fitting, 59–60, 59*t*, 60*f*, 61*t*

D

DE, 132–136
Demand control logic, 259–261, 260*f*
Deterministic methods, 161
Differential evolution algorithm, 103–106, 105*f*
  crossover, 106
  mutation, 105–106
  selection, 106
Different intervals, proposed method with, 182–185, 185*f*
Digital overcurrent relays, equations of, 49–50
Digital relays, functional structure of, 18–30, 18*f*
  dynamic performance, 27, 28–29*f*
  instantaneous OCR, 23–24, 23*f*
  inverse time relay, 24–27, 24*f*, 26*f*
  limitations of, 27–30, 29–30*f*
  phasor estimation, 19–23, 19*f*
Directional overcurrent relays (DOCR), 10–14, 11–12*f*, 269–274, 270–271*f*, 270*t*, 272*t*
  active networks, lack of sensitivity in, 235–238, 237*t*
    negative sequence relays, 236–238, 237*f*
  adaptive logic, current values used, 263
  adaptive pickup current, 240–244, 255–258, 256*f*, 258*f*
    relay setting groups, 241–242, 242*f*
    utility, microgrid connected to, 242–244, 242*t*, 243–244*f*
  CIGRs connection, 235–238
  continuous adaptive overcurrent, 245–250, 245*f*
    adaptive relay, control logic for, 249–250, 249*f*
    overcurrent relay, adaptive characteristic of, 246–249, 246–248*f*
    RMS, pickup current with, 250–258, 252–253*f*
  control logic, 255
  load demand, pickup current with, 258–262, 259*f*
    demand control logic, 259–261, 260*f*
    evaluation, 261–262, 261–262*f*
  microgrids (MG), 233
  sensitivity detection system, 238–240, 239*f*
Distance relay algorithm, 275–279, 276–277*f*, 279*f*
Distance relay coordination
  distance relay algorithm, 275–279, 276–277*f*, 279*f*
  DOCR, 269–274, 270–271*f*, 270*t*, 272*t*
  dynamic coordination, 287–291, 288–291*f*
  nonconventional time curves (NCC), 273–274, 273*f*
  optimization formulation for, 281–287, 282*f*
    results, 283–287, 286–287*t*
    test systems, 282–283, 283*f*
  protection purposes, signal processing for, 274*f*
  relay operation validation, 279–281, 280*f*
Distributed energy resources (DERs), 1
Distribution management system (DMS), 194
DOCR. *See* Directional overcurrent relays (DOCR)
Dynamic coordination, 287–291, 288–291*f*
Dynamic J (I) function, 141–157, 142*f*
  case studies, 148–157, 149*t*
  determination of, 153, 156*f*
  proposed model, 142–148, 143*f*, 145–146*f*, 148*f*
  proposed nonconventional curves, parameters for, 150–152, 152–153*f*, 154*t*
  protection devices, coordination results for, 154–157, 157*f*
  standard curves, protection coordination with, 152–153, 156*f*
  standard curves, transformer protection with, 149–150, 150*f*

E

Electrical systems, dynamic operating conditions, 296–300, 297*t*
  change of demand, 299–300, 300*f*

input/output of generators, 298–299, 298*f*
topological changes, 297–298, 298*f*

F

Finite impulse response (FIR), 275
Frequency change rate method (ROCOF), 211, 211–212*f*

G

Gauss-Newton method, 52
Genetic algorithm (GA), 268
Genetic operators, 91–96
crossover, 94–95
elitism, 93–94
multipoint crossover, 95
mutation, 95–96, 96*f*
population, 91
selection, 91–94, 93*f*
single-point crossover, 94–95
G-2 generator, output of, 185–187, 186*f*, 186–187*t*
Gray wolf optimization (GWO), 107–109, 108*f*, 132–136
archive integration, 112
block diagram, 110
DOCRs coordination, 109–110
enhanced hunting, 109–110
grid integration mechanism, 113
hunting, 108, 109*f*
leader selection integration mechanism, 113
prey encircling, 108–109
test systems and results, 113

I

Infeed in coordination, 342, 342*t*
Infinite impulse response (IIR), 275
Input/output of generators, 298–299, 298*f*
Intermittent generation sources, 213
Internal signal generation, 362–364, 362–364*f*, 364*t*
Interpolation, 50
Invasive weed optimization, 99–103
dispersion, 101, 102*f*
DOCR coordination, IWO for, 102–103
rolling down, 102
spreading, 101, 102*f*
Ion/lithium (Li-ion) batteries, 201
Islanding detection, 209–213, 210*f*
Island saving scheme, 213–216, 214–217*f*

L

Large-scale storage systems, technical characteristics of, 202*t*
Lead-acid batteries, 200
Least squares regression, 50
Li-ion technology battery, 201
Line 6-4, output of, 188, 189*t*
Load demand, pickup current with, 258–262, 259*f*
demand control logic, 259–261, 260*f*
evaluation, 261–262, 261–262*f*
Load flow and fault current analysis, 304
Low-voltage networks, 206

M

Meshed networks, 11–12
Metaheuristic algorithms
ant colony optimization (ACO), 81–82
biogeography-based optimization (BBO), 81–82
covariance matrix evolution strategy (CMA-ES), 81–82
crossover, 168
differential evolution algorithm, 103–106, 105*f*
crossover, 106
mutation, 105–106
selection, 106
genetic operators, 91–96
crossover, 94–95
elitism, 93–94
multipoint crossover, 95
mutation, 95–96, 96*f*
population, 91
selection, 91–94, 93*f*
single-point crossover, 94–95
gray wolf optimization (GWO), 107–109, 108*f*
archive integration, 112
block diagram, 110
DOCRs coordination, 109–110
enhanced hunting, 109–110
grid integration mechanism, 113
hunting, 108, 109*f*
leader selection integration mechanism, 113
prey encircling, 108–109
test systems and results, 113
initial population, 167–168
interval method, 166–174, 167*f*
invasive weed optimization, 99–103
dispersion, 101, 102*f*

Metaheuristic algorithms *(Continued)*
DOCR coordination, IWO for, 102–103
rolling down, 102
spreading, 101, 102*f*
metaheuristic optimization algorithms, 96–113
DOCR coordination, GA for, 98–99
GA, relay coordination using, 99
genetic algorithms, 97–99
mixed nonlinear programming (MINLP), 81
mutation, 168
relay coordination formulation, data preparation for, 83–90, 83*f*
constraints, 86–87
equality constraint, 86
inequality constraint, 86–87
initial population, 85–86
multiobjective function, 88–90
objective function, 87–90
random variables, 84–85, 85*f*
weighted objective function, 87–88
relay coordination verification, 169–174
selection, 168–169
self-adaptive differential evolution (SADE), 81–82
test case I, 114, 114*t*
test case II, 112*f*, 114–117, 116*t*
time intervals, proposed coordination method using, 167–169
Metaheuristic optimization algorithms, 96–113
DOCR coordination, GA for, 98–99
GA, relay coordination using, 99
genetic algorithms, 97–99
Microgrid (MG)
architecture of, 194–205
battery technologies, 200
microturbine model, 199
Monte Carlo analytical method, 197, 197*f*
Photovoltaic model, 198
renewable sources models, 195–205, 196–199*f*
battery management system (BMS), 201, 203*f*
BESS, power and energy of, 201–205, 204–205*f*
DOCRs coordination proposal, 217–228, 218*f*
academic test system, 219–228, 219*f*, 220–222*t*, 223*f*, 225*t*, 227–228*f*, 230*t*
large-scale storage systems, technical characteristics of, 202*t*
Li-ion technology battery, 201
protection, 205–216
frequency change rate method (ROCOF), 211, 211–212*f*
implementation, 210–211
islanding detection, 209–213, 210*f*
island saving scheme, 213–216, 214–217*f*
weak source contribution, 208–209, 208–209*f*
Microturbine model, 199
Mixed nonlinear programming (MINLP), 81
Monte Carlo analytical method, 197, 197*f*

N

Network topology, 303, 304*f*
Nickel/cadmium (NiCd) batteries, 200
Nine-bus IEEE system, topological changes to, 185–189
Nonconventional curves (NCCs), 7
Nonconventional DOCR based on the time interval criterion
conventional and interval coordination methods, 178–182, 178*t*, 180–182*t*, 181*f*, 183–184*f*
deterministic methods, 161
metaheuristic algorithms
crossover, 168
initial population, 167–168
interval method, 166–174, 167*f*
mutation, 168
relay coordination verification, 169–174
selection, 168–169
time intervals, proposed coordination method using, 167–169
novel method of, 163–166
potential applications, 161–162
proposed method
G-2 generator, output of, 185–187, 186*f*, 186–187*t*
line 6-4, output of, 188, 189*t*
nine-bus IEEE system, topological changes to, 185–189
time intervals coordination, 188–189, 190*t*
proposed method, optimization algorithm with, 176–178, 177*f*
proposed method with different intervals, 182–185, 185*f*
proposed optimization problem, 164–166
objective function, 164–165
restrictions, 165–166
time function, 165
results, 175–189
test systems, 174, 175–176*f*

time intervals, OCR coordination via, 163–166, 163*f*
Nonconventional time curves (NCC), 273–274, 273*f*
coordination by IWO, 127–132, 127–128*f*, 129*t*, 130–132*f*
DE, 132–136
defintion, 121–122
dynamic J (I) function, 141–157, 142*f*
case studies, 148–157, 149*t*
determination of, 153, 156*f*
proposed model, 142–148, 143*f*, 145–146*f*, 148*f*
proposed nonconventional curves, parameters for, 150–152, 152–153*f*, 154*t*
protection devices, coordination results for, 154–157, 157*f*
standard curves, protection coordination with, 152–153, 156*f*
standard curves, transformer protection with, 149–150, 150*f*
GWO, 132–136
phase and ground coordination, 137–141, 138–139*f*
academic test system, 138–139
results, 139–141
proposed general model, 122–127, 124*t*, 126*f*
test systems, 133–136, 133*f*, 134–135*t*, 136–137*f*
Nonlinear data fitting, 51

O

Online coordination system, special considerations for, 303–310
load flow and fault current analysis, 304
network topology, 303, 304*f*
parallel lines, 308
radial lines detection, 308–309
sensitivity analysis, 309–310, 310*f*
special considerations, 305–309
transformer protection, setting of relays for, 305–307
Online protection system, 300–303, 301*f*
coordination algorithm, 302–303
Operating time
validation, 336–337, 337*f*
verification, 347–354, 348–353*t*, 348*f*, 350*f*, 354–355*f*, 356*t*
Operation time analysis, 75, 75*f*
Optimization formulation, 281–287, 282*f*
problem, 281
results, 283–287, 286–287*t*
test systems, 282–283, 283*f*
Overcurrent coordination, fault location for, 2*f*
Overcurrent devices, coordination of, 31–34
Overcurrent relays (OCRs)
adaptive location, 76
adaptive operating time for
curve fitting program, 53–60, 54*f*
definition, 38–41
evaluation of, 41–43
implementation, 44–46, 45–46*f*
proposed relay, functional structure of, 43–46, 44*f*
time curve fitting algorithms, 46–53
adaptive relay, 70–71
conventional-adaptive-conventional relay set, 69–70
conventional relays, 70–71
coordinating relays
general procedure for, 7–9
coordination, 5–10, 6*f*, 66–71
digital relays, functional structure of, 18–30, 18*f*
dynamic performance, 27, 28–29*f*
instantaneous OCR, 23–24, 23*f*
inverse time relay, 24–27, 24*f*, 26*f*
limitations of, 27–30, 29–30*f*
phasor estimation, 19–23, 19*f*
distributed energy resources (DERs), 1
DOCR coordination, 10–14, 11–12*f*
general considerations, 14–17, 14*f*
meshed networks, 11–12
nonconventional curves (NCCs), 7
operation time analysis, 75, 75*f*
overcurrent coordination, fault location for, 2*f*
overview, 1–2
plug setting multiplier (PSM), 3
practical applications, 30–34
overcurrent devices, coordination of, 31–34
relay-fuse coordination, 32, 33*f*
relay-recloser coordination, 33, 34*f*
relay-relay coordination, 31–32
radial networks coordination, 7
recloser, 70–71
security, 10
sensitivity analysis, 10, 16*t*, 71
six-bus IEEE network system, 15*f*, 16
small generators, 12–13, 13*f*
time curve, 2–5, 3*f*

P

Parallel lines, 308
Phase and ground coordination, 137–141, 138–139*f*
  academic test system, 138–139
  results, 139–141
Photovoltaic model, 198
Plug setting multiplier (PSM), 3
Point of common coupling (PCC), 193, 210
Polynomial regression, 53
Potential transformer (PT), 274
Power wind model, 195
Proposed general model, 122–127, 124*t*, 126*f*
Proposed method, optimization algorithm with, 176–178, 177*f*
Proposed model, 142–148, 143*f*, 145–146*f*, 148*f*
Proposed nonconventional curves, parameters for, 150–152, 152–153*f*, 154*t*
Proposed optimization problem, 164–166
  objective function, 164–165
  restrictions, 165–166
  time function, 165
Protection devices, coordination results for, 154–157, 157*f*
Protection purposes, signal processing for, 274*f*

R

Radial lines detection, 308–309
Radial networks coordination, 7
Real time, acquisition of signals in, 346–354, 346*f*
Recloser, 70–71
  coordination, 342–344, 343*f*
Relay coordination formulation, data preparation for, 83–90, 83*f*
  constraints, 86–87
  equality constraint, 86
  inequality constraint, 86–87
  initial population, 85–86
  multiobjective function, 88–90
  objective function, 87–90
  random variables, 84–85, 85*f*
  weighted objective function, 87–88
Relay-fuse coordination, 32, 33*f*
Relay operation validation, 279–281, 280*f*
Relay programming, 330–332, 331*f*
Relay-recloser coordination, 33, 34*f*
Relay-relay coordination, 31–32, 340–342, 340–341*f*, 341*t*
Relay time curve fitting, 55
Remedial Action Scheme (RAS), 214
Renewable sources (RES), 193, 195–205, 196–199*f*
  interconnection of, 324

S

Security, 10
Self-adaptive differential evolution (SADE), 81–82
Sensitivity, 10, 16*t*, 355–359, 357–359*f*
  analysis, 71, 309–310, 310*f*
  detection system, 238–240, 239*f*
Signal file, 359–362, 360–362*f*
Six-bus IEEE network system, 15*f*, 16
Small generators, 12–13, 13*f*
Standard curves, protection coordination with, 152–153, 156*f*
Standard curves, transformer protection with, 149–150, 150*f*
States of charge (SoC), 201
Subharmonic currents, 339, 339*f*

T

Test systems, 133–136, 133*f*, 134–135*t*, 136–137*f*, 174, 175–176*f*, 313–314
Time curve, 2–5, 3*f*
Time curve fitting algorithms, 46–53
  digital overcurrent relays, equations of, 49–50
  equations of, 47–50
  time curve fitting program, 50–53, 51*f*
    nonlinear regression, 51–52
    polynomial regression, 53
Time curve fitting program, 50–53, 51*f*
  nonlinear regression, 51–52
  polynomial regression, 53
Time intervals
  coordination, 188–189, 190*t*, 324–325, 325*f*
  OCR coordination via, 163–166, 163*f*
Topological changes, 297–298, 298*f*
Transformer protection, setting of relays for, 305–307

V

Variable fault current, 41–42
Virtual relay design
  coordination examples, 339–344
    infeed in coordination, 342, 342*t*
    recloser-relay coordination, 342–344, 343*f*
    relay-relay coordination, 340–342, 340–341*f*, 341*t*
  definition, 334, 335*f*

internal signal generation, 362–364, 362–364*f*, 364*t*
laboratory test structure, 344–359, 344–345*f*
operating time verification, 347–354, 348–353*t*, 348*f*, 350*f*, 354–355*f*, 356*t*
real time, acquisition of signals in, 346–354, 346*f*
sensitivity, 355–359, 357–359*f*
modules of, 335–336
relay programming, 330–332, 331*f*
signal file, 359–362, 360–362*f*
stand-alone examples, 336–344
cold load current, 338–339, 338*f*
operating time validation, 336–337, 337*f*
subharmonic currents, 339, 339*f*
Virtual OCR, 332–336, 334*f*

W

Weak source contribution, 208–209, 208–209*f*
Wind generators, 214–215

FSC
www.fsc.org
MIX
Papier | Fördert gute Waldnutzung
FSC® C083411

Kolibri 360 GmbH
Ferdinand-Jühlke-Straße 7
99095 Erfurt, Deutschland
produktsicherheit@kolibri360.de